IP Communications and Services for NGN

IP Communications and Services for NGN

JOHNSON I. AGBINYA

CRC Press
Taylor & Francis Group
Boca Raton   London   New York

CRC Press is an imprint of the
Taylor & Francis Group, an **Informa** business

Auerbach Publications
Taylor & Francis Group
6000 Broken Sound Parkway NW, Suite 300
Boca Raton, FL 33487-2742

© 2010 by Taylor and Francis Group, LLC
Auerbach Publications is an imprint of Taylor & Francis Group, an Informa business

No claim to original U.S. Government works

Printed in the United States of America on acid-free paper
10 9 8 7 6 5 4 3 2 1

International Standard Book Number: 978-1-4200-7090-3 (Hardback)

Library of Congress Cataloging-in-Publication Data

Agbinya, Johnson I.
   IP communications and services for NGN / author, Johnson I. Agbinya.
      p. cm.
   Includes bibliographical references and index.
   ISBN 978-1-4200-7090-3 (hardcover : alk. paper)
   1. Digital communications. 2. Telecommunication systems. 3. Convergence (Telecommunication) I. Title. II. Title: Internet protocol communications and services for next generation networking.

TK5103.7.A394 2010
004.6'2--dc22   2009031626

Visit the Taylor & Francis Web site at
http://www.taylorandfrancis.com

and the Auerbach Web site at
http://www.auerbach-publications.com

Contents

List of Figures

List of Tables

List of Acronyms

1G: First Generation
1xRTT: Radio Transmission Technology (one frequency band)
2.5G: Second-and-a-Half Generation
2G: Second Generation
3G: Third Generation
3GPP: Third Generation Partnership Program
4G: Fourth Generation
AAAF: Foreign AAA Server
ACELP: Adaptive Code Excited Linear Prediction
ACN: Airborne Communication Node
ACR: Absolute Category Rating
ACUTA: Association for Communications Technology Professionals in Higher Education
ADC: Analog-to-Digital Converter
ADPCM: Adaptive Differential Pulse Code Modulation, Adaptive Differential PCM
ADSL: Asynchronous Digital Subscriber Line
AES: Advanced Encryption Standard
ALG: Application Level Gateway
AMR–WB: Adaptive Multi-Rate–Wide Band (coder)
ANNM: Asynchronous Subscribe & Notify Model
ANSI-41: American National Standards Institute-41
AOR: Address-Of-Record
AP: Access Point
API: Application Programming Interface
AR: Access Router
ASAP: Aggregate Server Access Protocol
ASCONF: Address Configuration
ASD: Answer-Signal Delay
ASNM: Asynchronous Subscribe & Notify Model
ATM: Asynchronous Transfer Mode

AUC: Authentication Center
AUXL: Auxiliary Link
B3G: Beyond 3G
BAHAMA: Broadband Ad-Hoc ATM local Area network
BER: Bit Error Rate
BGP: Border Gateway Protocol
BLER: Block Error Rate
BLOS: Beyond Line-Of-Sight
BN: Backbone Node
BRH: Bandwidth Request Header
BS: Base Station
BSC: Base Station Controller
BSS: Base Station Subsystem
BT: Base Transceiver Station
C: Client
CAPEX: CAPital EXpenditures
CCS: Call Control Server
CDMA: Code Division Multiple Access
CDMA2000: Code Division Multiple Access 2000
1xEV-DO: 1x Evolution Data Only
CDMAOne: Code Division Multiple Access One
CDR: Call Detail Record
CELP: Code-Excited Linear Prediction
CGF: Charging Gateway Function
CMS: Cryptographic Message Syntax
CN: Core Network, Correspondent Node
CO: Central Office
CoA: Care-of Address
COP: Code Of Practice
CoS: Class of Service
CR: Constraint-based Routing
CRD: Call-Release Delay
CR–LDP: Constraint-Based Label Distribution Protocol
CS: Circuit Switching, Class of Service
CS-ACELP: Conjugate Structure-Algebraic Code-Excited Linear Prediction
CS-CELP: Conjugate Structure Code-Excited Linear Prediction
CSCF: Call Session Control Function
CSRC: Contributing Source identifier
DAD: Duplicate Address Detection
DAR: Dynamic Address Reconfiguration
DARPA: Defense Advanced Research Projects Agency
DCR: Degradation Category Rating
DECT: Digital Enhanced Cordless Telephone

DES: Data Encryption Standard
DF: Delay Factor
DHC: Dynamic Host Configuration Protocol
DiffServ: Differential Services
DL: Downlink
DLCI: Data-Link Connection Identifier
DMOS: Degradation Mean Opinion Score
DMT: Discrete MultiTone (modulation)
DNIS: Dialed Number Identification Service
DNS: Domain Name Server
DSL: Digital Subscriber Line
DSP: Digital Signal Processor
DTLS: Datagram Transport Layer Security
DTMF: Dual-Tone Multi Frequency
E2E: End-to-End
EDGE: Enhanced Data rate for GSM Evolution
EIR: Equipment Identity Register
ENRP: Endpoint Name Resolution Protocol
ESP: Encapsulating Security Payload
ETSI: European Telecommunications Standards Institute
FCC: Federal Communications Commission
FDMA: Frequency Division Multiple Access
FEC: Forward Equivalence Class, Forward Error Correction
FER: Frame Error Rate
FoIP: Fax over Internet Protocol, Fax over IP
FTP: File Transfer Protocol
FTTC: Fiber-To-The-Curb
FTTH: Fiber-To-The-Home
FTTN: Fiber-To-The-Node
GEO: Geosynchronous Orbit
GGSN: Gateway GPRS Support Node
GMH: Generic MAC Header
GMM: GPRS Mobility Management
GMPLS: Generalized MPLS
GMSC: GSM Mobile Switching Center
GoS: Grade-of-Service
GPRS: General Packet Radio Service
GSM: Global Systems for Mobile Communications
GWO: Geostationary Earth Orbit
HA: Home Agent
HAE: High Altitude Endurance
HAWAII: Handoff Aware Wireless Access Internet Infrastructure
HDR: High Data Rate

HDTV: High-Definition Television
HE: Home Environment
HLR: Home Location Register
HoA: Home Address
HOL: Head-Of-Line
HR: Home Registrar
HSDPA: High-Speed Downlink Packet Access
HSS: Home Subscriber Server
HTTP: Hyper Text Transfer Protocol
I-CSCF: Interrogating CSCF
IEEE: Institute of Electrical and Electronics Engineers
IETF: Internet Engineering Task Force
IFP: Internet Fax Protocol
IGMP: Internet Group Multicast Protocol
IGRP: Interior Gateway Routing Protocol
IHL: Internet Header Length
IM: Instant Messaging
IMS: IP Multimedia Subsystem
IMT2000: International Mobile Telecommunications-2000
IN: Intelligent Network
IntServ: Integrated Services
IP: Internet Protocol
IPBS: Internet Protocol PBX
IP-PBX: Internet Protocol Private Branch Exchange
IPSec: IP Security
IPSL: Inter-Proxy Server Link
IPTV: Internet Protocol Television
IS-95B: Interim Standard 95 version B
IS-95C: Interim Standard 95 version C
ISDN: Integrated Services Digital Network
IS-IS: Intermediate System to Intermediate System
ISO: International Organization for Standardization
ISP: Internet Service Provider
ITU-T: International Telecommunication Union (Telecommunication Section)
LAN: Local Area Network
LD-CELP: Low Delay CELP
LDP: Label Distribution Protocol
LEO: Low Earth Orbiting (satellite)
LER: Label Edge Router
LIB: Label Information Base
LLC: Logical Link Control
LOS: Line-Of-Sight
LPC: Linear Predictive Coding (Coder)

LSP: Label-Switched Path
LSR: Label-Switched Router
LST: Laplace-Stieltjes Transform
MAC: Media Access Control, Message Authentication Code
MAN: Metropolitan Area Network
MANET: Mobile Ad-hoc NETwork architecture
MBN: Mobile Backbone Network
MBWA: Mobile Broadband Wireless Access
MDI: Media Delivery Index
ME: Mobile Equipment
MEGACO: Media Gateway Control Protocol
MGC: Media Gateway Controller
MGCF: Media Gateway Control Function
MGCP: Multimedia Gateway Control Protocol
MGW: Media Gateway
MH: Mobile Host
MIMO: Multiple-Input and Multiple-Output
MIP: Mobile IP Protocol
MIPL: Mobile IPv6 for Linux
MLR: Media Loss Rate
MLT: Modulated Lapped Transform
MMMUSIC: Multi-party Multi-Media Session Control
MMS: MultiMedia Service, Multimedia Messaging Service
MN: Mobile Node
MNP: Mobile Node Prefix
MNRU: Modulated Noise Reference Unit
MOS: Mean Opinion Score
MPDU: MAC Protocol Data Unit
MPEG: Moving Pictures Experts Group
MPEG2-TS: Moving Picture Experts Group2 Transport Stream
MPLS: Multi Protocol Label Switching
MP-MLQ: Multi-Pulse-Maximum Likelihood Quantization
MP-MLQ: Multi-Pulse-Multilevel Quantization
MR: Mobile Router
MRF: Media Resource Function
MS: Mobile Station
MSC: Mobile Switching Center
MSCTP, mSCTP: Mobile SCTP
MSDU: MAC Service Data Unit
MT: Master Terminal
MTU: Maximum Transmission Unit
m-WLAN: mobile Wireless Local Area Network
NAS: Network Access Server

NAT: Network Address Translation
NEMO: NEtwork in MOtion
NS: Name Server
NSS: Network and Switching Subsystem
NTSC: National Television Standards Committee
O&M: Operations and Maintenance
OAN: Optical Access Network
OFDM: Orthogonal Frequency Division Multiplexing
OLT: Optical Line Terminator
OMC: Operating network Management Center
OPEX: OPerational EXpenditures
OSI: Open Systems Interconnection
OSPF: Open Shortest Path First
OSS: Operations Support System
PAL: Phase Alternating Lines
PAN: Personal Area Network
PBX: Private Branch Exchange
PCM: Pulse Code Modulation
P-CSCF: Proxy-Call Session Control Function
PDA: Personal Digital Assistant
PDD: Post-Dialing Delay
PDU: Protocol Data Unit
PE: Pool Element
PLMN: Public Land Mobile Network
PLR: Packet Loss Ratio
PoC: Push-to-talk over Cellular
PON: Passive Optical Network
POP: Point of Presence
PPP: Point-to-Point Protocol
PS: Packet Switching
PSTN: Public Switched Telephone Network
PU: Pool User
PWT: Personal Wireless Telecommunication
QAM: Quadrature Amplitude Modulation
QoE: Quality of Experience
QoS: Quality-of-Service
RA: Router Advertisement
RADIUS: Remote Authentication Dial In User Service
RAN: Radio Access Network
RFA: Foreign Agent Router
RFC: Request For Comment
RLCAM: Radio Link Control in Acknowledged Mode
RNC: Radio Network Controller

RNS: Radio Network Subsystem
RP: Rendezvous Point
RPE-LPT: Residual Pulse Excited-Long Term Prediction
RR: Receiver Report
RRC: Radio Resource Control
RSerPool: Reliable Server Pooling
RSS: Received Signal Strength
RSVP: Resource Reservation Protocol
RTCP: Real Time Control Protocol
RTP: Real-time Transport Protocol
RTSP: Real Time Streaming Protocol
RTT: Return-Trip Time, Round-Trip Time
S/MIME: Secure Multi-Purpose Internet Mail Extensions
SA: Security Association
SAA: Stateless Address Autoconfiguration
S-CSCF: Serving CSCF
SCTP: Stream Control Transmission Protocol
SDP: Session Description Protocol
SG: Signaling Gateway
SGSN: Serving GPRS Support Node
SHA-1: Secure Hash Algorithm-1
SIGINT: Signal Intelligence
SIM: Synchronous Invite Model, Subscriber Identity Module
SIP: Session Initiation Protocol
SM: Session Management
SMB: Small and Medium-sized Business
SMS: Short Messaging Service
SMTP: Simple Mail Transfer Protocol
SNDCP: SubNetwork Dependent Convergence Protocol
SNMP: Simple Network Management Protocol (Proxy)
SNR: Signal-to-Noise Ratio
SONET: Synchronous Optical NETwork
SP: Service Provider
SPIT: SPam over Internet Telephony
SR: Sender Report
SS7: Signaling System Number 7
SSH: Secure Shell
SSL: Secure Socket Layer
SSN: Stream Sequence Number
SSO: Single Sign-On
SSRC: Synchronization Source
ST: Slave Terminal
STB: Set-Top Box

SVA: Simple VoIP Application
TCP: Transmission Control Protocol
TD-CDMA: Time Division CDMA
TDM: Time Division Multiplexing
TDMA: Time Division Multiplex Access
TD-SCDMA: Time Division-Synchronous Code Division Multiple Access
TE: Terminal Equipment
TIA: Telecommunications Industry Association
TLS: Transport Layer Security
TOS: Type of Service
TPC: Transmission Power Control
TU: Transaction User
UA: User Agent
UAC: User Agent Client
UAS: User Agent Server
UAV: Unmanned Airborne Vehicle
UCAN: Unified Cellular and Ad-hoc Network architecture
UDP: User Datagram Protocol
UE: User Equipment
U-HAAP: Undedicated High-Altitude Aeronautical Platform
UL: Uplink
UMTS: Universal Mobile Telecommunications System
URI: Uniform Resource Identifier
USIM: User Subscriber Identity Module
UTRAN: UMTS Terrestrial Radio Access Network, UMTS Terrestrial RAN
UWB: Ultra Wideband
VCI: Virtual Circuit Identifier
VDSL2: Very High Speed Digital Subscriber Line 2
VLR: Virtual Location Register
VOD: Video On Demand
VoIP: Voice over Internet Protocol, Voice over IP
VoWLAN: Voice over Wireless LAN
VPF: Voice Peering Fabric
VPI: Virtual Path Identifier
VPN: Virtual Private Network
VS: Video Server
VSELP: Vector Sum Excited Linear Prediction
WAN: Wide Area Network
WAP: Wireless Application Protocol
WCDMA: Wideband Code Division Multiple Access
WDM: Wavelength Division Multiplexing
WIBL: Wireless Inter-Base station Link
Wi-Fi: Wireless LAN

WiMAX: Worldwide Interoperability for Microwave Access
WISP: Wireless Internet Service Provider
WLAN: Wireless Local Area Network
WMAN: Wide Area Metropolitan Network, Wireless Metropolitan Area
Network
WPAN: Wireless Personal Area Network

Chapter 1

Access Networks

This chapter is an overview of different architectures of fixed, mobile and wireless networks, which provide solutions to support mobility for many users. Air interfaces such as GSM, GPRS, UMTS (cellular networks) and other wireless architectures that support mobility (ad-hoc networks). Each network architecture model is illustrated and explained. Then, the advantages and disadvantages of these network architecture models are discussed.

Public Switched Telephone Network (PSTN)

The public switched telephone network (PSTN) is the most familiar network for conversational voice. It is a copper-based wired time-division multiplexed (TDM) backbone network. Wired copper pairs are used to form the local loop. It is naturally very rugged and has for decades enjoyed several refinements and evolution. It was designed primarily for voice communications using the log-PCM (pulse code modulation) coding.

Speech coding for PSTN

Over the last three decades a great deal of progress has been made in speech compression leading to a plethora of speech compression standards for PSTN and mobile communication networks. In terms of the technical algorithms used, speech compression in general may be grouped into analytic, statistical and parametric methods. In this book however, for ease of analysis the compression methods are divided

1

into two broad groups, narrowband and wideband speech coding standards. The use of different speech bandwidths, varying sampling rates lead to different bit rates at the output of the coding algorithms. A narrowband voice signal is defined here as voice signal contains frequencies below about 3.4 KHz. Wideband speech signal provide larger bandwidth to accommodate for the finesse in the frequency content of speech and more specifically to accommodate for distinguishing fine female voice and for reproduction. As such wideband speech contains frequencies from 50 Hz to around 7 KHz. Compared with narrowband speech, wideband speech represents better voice quality content for both male and female and therefore the resulting coding standards are expected to have higher mean opinion scores (MOS). In practice wideband speech models female voices better. The frequency contents of female voices are clustered in the higher frequency bands above 3.4 KHz.

Since the early days of telephony narrow band speech has been traditionally sampled at 8 kHz. Doing so ensures that voice frequency component up to 4 kHz are accounted for through the Nyquist criterion. When each sample of narrow band speech is represented by eight bits, the resulting bit rate is 64 kbps and the coding standard is called pulse code modulation (PCM). The rational for the coding technique was to matches the capacity of copper and hence copper wires have been. The sound from a voice signal is an ear representation of the frequency spectrum of the signal. ITU-T maintains a record of speech standards and the PCM coding format was designated by the ITU-T as G.711. Since the coding standard afford clear distinction of speakers' voices and intonations, G.711 is often referred to as the toll quality standard for (telephone call quality) voice communication.

The oldest of the speech compression standards G.711 (alias PCM) has a mean opinion score of between 4.4 and 4.5 for just one round of coding. In PCM, speech is sampled at 8k Hz and each sample is represented as 8 bit value (64 kbps rate). Bandwidth-wise G.711 is inefficient because the channel is occupied by one user continuously until it is relinquished. In order to fit more voice calls onto one PCM channel and also maintain toll quality, several new voice compression standards were developed for the PSTN. One of such new standards uses adaptive differential pulse code modulation (ADPCM) technique and was standardized by CCITT as G.721 at a bit rate of 32 kbps. Improvements on G.721 came in the form of G.726 and G.727 standards at bit rates of 40, 32, 24 and 16 kbps. G.726 specified how either A-law or μ – law non-linear compressions could be used to adapt PCM to ADPCM at these data rates.

Other voice compression methods have been used as well. For example G.728 also known as LD-CELP (low delay-code excited linear prediction) uses 8 kHz sampling rate and results to data rate of 16kbps. This means 4 G.728 channels would fit into one PCM channel. The G.729 ITU-T standard is also based on LD-CELP technique resulting in 32kbps compression on 10ms frames and thus supports low transmission delays. It is therefore useful in application where low transit delay is essential such as in teleconferences. A variant of CELP called the conjugate-structure algebraic CELP (CS-ACELP) is supposedly a reduced complexity version

Table 1.1 **Parameters of Voice Codecs for PSTN**

Codec	Name	Frame Size/ Look Ahead (ms)	Complexity (MIP)	Bit Rate (kbps)	MOS
G.711	PCM	0.125	<<1	64	4.0+
G.721/726	ADPCM	0.125	1.25	32	~4.0
G.728	LD-CELP	0.625	30	16	3.9
G.729	CS-CELP	10/5	20	8	4.0
G.729A	CS-ACELP	10/5	12	8	4.0
G.723.1	ACELP/MP-MLQ	30/7.5	11	5.3/6.3	3.7/3.9

of CS-CELP and suitable for use in voice over IP (VoIP). In all the new coding standards, the main objectives are to reduce the bit rates a lot below PCM and to maintain toll quality. The MOS performance, bit rates and the complexities of these coding standards are shown in Table 1. It is noteworthy that the MOS is not an absolute rating value of a coding standard because it is a subjective measure and changes from test to test. Different values would therefore be obtained for a particular test speech file and different groups of test persons. Therefore values in Table 1 are to be seen as approximations.

The G.723.1 ITU-T standard is a dual rate codec at 6.3 kbps and 5.6 kbps and provides near toll quality speech. It is a highly bandwidth efficient codec designed originally for video telephony and also could be used in situations where excellent toll quality is not essential. The lower bit rate version uses ACELP and the higher bit rate version uses MP-MLQ (multi-pulse maximum likelihood quantization) techniques.

Tandem Coding and Transcoding

During voice communication across different types of networks and across several countries, calls often traverse several networks requiring re-coding or more than one coding of the voice information at intermediate networks [52]. This repeated coding is called tandem coding. Transcoding on the other hand is re-coding from one codec standard to another codec standard with the objective of not introducing any new losses. Thus for example, coding could be from ADPCM to LD-CELP from CS-CELP to MP-MLQ standards. Hence the objective is lossless re-coding to match a network using a different codec in the path of the voice communication. This situation may happen in inter-connectivity situations between countries using different communication systems.

The mean opinion score for G.711 for 8 tandem coding chain is greater than 4. Table 2 is a summary of the performance of the narrowband speech coding standards with tandem coding. Tandem coding requires repeated coding and decoding at the intermediate networks. The objective in tandem encoding is to minimize as

Table 1.2   Tandem & Transcoding Performance of Some Narrowband Codecs (adapted from [52])

Codec/No. of Tandem Coding	Name	No of Tandem Coding	MOS
G.711 x 4	PCM	4	>4.0
G.726 x 4	ADPCM	4	2.91
G.729 x 2	CS-CELP	2	3.27
G.729 x 3	CS-CELP	3	2.68
G.726 + G.729	ADPCM+CS-CELP	Transcoding	3.56
G.729 + G.726	CS-CELP+ADPCM	Transcoding	3.48

much as possible further coding losses in speech quality (beyond the previous coding losses) due to re-coding.

The results in Table 2 represent concerns when VoIP communication traverses several intermediate networks. The MOS drops dramatically for non-PCM coding standards. The last two rows show transcoding between two coding standards in different orders. They show that the speech quality received depends on the order of transcoding. For example, the speech quality is worse for transcoding from G.729+G.726 (MOS=3.48) but better when the transcoding order is reversed (MOS=3.56). This problem can occur with different (interconnectivities) terminations at different networks. Therefore a PSTN VoIP call using a G.726 codec terminated on a network using G.729 will sound better compared to when the reverse is the case. Hence termination and connectivity charges could reflect the type of coding standard being used by an operator or a voice over IP provider.

Wideband speech coding standards were proposed originally for video conferencing using H.323. They take input speech in the range 50 Hz to 7 KHz. The objective in wideband speech compression is to improve voice intelligibility, speaker identifiability and preservation of naturalness of voice. The most familiar example of wideband speech standard is the adaptive differential PCM (ADPCM). It segments speech into two bands, a narrow band and the higher band. ADPCM produces toll quality speech at 64 kbps, 56 kbps and 48 kbps. The advent of VoIP has led to considerable interest in ADPCM. Although voice clarity and intelligibility are important in modern IP phones, the need to support more channels means that G.722 codec due to its high bit rates is not always going to be the best choice. However, it also represents a problem for operators whose networks are unable to support its rate.

Adaptive Multi-Rate – Wideband (AMR-WB) Codec

Orthogonal transforms have also been used to implement wideband speech codecs include G.722.1 and G.722.2 standards. One of the major strengths of CELP is the application of analysis by synthesis method. This means CELP codecs try to synthesize speech and also look ahead in time expecting what the next speech frame ought to be. The G.722.2 standard uses adaptive code excited liner predictor (ACELP). G.722.2 is a multiple bit rate codec standardised by the 3GPP and adopted by the ITU-T. By nature it is designed as an adaptive multirate wideband (AMR-WB) speech coder and provides multiple bits rates at 6.6, 8.85, 12.65, 14.25, 15.85, 18.25, 19.85, 23.05 and 23.85 kbps. The coder works well for speech less so for music. This is understandable because high fidelity music requires higher sampling rates beyond the 7 kHz in use by wideband codecs. G.722.2 achieve toll quality performance at 23.05 kbps and higher. It has been shown to provide at 23.85 kbps rate an MOS of 4.5 for a French test and 4.2 at 12.65 kbps rate.

The G.722.1 uses another orthogonal transform, the modulated lapped transform (MLT) and uses a filter bank to decompose the speech signal into sub-bands. The reason for doing this is to enable more efficient processing and coding of the speech signal components per band and also to achieve variable bit rate. The level of finesses required for voice at high frequency means that a voice component in the high frequency band can be handled more efficiently by representing each frequency component with more bits than at low frequency. This extends the dynamic range and intelligibility of the high frequency components. For these reasons G.722.1 performs much better for music compared with other wideband codecs. G.722.1 results to two bits rates at 24 and 32 kbps. It has achieved a MOS of 4.1 at 24 kbps when tested with British English. The properties of G.722.1 and G.722.2 are shown in Table 3.

Narrowband AMR Coding in LTE-Advanced

AMR Codec is a patented codec standardized by ETSI and 3GPP for use in GSM, 3G (UMTS) and LTE-Advanced [53]. 3GPP [53] specifications provide details of AMR as used for implementing speech processing in GSM, 3G and LTE-Advanced phones. This uniformity of codec standard helps in maintaining comparative

Table 1.3 Characteristics of Wideband Speech Coding Standards for PSTN

Codec	Frame size/ Look Ahead (ms)	Complexity (MIPS)	Bit Rate (kbps)	Coding Method	MOS
G.722	0.125/1.5	10	48, 56, 64	Subband ADPCM	
G.722.1	20/20	<15	24 and 32	Modulated Lapped Transform	4.1
G.722.2	20/5	<40	6.6 to 23.85	ACELP	4.2–4.5

Table 1.4 Characteristics of AMR for GSM, 3G, LTE-Advanced

Codec	Frame Size/ Look Ahead (ms)	Network Conditions	Bit Rate (kbps)	Coding Method	MOS
AMR	30/0	Ideal	12.2	ACELP	4.14
AMR	30/0	Stressed	12.2	ACELP	3.79
AMR	30/5		10.2 to 4.75	ACELP	

speech quality across a wide array of mobile phone terminals from various vendors and equipment manufacturers.

In the analogue-to-digital part speech is sampled at 8 kHz and each sample is represented as 13 bit numbers giving a raw bit rate of 104 kbps. The finer resolution of the speech samples above 8 bits ensures better speech signal-to-noise ratio. 3 GPP calls this process "analogue–to–uniform digital conversion". The uniform format is represented as two's compliment numbers. The DAC performs the reverse operation of converting from 13-bit/kHz uniform PCM to analogue operation. The 3G standard is a bit flexible allowing terminal equipment manufacturers to decide how the A/D function is implemented, either by direct conversion to 13-bit uniform PCM or by first converting to 8-bit A-law or mu-law followed by an 8-bit to 13-bit conversion. AMR codec is based on CELP coding model. Seven rates are recommended for LTE-Advanced in the form of 12.2, 10.2, 7.95, 7.4, 6.70, 5.90, 5.15 and 4.75 kbps [53]. Subtle coding differences were recommended. At the 12.2 kbps rate, short term prediction or linear prediction (LP) are recommended. LP analysis is performed twice pre speech frame (30ms symmetric windows) using the auto-correlation method without lookahead. At the 10.2, 7.95, 7.4, 6.70, 5.90, 5.15 and 4.75 kbps rates, short term prediction or linear prediction (LP) are also undertaken once per speech frames which also are 30ms long but using asymmetric windows. A lookahead of 40 samples (5ms) is used during the auto-correlation computation. Detail explanations on how to achieve each of the above bit rates are given in [53].

Mobile Networks

Wireless networks have two different types of architectures; these are infrastructure-based and ad-hoc networks. Many wireless local area networks (WLAN) need clearly defined network architecture and established infrastructure to gain access to other networks. This also enables them to provide important features such as packet forwarding functions and medium access control (MAC). The access points provide the means for backhaul communication through the global Internet. Usually communication takes place between the wireless nodes and the access point and in ad hoc mode between the wireless nodes.

Ad hoc networking between access points from different manufacturers pose challenges as they are not developed for interoperability. Usually for this to occur and inter-working unit is required and is used to negotiate communication and protocol adaptation between the proprietary access points. However, ad hoc networks in general do not need any infrastructure to work. Each node can communicate with another node without the need to connect to an access point first [1].

Wireless Cellular Networks

First generation (1G) wireless cellular network was an analog system. It was mainly based on circuit-switched technology and designed for voice communication and not data. The second generation (2G) of wireless mobile networks was based on digital data-signalling and circuit-switching. The two popular versions are GSM and CDMAOne. The most popular of them is the Global Systems for Mobile Communications (GSM), which is based on simple signalling system 7 (SS7). The data rate is about 13.6 Kbps. GSM as a European standard is not only popular there but also in most developing countries. Both GSM and CDMAOne have undergone series of evolutions that were aimed at providing them the capabilities to carry data with enhanced data rates. The evolution path for GSM includes the so-called General Packet Radio Service (GPRS). It is generally accepted as a second and a half generation (2.5G) network and was based on digital transmission and packed-switching. It increased the data communication speed over 8 time slots to about 171 kbps. Enhanced data rate for GSM evolution (EDGE) was created as the necessary step towards the third generation (3G) networks and boasts a data rate of about 384 kbps. Although not a full-fledged 3G network, it served as an intermediate network. It is less-known compared to all the other GSM evolutions because most operators have not implemented it, skipping in favor of UMTS or CDMA2000.

GSM

One of the first mobile technologies that provide the base for next-generation networks is GSM. It is based on low-band digital data-signalling and is a combination of Frequency Division Multiple Access (FDMA) and Time Division Multiple Access (TDMA). GSM systems operate in the 900 MHz and 1.8 GHz bands around the world, with the exception of the Americas, where it operates in the 1.9 GHz band. GSM's architecture presents three interfaces: the air interface Um, the Abis interface and The A interface. These interfaces provide connections between GSM network components [2, 50].

As Figure 1 shows, the Um interface is used between the mobile station (MS) and the base station subsystem (BSS); the MS is a device used by each user to access the network. The communication is established wirelessly with the base transceiver station (BTS), which is one of the components of the BSS. The BSS provides the connection between mobile stations into a cell in which the node is located. The BSS also has a base station controller (BSC) component, which is responsible for controlling several base transceiver stations. The Abis interface is a standardized,

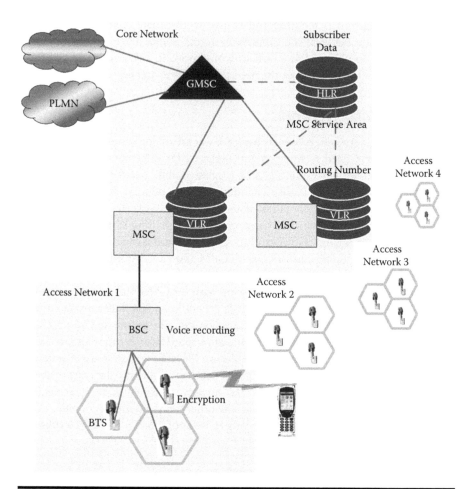

Figure 1.1   GSM system architecture.

open interface with a 16 Kbit/s user channel. It is used to connect the base station transceiver and the base station controller [2].

The 'A' interface is a standardized open interface with 64 Kbit/s user channel. This interface connects the BSC with the network and switching subsystem (NSS). This subsystem performs switching, mobility management, interconnection to other networks and system control functions. It has a GSM mobile switching centre (GMSC), home location register (HLR), virtual location register (VLR), authentication centre (AUC), equipment identity register (EIR) and operating network management centre (OMC). The MSC controls all connections via a separated network to/from a mobile terminal within the domain of the MSC. HLR is a central database containing user data; permanent and semi-permanent data of all subscribers assigned to the HLR. VLR is a local database for a subset of user data, including the data of all current users in the domain of the VLR. The AUC generates user-specific authentication parameters on the request of the VLR; authentication parameters are used for the authentication of mobile terminals and encryption of user data on the air interface within the GSM system. The EIR registers GSM mobile stations and user rights; therefore stolen or malfunctioning mobile stations can be located and sometimes even localised. The OMC presents different control capabilities for the radio subsystem and the network subsystem [2].

GPRS

The explosion of Internet usage started the demand for advanced wireless data communication services. However, 2G circuit-switch-based networks were too slow to support IP services. Therefore 2.5G systems were packet-based to increase the data rate up to 171 kbps. The most common of these generation technologies is the General Packet Radio System (GPRS) [50].

Figure 2 shows the GPRS architecture. This architecture was the intermediate step that was designed to allow the GSM network to provide Internet services before the deployment of full-scale 3G wireless systems. It presents two more components in comparison with GSM to provide an end-to-end packet transfer mode. These are the gateway GPRS support node (GGSN) and the serving GPRS support node (SGSN). The GGSN acts as a logical interface to external packet data networks; it is the MSC of the GPRS network. It performs functions such as protocol conversion between the IP backbone and BSS protocols, authentication of GPRS users, mobility management, routing of data to the relevant GGSN, interaction with the NSS via the SS7 network, collection of charging data records, which are related to GPRS calls, and traffic statistics collection. The SGSN is responsible for the delivery of packets to the MSC within its service area. It acts as a router to the external network; it functions are to route mobile-destined packets coming from external networks to the relevant SGSN, to route packets originating from a mobile to connect with an external network, to allow an interface to the external IP networks, to collect charging data and traffic statistics, and allocating dynamic or static IP addresses to mobiles.

This network is a packet network integrated with the GSM that provides a wireless IP connectivity. Theoretically the throughput of this network is 171 kbps, but realistically it is 54 kbps. Therefore, this network can provide only some Internet services, not those that require high bandwidth, such as multimedia services [3].

CDMA Cellular Networks

The evolution steps for CDMAOne are very different from those of GSM. These networks use CDMA as the air interface making it easier to upgrade them to 3G compared to rival systems based on FDMA/TDMA (eg. GSM). The basic CDMA system offers voice at 14.4 Kbit/sec data rates, which facilitates between 15 and 20 users. An upgrade called IS-95B offers data rates of up to 115 Kbits/sec, which means

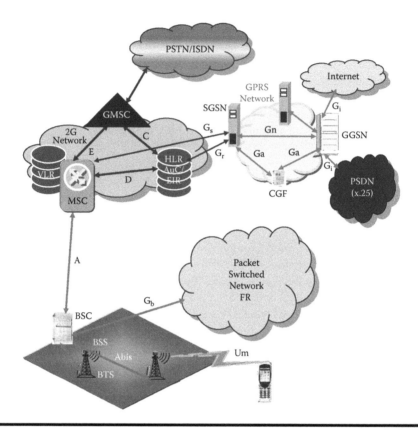

Figure 1.2   GPRS architecture. (From Shimoe, T. and Sano, T., *FUJITSU Sci. Tech. J.*, 134-139, December 2002. With permission.)

only two users per channel. In practice this comes to about 14.4 kbps maximum data rate and circuit switched, using voice-activated digital coding [4]. IS-95c is the evolution of IS-95B (standard ANSI-41) to offer 64 kbps and was discontinued.

Performance of IS-95B is comparable to that of GSM. Its authentication and security are as good as GSM's. Voice quality is high for light loads; however, when the number of users increase, the quality of voice degrades due to 'cell breathing'. All CDMA networks suffer from 'cell breathing' effect. This means that as the number of users increase, the cell coverage area reduces. Therefore at peak traffic period or regions, the system degrades because the cell range shrinks with poor receptions. However, for low traffic times and regions, reception is good. This means there could be a lack of consistency in service quality. To evolve CDMA networks towards the IMT2000 target of 2 Mbits/sec, CDMA systems use more codes, different modulation schemes, and wider bandwidths.

The 1xRTT (radio transmission technology) standard (alias, 1x multi-channel or 1xMC) is CDMA2000 at a carrier frequency of 1.25 MHz. It reuses the frequencies used by cdmaOne except that the number of Walsh codes used for spreading is doubled. The theoretical capacity of the network is 384 kbps (equivalent to that of EDGE) and equivalent to the pedestrian rate of UMTS. The practical data rate is much lower than this for a single user. In a nutshell:

$$1xRTT = IS\text{-}2000 = 3G1X = CDMA2000\ 1X$$

Some articles indicate that the theoretical data rate is actually 144 kbps and Nokia goes further to say that the achievable data rate is 76.8 kbps. Since cell capacity is 170 kbps, only 1 or 2 users can realistically use a cell at a time. The transmit power level is 200 mW. 1xRTT supports mobile IP. The reality is that, cdma2000 1X is equivalent to GPRS and 1xRTT networks support IP communications.

The 3xRTT CDMA standard (alias, 3x multichannel or 3xMC) is CDMA2000 with three times the carrier bandwidth (3x1.25 MHz) 3.75 MHz. It uses the same technology as 1xRTT and wider frequency channels. This means at least three times the data rate of 1xRTT can be achieved with 3xRTT. This means

$$3xRTT = IS\text{-}2000 = 3G3X = CDMA2000\ 3X$$

The theoretical data rate is about 1117 kbps. In practice the data rate is much less than this. Existing 1xRTT networks can be upgraded to 3xRTT to be compliant with UMT2000 requirements. The intention is that 3xRTT will inter-operate with WCDMA.

The 1xEV (EV = Evolution) solutions is the second step in the evolution of CDMA networks. This comes through the basic thrust for development efforts in CDMA2000 in the radio interface. Usually, the evolution occurs in the air interface in the form of changing the modulation scheme and increasing the number of Walsh codes (Spreading functions). In addition redundancies were built into the technologies up to 1xRTT to achieve real time to handle both voice and data. By removing the redundancies, further increase in data rates can be achieved. Qualcomm claims its high data rate (HDR) system can offer packet switched data rates for downlink at 2.4 Mbps and 144 kbps uplink. The 3GPP2 standardised this version and renamed it CDMA2000 1xEV-DO (1x evolution data only), or IS-856. This system cannot offer voice (DO for data only), therefore, voice will have to be offered by 1xRTT or cdmaOne. There are two advantages of 1xEV-DO over 3xRTT. First, using narrower frequency channels of 1.25 MHz reduces the cost of spectrum and building the network. Second, a larger amount of data can be carried if it were as in 3xRTT. This system will operate as an overlay, on top of the voice service.

Motorola followed a similar path to that of Qualcomm in evolving cdmaOne to CDMA2000, a packet switched network. Motorola developed a derivative of 1xRTT version of CDMA2000 and called it 1xTREME. They changed the modulation to achieve higher data rate. Unlike the HDR system, 1xTREME is designed to handle both voice and data. Motorola and Qualcomm have now decided to make concessions, the basis for the 1xEV-DO standard.

UMTS Architecture

The third generation (3G) cellular networks are based on digital signalling and support both packet-and circuit-switched networks. The most common networks are the Universal Mobile Telecommunications System (UMTS) (in Europe and Japan) and cdma2000 (in North America). 3G networks increase the speed to between 384 Kbps and 2 Mbps maximum [11,20].

3G mobile networks cover the weak points of 2G and 2.5G. They consist of multi-vendor networks and advanced standardisation of the interface between the functional elements of networks. 3G also has mobile network system specifications based on the GSM communications and GPRS, making it compatible with those networks. Even though 3G improves the 2G and 2.5G capabilities; poor channel quality and low data-rate connection with the base station are still issues. The two most common 3G technologies are the Universal Mobile Telecommunication Services (UMTS) in Europe and Japan and the Code Division Multiple Access 2000 (cdma2000) in North America.

The main elements in the UMTS network are the base station and switches. This network consists of a Radio Access Network (RAN) and a core network. The RAN architecture is divided into two major parts: the air interface and the UMTS Terrestrial Radio Access Network (UTRAN). The network architecture is based on the GSM/GPRS architecture. The components can also be classified into three main groups according to the part of the network where they work. First, the User Equipment UE, which has three main components: the Mobile Terminal (MT) or Mobile Equipment (ME); the user subscriber Identity Module USIM, which is a component that equips the MT; and the Terminal Equipment (TE). The UE is connected to the UTRAN using the air interface Uu. Second, the UTRAN presents two main components: the Radio Network Controller (RNC) and the Node B or base station. Third, the core network which consists of a packet-switched domain. This includes the 3G SGSNs and GGSNs which provide the same functionality that they provide in GPRS systems. Charging for services and access is carried out through the Charging Gateway Function (CGF), which also is part of the core network [4,50].

UTRAN consists of one or more Radio Network Subsystem (RNS). Each RNS consists of many RNCs. Each RNC is connected to other RNCs via the Iur interface. Each RNC is connected to the core Network via an Iu interface, which can be: Iu-CS which connects the RNC to the circuit-switching part of the CN, or Iu-PS, which connects the RNC to the packet-switched part of the CN. Additionally, each RNC consists of many Radio Network Subsystems (RNS), and each RNC is connected to many base stations, which are known as Node Bs [4].

Each Node B performs radio transmission and reception with cells. Its main task is to transmit data from and to the UE via the Uu air interface and the RNC via the Iub interface. This task includes Forward Error Correction (FEC); rate adaptation; W-CDMA spreading and de-spreading; QPSK modulation; handover and macro diversity for cells that belong to the same Node B; the Node B measures and reports on the quality and strength of connection and determines the frame error rate (FER) as a measured report to the RNC for cells that belong to different Node Bs, but are controlled by the same RNC. The node B also allows the UE to adjust its power using downlink (DL) transmission power control (TPC) commands via the inner-loop power control on the basis of uplink (UL) TCP information [4].

The RNC corresponds to the GSM BSC. Its functions are to perform the Radio Resource Control (RRC) protocol that defines the messages and procedures between UE and UTRAN. The main tasks of RNC are: to control logical resources provided by the Node Bs, to be responsible for the layer 2 processing of user data, closed-loop power control, handover control, admission control, code allocation, packet-scheduling, and macro-diversity combining/splitting over a number of Node Bs [4].

Finally, the core network consists of two separate but parallel networks: the switched service domain and the packet-switched service domain. The switched service domain focuses on the MSC. This domain is derived from the GSM infrastructure and thus shares many of its characteristics. It also enables GSM operators to share network infrastructure in the early stages of UMTS rollout. The packet-switched service domain focuses on the GSNs and uses IP to transport data traffic.

The core network (CN) consists of the following components, whose elements are: 3G MSC/VLR (also called wideband MSC), SGSN or 3G SGSN, GGSN or 3G GGSN, GMSC, SMSC (SMS), databases (HLR, VLR, EIR), and Charging gateway (CG). 3G MSC/VLR is a new implementation of the Mobile Switching and Visitor Location Register that combines the functionality and services as both the switch and a database. The MSC is used to switch the circuit-switched data, which includes controlling call signalling and coordinating the handover procedures between UTRANs. The 3G MSC has also been provided with both switching and packet capabilities. The VLR keeps a copy of visiting users and service profiles [4].

Hybrid Networks

This network uses fixed stations communicating between themselves via wireless Inter-base station links (WIBLs) to provide network functionality to mobile stations. As Figure 4 demonstrates, the mobile stations have one wireless hop to connect to a base station and this has a multi-wireless hop to connect to another. In this network, the mobile station can freely roam from cell to cell, without the need to belong to a permanent provider. Therefore, many SPs can be involved while providing a service [10]. In this network architecture, the cost is reduced because users can act as wireless network operators as well. Some users of the current Internet may install base stations for their needs. Those base stations can be interconnected to form a wide area network [11].

Cell-hopping architecture is a wireless multi-hop packet switching cellular network that works in the unlicensed spectrum. In this network the base stations form a coreless infrastructure (no core network), which is formed for wireless access

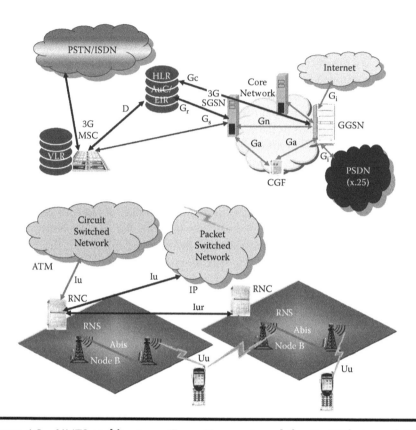

Figure 1.3   UMTS architecture. (From Hassan, J. and Jha, S., 28th Annual IEEE International Conference on Local Computer Networks (LCN'03), 2003. With permission.)

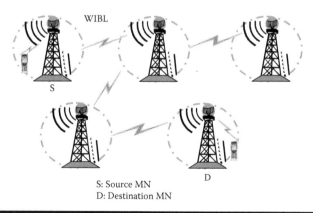

Figure 1.4   Cell-hopping architecture. (From Hassan, J. and Jha, S., 28th Annual IEEE International Conference on Local Computer Networks (LCN'03), 2003. With permission.)

points that do not perform switching functions. The network is formed by the BSs, which are responsible for routing and switching the packets by using WIBLs. There is no necessity for a home network in a cell-hopping architecture, because there is no core network [11].

In this network architecture, the access points are not mobiles. The base stations can thus easily save the routing information. There is no core backbone network. Therefore, the BSs have to provide routing and switching functionalities and act as local databases in order to save the information of the mobile stations under their coverage [10].

When a packet needs to be sent, it is sent to its host base station for forwarding. The base stations execute hop-by-hop routing of the packet, while all devices register themselves with the nearest BS [11].

Figure 5 illustrates the differences of cell-hopping architecture from other current cellular networks' infrastructure, such as GSM. This network does not have

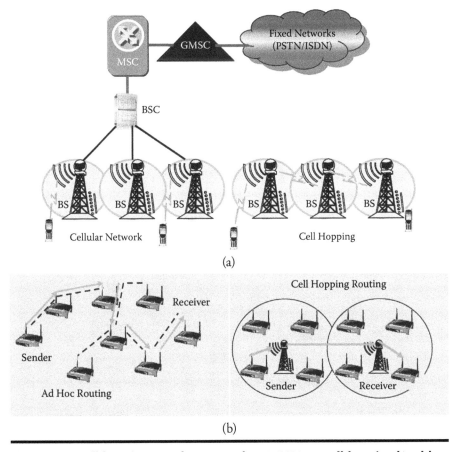

Figure 1.5 Cell hopping vs. other networks; a) GSM vs. cell hopping b) ad-hoc network vs. cell hopping. (From Hassan, J. and Jha, S., *Wireless Communications,* 16–21, October 2003. With permission.)

hierarchical entities as the MSCs and the BSCs. The BSs communicate and cooperate with each other to forward packets hop-by-hop. The figure also shows the differences from ad-hoc networks, which do not have cells managed by a BS [11].

Unified Cellular and Ad-hoc Network Architecture (UCAN)

The UCAN is proposed for enhancing cell throughput of networks as 3G, while maintaining fairness. In UCAN, a mobile client has both a 3G cellular link and IEEE 802.11 based peer-to-peer links. The 3G base station forwards packets from destination clients with poor channel quality to proxy clients with better channel quality. Then, the proxy clients use an ad-hoc network formed from other mobile clients and IEEE 802.11 wireless links to forward the packets to the destination; this improves the throughput [5].

High Data Rate (HDR) and IEE802.11 technologies were chosen because of their support for high data rate services and their popularity. This architecture is based on the idea of using the advantages of IEEE 802.11 interfaces to improve 3G wide area cell throughput. This design was based on three main observations. First, to provide the preferred anywhere always-on connection to provide Internet applications; it is necessary to manage the infrastructure of a wide-area cellular network. Second, given the popularity of 802.11b, this enables devices in 802.11b ad hoc mode to offer peer-to-peer communication, which is cost-effective. Finally, the IEEE 802.11b standard permits high bandwidth wireless communication up to 11 Mbps [5]. Figure 6 presents the network architecture. Some of those mobile devices associated with the HDR base station may have their interfaces in dormant mode, while other devices may be actively receiving data packets from the Internet via HDR downlink. Clients monitor the pilot bursts of the HDR downlink to estimate their channel conditions. At the same time, the devices turn on their IEEE 802.11b interfaces in ad-hoc mode, and run UCAN protocols. In the case of low

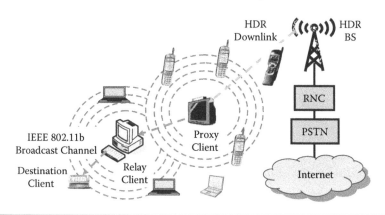

Figure 1.6   UCAN architecture. (From Luo, H., Ramjee, R., Sinha, P., Li, E. and Lu, S., *MobiCom' 03*, September 2003. With permission.)

HDR channel rate, the client stops transmitting directly and starts to transmit the frames to a proxy client with better channel rate. Then, the frames are passed on through IP tunneling, using the bandwidth IEEE 802.11 links [5].

Mobile Ad-hoc Network Architecture (MANET)

A MANET is a set of wireless nodes that can communicate with each other and move at the same time, without the need of support from any centralised administrator or wired infrastructure managing the communication between the nodes [21]. Therefore, MANETs are infrastructureless self-organised wireless networks. Each node can have routing capabilities while moving, to be able to forward the packets without the need to send the packet to a base station. It requires an auto-configuration protocol to permit dynamic assignment of nodes' network addresses [13]. It is a flexible network and conveniently deployed in almost any environment. In the network, hosts may connect in a multi-hop manner. This kind of network could present an enormous advantage in many cases such as battlefields, disaster areas, and outdoor congregations [12].

There are many ad-hoc network architecture configurations in current literature. These combine ad-hoc network architecture with other technologies' architectures in order to provide mobility. These architectures are presented in the following sections.

Multi-hop Mobile Ad-hoc Networks

This network forms part of the MANET mobile ad-hoc network family, which has become quite popular recently. These networks are formed with a set of mobile hosts which communicate between themselves indirectly through a sequence of wireless links without the need to pass through a base station. Here each mobile host serves as a router. Figure 7 represents an example of a typical MANET network scenario [15].

A Two-tier Heterogeneous Mobile Ad-hoc Network Architecture

This architecture presents a heterogeneous MANET network, which can support Internet access. It is a two-tier heterogeneous Mobile Ad-hoc Network Architecture, in which the lower tier of the network consists of a set of mobile hosts; each host is equipped with an IEEE 802.11 wireless LAN card. This card allows connection to the Internet and to handle the network partitioning problem. Then, the high tier is formed for a subset of mobile hosts and gateways, which can access the cellular network. The high tier is heterogeneous because it can support various network interfaces in the gateway hosts, such as IEEE 802.11 cards, PHS handsets, or GPRS handsets. Each interface supports different bandwidths and latency. Additionally, by adopting the cellular internetworking interfaces on the higher tier, the mobility of the network will not be reduced, and it will also improve connectivity to the Internet [12].

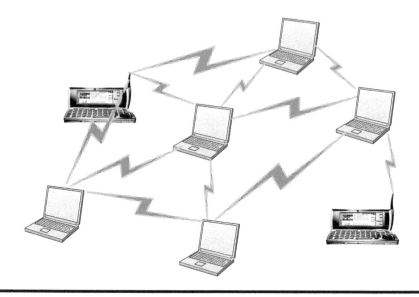

Figure 1.7   Example of mobile hoc network (MANET) [15].

The key issue in the proposed network architecture is how to use the bandwidths provided by the higher tier gateways efficiently. First, the lower tier presents a higher bandwidth than the higher tier, which is conformed by the cellular network. Second, the traffic concentration effect can saturate the former because the higher tier interface will have to serve multiple lower tier hosts. Therefore, here a load balance routing from hosts is necessary to share the gateways [12].

This network considers a set S of mobile hosts, which are equipped with a wireless 802.11 card. These hosts communicate to each other in a multi-hop manner working in ad hoc mode; they form the lower tier of the two-tier MANET. As can be seen from Figure 8, the higher tier can be a mixture of PHS handsets, GPRS handsets or IEEE 802.11 interfaces. Therefore, the higher tier is heterogenous in that sense. A mobile host without a higher tier link can access the Internet through one or more low-tier links leading to a gateway host. Therefore, the Internet services are extended through ad-hoc links [12].

Multi-layered Wireless Network Using Ad-hoc Network

This network uses a mobile radio communication system with a multilayered wireless network that uses ad-hoc networks. The major purposes of this proposed network are to obtain anti-shadow fading characteristics and achieve high user capacity [7].

Figure 9 shows the architecture of a three layered wireless network. Here, a 'wireless ad-hoc network is combined with the conventional wireless communication network such as a cellular network' [7]. The user terminals form a wireless

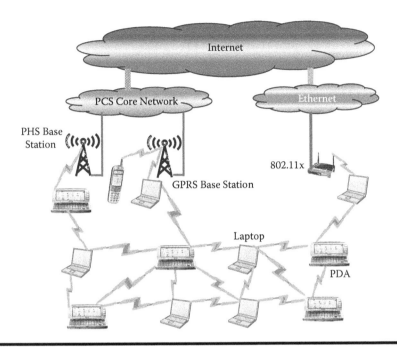

Figure 1.8 Example of the two-tier heterogeneous mobile ad-hoc network architecture [12].

ad-hoc network from the lower layer, L1. Each terminal has coverage of at least a few tens of metres. Here, each ad-hoc wireless network consists of one master terminal (MT) and several slave terminals (STs). A master terminal can transfer information to another master terminal in different wireless networks or to a slave terminal in the same ad-hoc wireless network. The slave terminals can access another slave terminal in the same wireless network under the master terminal control. As Figure 9 shows, each master terminal manages devices in a small coverage area; this is called a pico-cell network. This network layer is then managed for another layer, L2, formed for base stations (BSs) where each BS covers an area that group N out of M (M = 1, 2, 3, …) master terminals in L1. Each BS's coverage is several hundred metres (micro-cell). Then, the L3 network covers several kilometres (macro-cell). It is used when, due to shadow fading or other factors, some MSs cannot access the BS. The BS can then access the base stations in the upper layer L3 [6].

Multi-tier Mobile Ad-hoc Network

The ad-hoc networking (MANET) has received considerable attention recently; most of this attention has focused on single-tier and homogeneous networks (for example, the

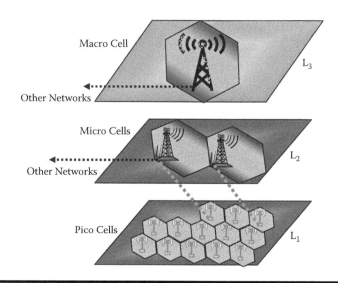

Figure 1.9   Multi-layered wireless communication network architecture [7].

same radio for every node). However, an investigation has been conducted on heterogeneous and multi-tier MANETs. Multi-tier MANETs present a 'coverage asymmetric' with much more area coverage by airborne nodes compared to ground nodes [21].

Figure 10 shows an overall Airborne Communication Node (ACN) network architecture. The figure presents a multi-dimensional network architecture, which has various layers to communicate and to cover different areas of the terrestrial surface. This network will use an undedicated high Altitude Aeronautical Platform (Undedicated-HAAP), which uses undedicated aircraft to support mobile coverage. The aircraft acts as satellite mobile coverage to provide city-wide high-speed coverage. This scheme uses licensed spectrum in the range 47.2–48.2 GHz. The different layers are determined according to distances from the terrestrial surface. These layers are: space layer, airborne layer and terrestrial layer [11]. However, not much research has been conducted on this kind of architecture for a commercial use; it has been conducted mostly for military uses. Different entities have presented ideas for finding an architecture solution to this network. The principal idea is explained as follows.

The Airborne Communications Node (ACN) program will enable an autonomous communication infrastructure that provides guaranteed communications, situational awareness and signals intelligence (SIGINT). It is a Mobile Backbone Network (MBN), which will basically present a hierarchical network in which the nodes will connect to each other at different earth levels; this means that a ground vehicle can create a connection with an aircraft node. In other words, ACN payloads can be integrated on platforms ranging from High Altitude Endurance (HAE) unmanned airborne vehicles (UAV) (for example, Global Hawk) to ground vehicles.

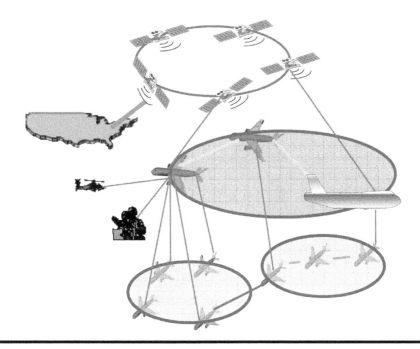

Figure 1.10 Multi-tier network architecture. (From U.S Air Force, "Airborne Networking," The MITRE Corporation, 2004. With permission.)

The ACN on an HAE platform will provide wide-area wireless communications and SIGINT services throughout the world. It will establish an early robust airborne infrastructure for intra-theatre line-of-site (LOS) and reach back beyond line-of-site (BLOS) situations. In this network architecture, the backbone nodes (BN) will have the ability of forming a multilevel backbone node network with long-range coverage radios. Figure 10 illustrates a multilevel hierarchy where one level supports ground communications; the next levels support UAV communication and satellites communication in order to provide higher-level and broader reach connections. The figure also shows that all layers are interconnected between themselves, in order to forward communication between the different layers [24,25].

X-hoc Network.

An ad-hoc network is composed of a set of nodes that interconnect to support some common objectives. The principal idea of an X-hoc network model is to consider a set of heterogeneous ad-hoc networks, in which some nodes need to communicate with other nodes located outside the boundaries of the ad-hoc network in which they are located. This architecture proposes a unique identifier for a set of ad-hoc networks ANi3 [21].

Figure 11 depicts an X-hoc network structured by five ad-hoc networks, AN1, AN2, AN3 and AN4 and an ad-hoc network that has a group of gateways that permit inter ad-hoc communications. Here the set of ad-hoc networks is dynamic;

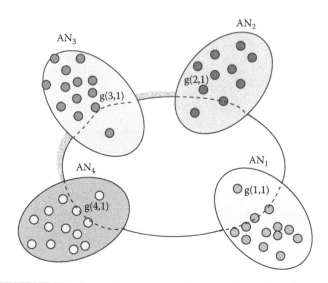

Figure 1.11   An X-hoc network. (From Ryu, B., Andersen, T., Elbatt, T. and Zhang, Y., IEEE, 2003, 1280–1285. With permission.)

there is no control for an ad-hoc network to appear or disappear. However, each node belongs to at least one ad-hoc network at any time [22].

Metropolitan Area Multi-tier Wireless Ad-hoc Network Architecture

The proposed system, named Ad-hoc City is a multi-tier wireless ad-hoc network architecture for general-purpose wide-area communications. The backbone network is itself also a mobile multi-hop network, composed of wireless devices mounted on mobile fleets. In an ad-hoc network, the mobile nodes self-organise to form a network without the need of infrastructure as a base station or access point. Each mobile node acts as a router forwarding packets from one node to another. It is a scalable multi-tier ad-hoc network architecture consisting of a small number of wired stations and a mobile multi-hop wireless backbone, which serves mobile users over a metropolitan area. The backbone nodes are implemented on mobile fleets, which cover the city area in space and time [6].

In this architecture, the buses represent the mobile nodes forming the wireless mobile backbone of the Ad-hoc City architecture. This is proposing a hybrid system within a single base station cell. The nodes may roam between different cells on a regular basis, and they do not have a home cell. The backbone in this architecture is composed of wireless devices mounted on mobile fleets, such as buses or delivery cars within a city covering the area in time and space. It can be organised in a multi-hop wireless ad-hoc network backbone, which can provide network access and general communication services throughout the city. In this architecture, several fixed base stations, with which nodes can communicate over multi-hop paths, are located throughout the metropolitan area. The individual mobile users can communicate over the network using different devices to connect to the Internet or communicate between them. Figure 12 depicts an example of these networks [6].

Figure 12 shows two types of mobile nodes in the architecture: personal mobile nodes such as laptops and PDAs; and network mobile nodes, which form the mobile backbone of the networking infrastructure. Network mobile nodes route packets between personal mobile nodes within the ad-hoc network; and between personal mobile nodes and the Internet, through a small set of fixed base stations [7].

In this architecture, the base station serves the mobile nodes within a topological cell. The cell coverage is determined by the number of wireless hops from the base station. Here the packets are routed along a possible multi-hop route between network mobile nodes to the nearest base station to the node or routes into the Internet. If the source and destination are both in the same ad-hoc network the packets will not need to be forwarded to a base station [6].

Wireless Moving Networks Architectures

This network design is an experimental ad-hoc network to be arranged on moving platforms. In this network, the terminals operate as moving wireless LANs carried by moving platforms. However, each mobile wireless LAN communicates with a global fixed network wirelessly through an anchor or master (M) node. Here the

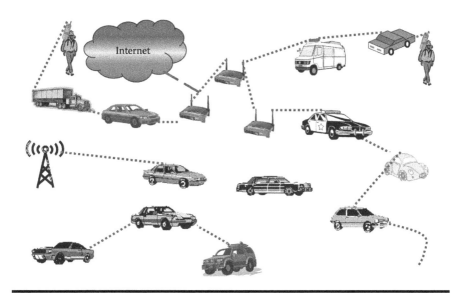

Figure 1.12   An Ad-hoc City architecture [6].

slave nodes and their master nodes retain a wireless LAN structure in a mobile scenario such as buses, trains, ships and airplanes [17].

The base station on the moving platform manages the local communication of the terminals on the platform and the external communications with remote base stations. This network is a hybrid of ad-hoc networks and WLANs. The terminals on the moving platform do not need to communicate with each other directly; they can use the supervision of the base station to which they all have access [17].

In the proposed network solution there can be two models: one static and the other dynamic. The static model is formed from a mobile WLAN (m-WLAN), which has 'n' static nodes which are connected to a supervisory mobile node (anchor node), as can be seen from Figure 15. The supervisory node is moving in the network and terminals are connected to it. These terminals can be considered with zero relative velocity respective to that node. Therefore, the overall resulting network is a connection of multiple 'N' supervisory mobile nodes where the 'static' nodes in each m-WLAN communicate through each corresponding supervisory node (anchor nodes). The static nodes (slaves) are randomly distributed in their resident platform [17].

In this version of dynamic mobile network (DMN), the mobility of the nodes is generally considered. As can be seen in Figure 15, if the slave nodes and the M node move with a different velocity vj, then the m-WLAN moves at average 'v' where v is: [17].

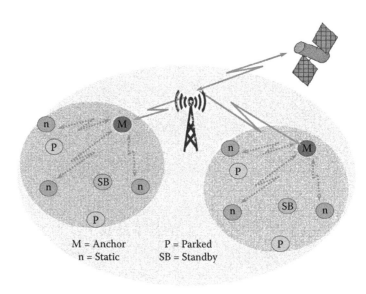

Figure 1.13 Moving WLAN linked to a cellular network. (From Agbinya, J. I., *Design Concepts: Wireless Moving Networks (WMN),* **University of Technology, Sydney, Australia, 2003, 1–6. With permission.)**

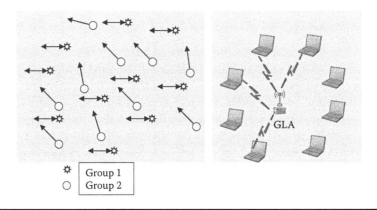

Figure 1.14   Overlap mobility model (a) and each group model (b).

Wireless Ad-hoc Network with Group Mobility

This network does not have a fixed backbone and the mobile stations move randomly. An ad-hoc network is a dynamically re-configurable wireless network without a fixed infrastructure. Ad-hoc networks have no fixed base station. The ad-hoc networks maintain a dynamic interconnection topology between mobile users. In

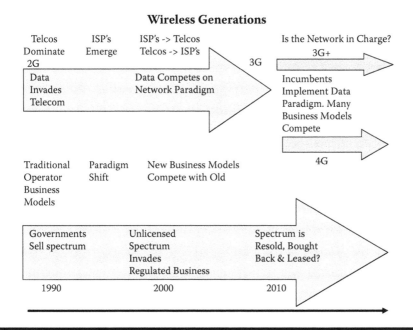

Figure 1.15   Wireless generation and tendency. (From Charas, P., "Peer-to-Peer Mobile Network Architecture,"IEEE, 2002, 55–61. With permission.)

ad-hoc networks the communication is point-to-point instead of among groups, based on individuals [19].

This architecture is based on the broadband ad hoc ATM local area network (BAHAMA) mobility model and uses IP protocol. The BAHAMA and Monarch project is a group mobility model researched by DARPA. BAHAMA uses a multi-channel scheme and consists of portable base stations and mobile nodes in a wireless ad-hoc network. The group mobility model is based on the overlap mobility model. Here each mobile base station is the leader of each group (GLA) and has random mobility. Several mobile nodes are included in each mobile base station, which combines switching capabilities with wireless access interfaces. The BS also communicates with other mobile nodes or mobile base stations via a wireless link. The mobile nodes with each group move with GLA [19].

As can be observed from Figure 16, the group mobility model in this architecture is based on the overlap mobility model. Different groups carry different tasks over the same area; the distinct requirements of each task cause the differences between mobility patterns. Each group in the network consists of one mobile base station and mobile nodes [19].

From Figure 16, the neighbouring coverage areas of a GLA overlap with each other, ensuring connection when a mobile node and GLA move from one coverage area to another. Each GLA has a set of channels allocated to it. A handoff occurs in this network when a GLA with ongoing calls of several nodes moves from one coverage area to another [19].

Advantages and Disadvantages of Previous Network Architecture Models

The rest of this section will present an analysis of the advantages and disadvantages of all the previously presented architectures. It examines the positive and negative aspects of the second-, second-and-a-half and third-generation networks: GSM, GPRS and 3G. GSM is based on ISDN signalling. This and other network factors limit the data communication speed in the second-generation mobile networks. Also, there is a limitation to provide IP packet services at high speed. 3G mobile networks cover the weak points of 2G and 2.5G. They consist of multi-vendor networks and advanced standardisation of the interface between the functional elements of networks. 3G also has mobile network system specifications based on the GSM and GPRS communications. 2G and 2.5G networks, such as GSM and GPRS, cannot provide ultra-wideband services. These architectures were not designed to carry traffic for services that require high-band capabilities, such as multimedia [8]. Although 3G improves these aspects, it presents other problems such as poor channel quality and low data-rate connection with the base station. To maintain fairness, 3G's services are provided with low data-rate to the users, but it carries, as a consequence, a reduction of the cell's aggregate throughput [5].

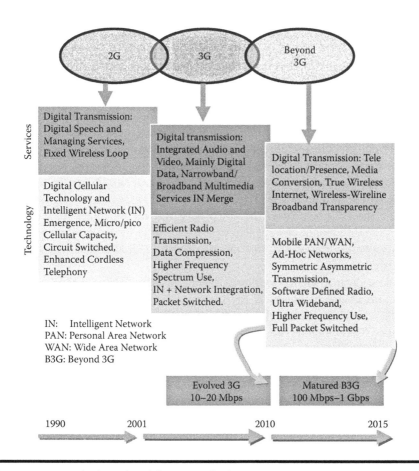

Figure 1.16   Evolution of mobile networks from 2G to B3G [42].

In the 1990s competition began between the classical Telcos and ISPs. 3G became a 3G IP project under pressure from the rampant success of the IP paradigm, while the arrival of the IEEE 802.11 business model increased the pressure on 3G business [18]. 3G base station forwards packets from destination clients with poor channel quality to proxy clients with better channel quality. Then, proxy clients use ad-hoc network composed of other mobile clients and IEEE 802.11 wireless links to forward packets to the appropriate destination. In this form the throughput is improved [5].

Mobile cellular networks contribute with anywhere everywhere access networks. However, cellular networks have some limitations, such as complex construction requirements, use of licensed frequency spectrum, and single ownership of the network. These factors result in very expensive networks [11].

Cell-hopping requires less infrastructure than traditional cellular networks, yet provides superior routing scalability compared to emerging infrastructureless ad-hoc networks. Cell-hopping architecture also provides a low-cost network, which allows users to play as wireless network operators by installing a low-cost base station suiting their needs. Then, interconnections between them will form the backbone network. In this form, the network permits the users to conduct their own business without worrying about switching or routing, unlike MANETs. Cell-hopping architecture also offers new business opportunities and enables low-cost basic services for common users. The small operators can receive a part of the revenue from the wireless provisioning business. However, much research needs to address cell–hopping, such as location management. Moreover, because there are many operators in the network without any central administration authority, charging will be a complex issue [11].

While the UCAN reduces the complexity and the reliability of 3G, it also increases the throughput of wide area wireless networks by using ad-hoc local area wireless networks. It uses cards that integrates both IEEE 802.11b and 3G wireless interfaces [5].

While wide-area wireless networks operate in infrastructure mode with fixed base stations serving mobile users, local area wireless networks can work in ad-hoc mode. Here, mobile clients relay packets for each other over multi-hop wireless links. The UCAN architecture, while increasing throughput, reduce overhead on the 3G uplink. However, the energy consumption of the protocol used is higher than for on-demand protocol [5].

The multi-layered ad-hoc wireless network's purposes are to obtain anti-shadow fading characteristics and to disperse the traffic in a service area [7].

The main advantage of ad-hoc wireless networks is the ability to make easy connections with small transmission power. Narrow coverage wireless communications can provide high user capacity by distributing the traffic in the service area. In other words, it is used to obtain anti-shadow fading and high user capacity [7].

The mobile ad-hoc network (MANET) can be flexibly and conveniently deployed in almost any environment without the need for fixed stations' network infrastructure base. MANET network architecture can have applications in situations like battlefield, disaster areas, festival ground, outdoor assemblies and rescue actions, where networks need to be deployed immediately, but the base station or fixed infrastructure is not available. For example, in an earthquake disaster all the stations may be down if there is no electricity. In this scenario, a MANET driven by battery power can be used to set up a network environment [12,15].

MANETs address some of the cellular mobile network limitations. However, the absence of any dedicated fixed switching/forwarding agent, and the uncertain and frequent mobility of a node are aspects which make ad-hoc network unstable. Even if the end node is not moving, it can also remain unstable because of the intermediate node movement [11].

MANET multi-hop allows the mobile terminals within a cell to take part in the routing process. Some authors think that throughput can increase in this way.

However, this scheme still depends on infrastructure-based cellular networks, and involves user devices in packet forwarding [11].

A MANET is usually assumed to be homogeneous, whereby each mobile node shares the same radio capacity. However, a homogeneous ad-hoc network suffers from poor scalability. Resent research has demonstrated its performance bottleneck. Poor scalability is due to the fact that in ad-hoc networks, most bandwidth of a node is consumed by forwarding packets. This is further aggravated by heavy routing overhead when the network size is large. Then, to achieve scalability a physically hierarchical ad-hoc network is a promising alternative [25].

A heterogeneous MANET, by adopting cellular networking interfaces on the high tier, would not reduce the mobility of the MANET. It will improve the MANET's connectivity to Internet. The cellular networks have much wider coverage and longer transmission distances, but much lower communication bandwidths, compared to IEEE an 802.11-based network. Therefore, it will be advantageous to combine these strengths in a heterogeneous network. This is the case in two-tier heterogeneous MANETs, in which the usefulness of ad-hoc networks is improved in both conquering the network portioning problem and improving its Internet access capability [12].

Here it is important to note that hierarchical large-scale ad-hoc networks can use different types of multi-tier; one uses a multi-tier architecture that combines various base stations at different levels from the ground (for example, cars communicating with aircraft). The other architecture uses different types of radio capabilities at different layers. In such a structure, nodes are first dynamically grouped into multi-hop clusters. These also can be combined in one network.

When a multi-tier network is built at different levels from the ground, many advantages can be listed, as the communication can be accessed from any place on the earth, and the network can also be used in case of a catastrophe or in a battlefield. However, there are many issues to examine, such as routing, management of the network, technology, requirement of the devices and methods of node localisation. Also, if the network is limited to undedicated HAAP the coverage is restricted to the area near to the path. In addition, the use of licensed bandwidth and aeroplanes as base stations is going to result in an expensive network that will prohibit small businesses from participating in it [11].

The X-hoc network purposed architecture could present a flexible and adaptable communication service which is capable of exploiting all possible resources to which a node is connected. The solution to this architecture encompasses high-level resource-sharing, user references, combinations of such resources as well as internetworking issues. Additionally, given both the current and expected widespread coverage of wireless technologies and ad-hoc networks, it is fairly easy to devise cases in which two nodes want to communicate. Furthermore, an increasing number of devices are released equipped with distinct wireless technologies that can be mapped into gateway nodes according to the X-hoc models. This network model's purpose is to explicitly deal with the coexistence of ad-hoc networks built by distinct devices supporting distinct technologies and aiming to exchange messages. However, X-hoc represents a really difficult task. Moreover, X-hoc communications should be further refined in terms of standard means to allow heterogeneous gateways to communicate among them. This is mainly a technological issue [22].

Metropolitan-area multi-tier wireless ad-hoc network architecture presents some advantages, as vehicles are used as the backbone of the network. They cover the area of a city in both space and time; these vehicles are buses or delivery vehicles within the city. In this manner, by using the fleets as a network backbone, the network infrastructure can be simplified. Additionally, the effect occasioned for any physical failure of any backbone node is lessened, as the vehicles in the area will cover any gaps caused by the failure. Results from simulation show that the ad-hoc city architecture is viable, and with some optimisation, this network could provide a good performance in a real deployment [6].

A network infrastructure consisting of infostations operating in the unlicensed spectrum bands have been proposed. Infostations do not provide continuous coverage, but provide hot spots of high-speed coverage; the users have to use the service of the cellular networks from the major providers [11].

Table 1 presents a comparison of some of the network approaches that have been related in this document.

A mobile IP is a network based on the traditional Internet. Therefore, it is compatible with the fixed Internet network. It allows mobility in the Internet environment. It can also provide all Internet actual services by using the same protocols. Additionally, mobile IP supports the mobility of terminals in a simple and scalable manner. It is well suited for macro mobility but not for micro mobility. It treats macro mobility and micro mobility in the same way. Therefore, the total signalling costs in mobile IP are enormous if the distance between home and the foreign agent is large [16]. Another disadvantage is that mobile IP does not provide sufficient mobility management compared with cellular networks. Mobile IP also requires a lot of software work; this makes its work slow. Mobile IP is also not very efficient. It increases congestion on the fixed Internet. This is because each mobile IP terminal's home and foreign agents need to constantly communicate to exchange information.

Combinatorial mobile IP improves some of the mobile IP disadvantages; it separates not only micro mobility from macro mobility, but also active mode from idle mode. Therefore, it reduces the total signalling cost efficiently [16].

Wireless moving networks present many applications. There is the case of a railway that carries many passengers per day. Therefore this huge customer base could use a moving network deployed on trains. The same could be applied in the case of buses [17]. However, more research work is necessary to solve all the requirements of providing broadband services on moving nodes. One of these is

Table 1.5 Comparison of Various Approaches

Technology	Frequency	Base Station	Switching Center	Hop	Self-Organizing	Scale
Ricochet	Unlicensed	Access points	Yes	Multi	No	Medium
Multihop cellular	Licensed	Required, fixed	Yes	Single/multi	No	Large
Infostations	Unlicensed	Required, fixed	No	Single	No	Small
CarNet	Unlicensed	No	No	Multi	Yes	Large
Terminode	Unlicensed	No	No	Multi	Yes	Large
U-HAAP	Licensed	Has, mobile	No	Single	No	Small

Source: Hassan, J. and Jha, S., *Wireless Communications*, October 2003, 16–21. With permission.

to solve the problem of sending data while the moving node is travelling at high speed.

A wireless ad-hoc network with group mobility improves wireless ad-hoc networks because it does not have a fixed backbone network and the base stations move randomly. The conventional channel allocation cannot efficiently predict mobility and it is not possible to support a burst of handoff traffic due to group mobility. A wireless ad-hoc network with a group mobility channel allocation scheme solves the problem. It efficiently supports a burst handoff using a guard channel and hello message. However, this proposed network brings an overhead of replicating services, which is inevitable when we demand continuous streaming services in a network like this, as previously described in this thesis [19].

Overview of Network Technology

Technologies such as Digital Enhanced Cordless Telephones (DECT), Ultra Wideband (UWB) and different versions of IEEE standards for wireless environments (802.11, 802.11a and 802.11g), are examined. However, this thesis focuses on those technologies that can be used in a mobile broadband environment (3G, B3G, 802.11x, 802.16e and 802.20).

Examining different generations in the mobile environment, Wide Area Network (WAN) technologies used on networks such as General Packet Ratio Service (GPRS) offer an average throughput of 19Kbps. Then, speeds of 100–130 Kbps were achieved through technologies such as Enhanced Data Rates for Global Evolution (EDGE). Third generation technologies such as Code Division Multiple Access (CDMA) can support speeds of 300–400 Kbps [40].

This chapter examines the most common wireless technologies and will discuss the possibility of using those technologies on a mobile wireless broadband environment in which the nodes are moving at the velocities of cars and trains. Different wireless technologies such as Digital Enhanced Cordless Telecommunications (DECT), Ultra wideband (UWB), 3G and different versions of 802 such as Wi-Fi (802.11) and WiMax (802.16), Mobile-Fi or MBWA (802.20) are considered. This chapter studies the pros and cons of each technology when it is used in a mobile wireless broadband environment.

Digital Enhanced Cordless Telecommunications (DECT)

DECT is a flexible digital radio access standard for cordless communications. It can be used to design and build a whole range of digital low-power wireless communication systems, from the residential cordless telephone to the corporate multimedia digital wireless office with interconnections to digital cellular systems. The U.S. equivalent is Personal Wireless Telecommunications (PWT). DECT and PWT are described in two standards (two series of standards) published by the European

Telecommunications Standards Institute (ETSI) and the U.S. Telecommunications Industry Association (TIA), respectively [29].

DECT is a low-power two-way digital wireless communications system. Low RF power means that the normal distance of any one handset from its associated base station may range up to about 200 metres. In some applications, it is extended by using directional antennae [30]

DECT provides voice, fax, data and multimedia traffic, wireless local area networks and wireless PBX. It plays an important role in communications developments such as Internet access and inter-working with other fixed and wireless services such as ISDN and GSM. It uses TDMA (Time division Multiple Access), with its low radio interference characteristics, and provides high system capacity [30].

DECT plays an important role in worldwide standards for short-range cordless mobility. It is easy to adapt for different frequency allocations [31]. Its key characteristics are:

Table 1.6   DECT Features

Characteristic	DECT
Frequency	1880–1900 MHz
Capacity	High traffic capacity up to 10.000 Erlang per km² per floor
Data Service	From 24 kbits/s up to 552 kbit/s (2 Mbits/s in preparation)
Coverage	Pico cellular coverage <300-m cells up to 3–5 km

After studying the DECT technology and examining its features, it can be said that although DECT supports wireless applications, this technology cannot be used to support mobile wireless broadband network applications. First, it does not support mobility. Second, its range of coverage is quite limited for covering large areas, such as a whole city, and it will require a lot of handover mechanisms to cover big areas while controlling the movement of vehicles. Finally, its data rate is not sufficient to provide all broadband services. For instance, VoIP requires at least 64 Kbps up and down-load.

Ultra-wideband (UWB)

UWB technology offers a solution for the bandwidth, cost, power consumption, and physical size requirements of next-generation devices. UWB complements existing longer radio mobile wireless technologies by providing high-speed interconnections across multiple devices and PCs within the digital home and office environments. For instance, this emerging technology provides high bandwidth to create high-speed WPANs that can connect devices throughout home and office. UWB is emerging as a solution for the IEEE 802.15.3a (TG3a) standard. The data rate is greater than 110 Mb/s to satisfy multimedia industry needs for wireless personal area network (WPAN) communications [35–37].

UWB (Ultra-Wideband) has been allocated 7.500 MHz of the spectrum in the 3.1 to 10.6 GHz frequency band by the Federal Communication Commission (FCC). The FCC defines UWB as any signal that occupies more than 500 MHz bandwidth in the 3.1 to 10.6 GHz band. The main limitations are provided by the low-power spectral density of 41 dBm/MHz (dramatic channel capacity at short range that limits interference and the fact that the transmit signal must occupy at least 500 MHz at all times) [35,36].

UWB systems use modulations techniques, such as Orthogonal Frequency Division Multiplexing (OFDM), to occupy these extremely wide bandwidths. IEEE 802.15.3a is being developed for high bit rate PAN applications, and UWB is the most promising technology to support the stringent requirements of 110, 200 and 480 Mb/s. The range for 110 Mb/s is 30 feet and for higher bit rates, it is reduced. The table below summarises UWB features [36].

For low data-rate systems, under 2 Mb/s the range can be extended to 100 to 300 metres. However, this is still a very small range for mobile broadband applications which want to be supported in the network architecture presented in Chapter 2. UWB is a short-range technology, therefore is ideal for WPANs. UWB complements existing longer radio technologies such as Wi-Fi, WiMAX, and cellular communications. UWB is suitable for providing the needed low cost, low power consumption, high bandwidth interface solution for transmitting data from host devices to devices in a limited area up to 10 or 30 metres [35], [37] and [39].

After studying the UWB technology and examining its features, it can be seen that although UWB supports wireless applications and high data rate, this technology could not be used to support mobile wireless broadband network applications in a large range environment. In order to do that, it will be necessary to combine the technology with other technology to provide longer range coverage. Additionally, it does not support mobility to allow nodes to move at high speed; as mentioned before, UWB is ideal for use in an office or home wireless broadband environment.

Table 1.7   UWB Features

Characteristic	UWB
Frequency of operation	3.1 to 10.6 GHz
Bandwidth	7.500 MHz
Bit rate	110 and 200 Mb/s
Range	30 and 12 ft
Power consumption	100 and 250 mW

Source: Intel in Communications, 2004, 1–8. With permission.

3G Technology

The third generation (3G) transformed wireless communications into on-line, real-time connectivity. The trend of introducing higher data rates over mobile wireless networks ended with the introduction of third-generation 3G systems by the International Telecommunication Union (ITU). ITU started the IMT-2000 project, the technical framework for 3G, whose main goal was to establish a global standard for the next generation of mobile communications. It supports 144 Kbps for voice and data services at high-velocity mobility, 384 Kbps for low-speed mobility and 2 Mbps for fixed location terminals. The frequency bands allocation are 1885–2025 MHz and were set aside for the IMT-2000 project.

This standard's aim is to harmonise worldwide 3G systems and it has grouped five standards together. These are:

- W-CDMA
- CDMA2000
- TD-CDMA/TD-SCDMA
- DECT
- UWC-136.

The table below shows the main features of 3G technology:

Table 1.8   3G Features [34]

Characteristic	3G
Frequency of operation	Below 2.7 GHz
Bandwidth	Each channel <5 MHz
Bit rate	144 kbps (120 km/h) up to 2 Mbps (10 km/h)
Mobility support	Up to 300–500 km/h

It is important to note that 3G networks are demonstrating an evolution to the next generation of technologies. This means that the current 3G technology will be improved. In general, the mobile cellular networks have presented an evolution, as the figure below shows. The figure demonstrates that 3G networks are having an internal evolution, and this is called Beyond 3G (B3G). This technology's evolution will provide more coverage in high populated areas to carry more traffic and provide more and better services to the user [42].

To improve the data rate and increase spectral efficiency, two main approaches are being used. The first one is High Speed Downlink Packet Access (HSDA), which is based on a single carrier modulation. It aims to provide rates of up to 10.8 Mbps per carrier. The second option is to use Orthogonal Frequency Division Multiplexing, which is a multiple carrier modulation technique. It aims to provide rates of more than 20 Mbps. However, this frequency modulation already exists on other standards as 802.11, 802.16 and 802.20, whose adaptation to cellular environment is possible in the medium term [43].

After studying 3G and examining the B3G technology, it can be seen that 3G or the evolution of 3G can be used as technology support on the proposed network architecture. 3G provides mobility and a high data rate. However, its data rate is reduced when the velocity is increased, and it will be a problem in the proposed network architecture because the speed on trains is high and the technology will decrease its performance at these velocities.

The beyond 3G can improve the actual 3G limitations and it could be an excellent technology solution to use on the proposed network architecture. However, this technology has not been used yet and more studies are necessary before the technology is ready to be used. But the network architecture solution proposed in Chapter 2 will not be possible if there is not a technology that can support its requirements, such as velocity, coverage and bit rate. Therefore, it is necessary to study the technologies that are currently being improved to see where improvements are occurring, and to determine whether these future improvements will satisfy the technical requirements for the proposed network architecture.

802.11, 802.16 and 802.20

802.11, 802.16 and 802.20 are IEEE standards, which are applied to wireless interfaces for radio frequency transmissions. These standards have been applied in personal area networks (PAN) with a 10-metres range, local area networks (LAN) with a 100 metre-range, and metropolitan area networks (MAN) with a 1,000–metre- or more range. The figure below shows the range of coverage of 802 standards [47].

802.11 (Wi-Fi)

This technology is an IEEE standard which provides a wireless solution for fixed wireless networks in a Local Area Network (LAN). It provides a best effort service only. Therefore, it is very difficult to provide quality of service (QoS) guarantees for throughput-sensitive

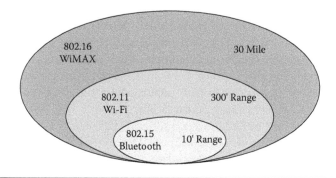

Figure 1.17 IEEE standards ranges. (From Charas, P., "Peer-to-Peer Mobile Network Architecture,"IEEE, 2002, 55–61. With permission.)

and delay-sensitive multimedia applications. It is a challenge to use IEEE 802.11 to meet multimedia services and applications demands over long distances, due to the lack of quality support. Additionally, IEEE 802.11 versions do not provide mobility and their coverage range is limited to short distances. Therefore, it cannot be used as the main technology solution (backbone) for the proposed architecture [44].

The following table presents the main aspects and capabilities of different versions of 802.11 technologies.

Table 1.9   802.11 Standard

Task Group	Responsibility
802.11a—OFDM in 5 GHz Band	Specification enabling up to 54 Mb/s to be achieved in the 5 GHz unlicensed radio band by utilizing OFDM.
802.11b—HR/DSSS in 2.4 GHz Band	Specification enabling up to 22 Mb/s to be achieved in the 2.4 GHz unlicensed radio band by utilizing HR/DSSS.
802.11c—Bridge Operation Procedures	Provides required information to ensure proper bridge operations, which is required when developing access points.
802.11d—Global Harmonization	Covers additional regulatory domains, which is especially important for operation in the 5 GHz bands because the use of these frequencies differs widely from one country to another. As with 802.11c, the 802.11d standard mosty applies to companies developing 802.11 products.
802.11e—MAC Enhancements for QoS	Covers issues of MAC enhancements for quality of service, such as EDCF service differentiation and hybrid coordination function (HCF).
802.11f—Inter Access Point Protocol (IAPP)	Provides interoperability for users roaming from one access point to another of different vendor.
802.11g—OFDM in 2.4 GHz band	Specification enabling high data rates (36 or 54 Mb/s) to be achieved in the 2.4 GHz unlicensed radio band.
802.11h—Dynamic Frequency Selection (DFS)	Dynamic channel selection and transmission power control.
802.11i—Security	Specification for WLAN security to replace the weak Wired Equivalent Privacy (WEP).

Source: Zhu, H., Chlamtac, I., and Prabhakaran, B., *IEEE Wireless Communications,* August 2004, 6–14. With permission.

802.11a applications in hot spots and 802.11b/g present new opportunities for companies. Their limitations include range of coverage and access methods. This technology competes with DECT and UBW technologies for multimedia application over short distances. Consequently, in order to provide a solution for the proposed architecture in Chapter 2, these technologies need to be combined with a technology that provides a solution in a mobile broadband large-area environment (backbone technology).

802.16 (WiMAX) and 802.20 (MBWA)

The IEEE 802.16 technology (Worldwide Interoperability for Microwaves Access, WiMAX) is a wireless standard developed by the working group 16 of the Institute of Electrical and Electronics Engineers (IEEE). It is specialised for point-to-multipoint broadband wireless access. It released its first standard in December 2001 which addresses systems operating between 10 GHz and 66 GHz frequency bands. In January 2003, 802.16a was approved, which addresses systems operating in the spectrum between the 2 GHz and 11 GHz frequency bands. It was limited to line-of-sight (LOS) operations to the base station. Then, IEEE standard 802.16a-2004 addressed only fixed systems. IEEE 802.16e adds mobility components to the standard. It was closely followed by the IEEE 802.20 Mobile Broadband Wireless Access (MBWA) technology, which was approved by IEEE Standards Board on December 11, 2002. Its aim was to prepare a formal specification for a packet-based air interface designed for multimedia IP-based services [46, 47].

The table below summarises 802.16 standards:

Table 1.10   WiMAX Standards

	802.16	*802.16a*	*802.16e*
Date Completed	December 2001	January 2003	2Q 2005
Spectrum	10 to 66 GHz	<11 GHz	<6 GHz
Operation	LOS	Non-LOS	Non-LOS/Mobile
Bit Rate	32–134 Mbps	Up to 75 Mbps	Up to 15 Mbps
Range of Coverage	1 to 3 Miles	3–5 Miles	1–3 Miles

Source: Richards, J., and Rupy, K., *WiMAX: The Next Big Step in Wireless Broadband? A Technical and Regulatory Overview,* Keller and Heckman LLP, November 2004, 1–6. With permission.

Apart from the 802.11 standards, the new WiMax (802.16x) series promises metropolitan wireless access networks with a much greater coverage range. It is truly broadband and can be used to create a wireless backbone. It promises a range of up to 50 km and a bit rate of up to 70 Mbps in addition to QoS. This means that 802.16 can be used to create a fixed wireless access backbone that could support hundreds of fixed subscriber stations. Indeed, a network of hotspots can be created. In future, 802.16e will support mobile wireless transmission directly to end-users in a manner similar to GPRS and 1xRTT (radio transmission technology). WiMAX will rival DSL and overlay the much needed incentive for a truly broadband wireless access with large coverage ranges. 802.20 is following closely and promises to better the performance of 802.16. 802.20 will address wide-area wireless networks specifically including mobility. The IEEE group working on 802.20 technology specifies a physical and medium access control layer of the air interface, operating bands below 3.5 Ghz and allowing the users to transmit data with a peak data rate of over 1 Mbps. It supports various vehicular mobility classes at up to 250 km/h [14, 48] and [45, 46].

The mission of IEEE 802.20 is to provide the specification for a fully mobile broadband solution using an efficient transport of IP-based services. It attempts to enable worldwide deployed multi-vendor mobile broadband wireless access networks. This specification fills the performance gap between the high data-rate low mobility (small vehicular speed at around 60 km/h in urban areas) services currently developed in 802 and the high-mobility cellular networks. The technology specifications are summarised in the table below [43]. At low speeds Doppler can be neglected as it has limited effects on the signal. However at higher speeds, it can significantly modify the transmitted signal on reception.

WiMAX and MBWA do not conflict with WiFi but complement it. The logical link controller is the same for both technologies; they can be both bridge and router to them. WiMAX or MBWA can be used as a backbone for WiFi to connect to the Internet and provide a wireless extension to cable and DSL for the last mile broadband access.

WiMAX is considerably different from WiFi. In WiFi the user need to use an Access Point (AP) to access the medium on a random basis. On the other hand, WiMAX and MBWA are for scheduling MAC, whereby the user requests when it starts the communication, after, the user is allocated a time slot by the base station.

Table 1.11   MBWA Solution Characteristics

Characteristics	Value for 1.25 MHz	Value for 5 MHz
Mobility	Up to 250 km/h	
Peak user data rate (Downlink DL)	>1 Mbps	>4 Mbps
Peak user data rate (Uplink UL)	>300 kbps	>1.2 Mbps
Peak aggregate data rate per cell (DL)	>4 Mbps	>16 Mbps
Peak aggregate data rate per cell (UL)	>800 kbps	>3.2 Mbps
Spectrum (maximum operating frequency)	<3.5 GHz	

Source: Klerer, M., *IEEE-SA Standard Board,* March 2003. With permission.

This time slot can remain assigned to the subscriber and other users cannot use it; they have to request a time slot assignation. Additionally, WiMAX and MBWA are much more time-efficient than 802.11; their 50 km coverage can cover a metropolitan area, creating a wide area metropolitan network (WMAN) and it will allow mobility [46].

Summary

This chapter provided an overview of network architectures. From the traditional fixed network architecture, it expanded the scope to include hierarchical mobile network architectures and then ad-hoc networks (single-tier and multi-tier). It also presented various architectures for moving wireless networks and showed that mobility adds a new dimension to link stability. The chapter concluded by discussing WiMAX in its various forms and the IEEE 802.16 standards.

Practice Set: Review Questions

1. The carrier bandwidth of 3xRTT is:
   a. 3 MHz
   b. 3.75 MHz
   c. 300 kHz
   d. 1.25 MHz
   e. 5 MHz
2. The carrier bandwidth of WCDMA is:
   a. 3 MHz
   b. 3.75 MHz
   c. 0200 kHz
   d. 1.25 MHz
   e. 5 MHz
3. Speech sampled at 8 kHz and quantized at 8 bits per sample has a bit rate of:
   a. 13 kbps
   b. 16 kbps
   c. 64 kbps
   d. 8 kbps
   e. 32 kbps
4. Speech sampled at 6 kHz:
   a. Will have intersymbol interference
   b. Will lose the high-frequency components
   c. Will not represent voice frequencies well
   d. Has disobeyed the Nyquist criterion
   e. All of the above

5. Toll-quality speech means:
   a. Voice quality as used in toll gates
   b. Shows a heavy toll on frequency content
   c. Sounds like the actual voice of the person when sent over a phone line
   d. Has no real value to voice signal representation
   e. All of the above
6. The radio network in a GSM system consists of:
   a. Radio stations operated by many vendors
   b. Base stations that can transmit and receive voice signals
   c. Large filters for set up on base stations
   d. A network of all radio stations in a city
   e. None of the above
7. 1xRTT has the following channel bandwidth:
   a. 1 MHz
   b. 4 MHz
   c. 1.25 MHz
   d. 1.52 MHz
   e. All of the above
8. The channel bandwidth of 4xRTT is equivalent to:
   a. 4 MHz
   b. 8 MHz
   c. 6 MHz
   d. 5 MHz
   e. None of the above

Exercises

1. Speech is sampled at 8 kHz and encoded using a frame size of 128 samples. How many seconds is this frame size?
2. If each sample is quantized and represented with 8 bits, what is the uncompressed bit rate?
3. Is there anything to be gained by representing each of the speech samples with 16 bits instead of 8 bits? Justify your answer.
4. Which sampling rate is likely to represent a female voice better, at 8 kHz or at 16 kHz?
5. The mean opinion score of a speech encoder is given as 4. What does this mean?
6. What is toll-quality speech?
7. List and explain all the GSM interfaces.
8. What is a GSM time slot? How long is a time slot?
9. Distinguish between the GSM and GPRS

10. If the data rate per time slot for GSM is 14.4 kbps, what is the expected theoretical maximum data rate for GPRS?
11. List and explain all the GPRS interfaces.
12. What are the differences between IS-95a and IS-95B?
13. What is 1xRTT? What is the channel bandwidth of 1xRTT?
14. In your view, is 3xRTT a 1G, 2G, or 3G type of network?
15. What is CDMA2000 EV-DO? Mention its advantages over 3xRTT.
16. List the main network elements (hardware) used in the radio network of WCDMA. What are the functions of the radio network section of WCDMA?
17. List all the interfaces used in a typical UMTS mobile network.
18. Is MANET a fixed or mobile network? What is it?
19. What is an ad-hoc network?
20. Distinguish between the IEEE 802.11a, 802.11b, and 802.11g.

References

[1] Shiller, J. H., Mobile Communications, 6th edn, Addison-Wesley, Harlow, 2003.
[2] Cai, J. and Goodman, D. J., 'General Packet Radio Service in GSM', IEEE Communication magazine, October 1997, pp. 122–131.
[3] Depaoli, R. and Moiso, C., 'Network Intelligence for GPRS', IEEE, 2001, pp. 4–8.
[4] Sandrasegaran, K. Class notes wireless networking technology at University of Technology, Sydney, 2004.
[5] Luo, H., Ramjee, R., Sinha, P., Li, E. and Lu, S., 'UCAN: A Unified Cellular and Ad-hoc Network Architecture', MobiCom' 03, September 2003, viewed September 2004, <http://www.bell-labs.com/user/ramjee/papers/ucan.pdf >.
[6] Jetcheva, J. G., Chun Hu, Y., Chaudhuri, S. P., Saha, A. K. and Johnson, D. B., 'Design and Evaluation of Metropolitan Area Multitier Wireless Ad Hoc Network Architecture', Fifth IEEE Workshop on Mobile Computing Systems & Aplications (WMCSA 2003), 2003, viewed September 2004, <http://monarch.cs.rice.edu/monarch-papers/wmcsa03.pdf >.
[7] Ogose, S., Sasaki, S. and Itoh, K., 'Mobile Communications Systems with Multi-layered Wireless Network using ad hoc Network', IEEE, 2003, pp. 695–698.
[8] Mathur, A.K., 'The speed demon: 3G telecommunications a high-level architecture study of 3G and UMTS', IBM, March 2002, viewed September 2004, <http://www-106.ibm.com/developerworks/wireless/library/wi-speed/?article=wir>.
[9] Shimoe, T. and Sano, T., 'IMT-2000 Network architecture', FUJITSU Sci. Tech. J., December 2002, pp. 134–139.
[10] Hassan, J. and Jha, S., 'Reducing Signaling Overhead in Cell-Hopping Mobile Wireless Networks using Location Caches', 28th Annual IEEE International Conference on local computer networks (LCN'03), 2003.
[11] Hassan, J. and Jha, S., 'Cell Hopping: A Lightweight Architecture for Wireless Communications', Wireless Communications, October 2003, pp. 16–21.
[12] Huang, C. F., Lee, H. W. and Tseng, Y. C., 'A Two-Tier Heterogeneous Mobile Ad hoc Network Architecture and Its Load-Balance Routing problem', MOE Program, 2003, viewed October 2004, <http://www.csie.nctu.edu.tw/~yctseng/papers.pub/mobile54-gw-vtc2003fall.pdf >.

[13] Weniger, K., Zitterbart, M. and Karlsruhe, U., 'Address Autoconfiguration in Mobile Ad Hoc Networks: Current Approaches and Future Directions', IEEE Network, July/August 2004, pp. 6–11.

[14] Navarrete, G. P., Sandvik, S. I. and Agbinya, J. I., 'Architectures of Moving Wireless Networks', Preliminary program of the 3rd workshop on the internet, telecommunications and signal processing (WITSP'04), University of Technology Sydney, Adelaide, Australia, December 2004, p. 6.

[15] Tseng, Y. C., Li, Y. F. and Chang, Y. C., 'On Route Lifetime in Multihop Mobile Ad Hoc Network', IEEE Transactions on mobile computing, Vol. 2, No. 4, October–December 2003, pp. 366–376.

[16] Choi, T., Kim, L., Nah, J. and Song, J., 'Combinatorial Mobile IP: A New Efficient Mobility Management Using Minimized paging and local Registration in Mobile IP Environment', Kluwer Academic Publishers, October 2004, pp. 311–321.

[17] Agbinya, J. I., 'Design Concepts: Wwireless Moving Networks (WMN)', University of Technology, Sydney, 2003, pp. 1–6.

[18] Charas, P., 'Peer-to-Peer Mobile Network Architecture', IEEE, 2002, pp. 55–61.

[19] Seo, J. H. and Han, K. J., 'Channel Allocation Scheme for Handoff Control in Wireless Ad Hoc Network with Group Mobility', Kluwer Academic Publishers, March 2004, pp. 273–285.

[20] Ramos, R. E. and Madani, K., 'A Novel Generic Distributed Intelligent Re-configurable Mobile Network Architecture', VTC-Spring Conference, Westminster University, May 2003, Rhodes, Greece, pp. 139–144.

[21] Ryu, B., Andersen, T., Elbatt, T. and Zhang, Y., 'Multi-Tier Mobile Ad Hoc Networks: Architecture, Protocols, and Performance', IEEE, 2003, pp. 1280–1285.

[22] Patini, M., Beraldi, R., Marchetti, C. and Baldoni, R., 'A Middleware Architecture for Inter ad-hoc networks Communication', *Fourth International Conference on Web Information Systems Engineering Workshops (WISEW'03)*, **IEEE, 2003, Roma, Italy, pp. 201–208.**

[23] U.S AIR FORCE, 'Airborne Networking', The MITRE Corporation, 2004, pp.1–20.

[24] 'Airborne Communication Node (CAN/AJCN)', *Advanced technology office*, viewed 19 October 2004, <http://www.darpa.mil/ato/programs/acn.htm>, 2003.

[25] Xu, K., Hong, X. and Gerla, M., 'Landmark Routing in Ad Hoc Networks with Mobile Backbones', *Journal of Parallel and Distributed Computing*, vol. 63, no. 2, February 2003, pp. 110–122.

[26] William, S., 'Mobile IP', The Internet Protocol Journal, vol. 4, no. 2, June 2001, pp.1–32.

[27] 'PLMN Switching network elements in GSM', Understanding communications, Ericsson, November 2004, viewed August 2004, <http://www.ericsson.com/support/telecom/part-d/d-3-2.shtml>.

[28] Forouzan, B. A., 'TCP/IP Protocol Suite', 2nd edn, McGraw-Hill, New York, US, 2003.

[29] Phillips, J. A. and Namee G. M., 'Personal Wireless Communication with DECT and PWT', Artech House Publishers, Boston, London, 1998.

[30] 'DECT Technology' *TDC Ltd*, January 25 2005, viewed February 2005, <http://www.tdc.co.uk/dect/>.

[31] Kleindl, G., 'DECT™ Summary', *ETSI organization*, 2005, viewed February 2005, <http://portal.etsi.org/dect/summary>.

[32] 'DECT and Health – Facts and Figures', *DECT forum*, 2004, viewed January 2005, <http://www.dect.ch/>.

[33] 'DECT in Europe and around the world', *LAND MOBILE*, 2004, viewed January 2005, <www.landmobile.co.uk>.

[34] 621.38456

[35] 'Ultra-Wideband (UWB) Technology' *Technology and research at Intel*, viewed February 2005, <http://www.intel.com/tecnology/comms/uwb/>.

[36] Aiello. G. R. and Rogerson G. D.,'Ultra wideband wireless systems', *IEEE microwave magazine*, June 2003, pp. 36–47.

[37] 'Ultra Wideband (UWB) Technology enabling high speed wireless personal area networks' *Intel in communications*, USA, 2004, pp. 1–8.

[38] Fontana R.J, 'Recent Developments in Shot Pulse Electromagnetic or UWB the Old-fashioned way?' *Multispectral solutions*, Inc., March 2004.

[39] 'Ultra Wideband (UWB) Technology enabling high speed wireless personal area networks' *Intel Corporation*, USA 2004, pp. 1–4.

[40] 'Understanding WiMAX and 3G for Portable/Mobile Broadband Wireless', *Intel technical white paper on broadband wireless Access*, December 2004, pp. 1–14.

[41] Hurel. J. L, Lerouge. C., Evci. C. and Gui. L, 'Mobile Networks Evolution: from 3G onwarda', *Alcatel Telecommunications Review*, 2003, pp. 1–9.

[42] Daly. B, 'Evolution to 3G: Where We've Been, Where We're Going' *3G information – technology overview*, viewed April 2005, <http://www.iscoint.com/3g.html>.

[43] Klerer. M, 'Introduction to IEEE 802.20', IEEE-SA Standard Board, March 2003.

[44] Zhu. H. Chlamtac. I. and Prabhakaran. B, 'A Survey of Quality of Service in IEEE 801.11 Networks', IEEE wireless communications, August 2004, pp. 6–14.

[45] 'Understanding the 802 standards: 80.16, and 802.20^{TM}' 802 IEEE Training for telecommunications Research Associates (TRA), 2004, viewed on 17 November 2004, <http://www.tra.com/pages/802ieee.cfm/site=TRA>.

[46] From Wikipedia 'WiMAX', viewed on 19 March 2005, <http://en.wikipedia.org/wiki/Wimax>.

[47] Richards. J. and Rupy. K, 'WiMAX: The Next Big Step in Wireless Broadband? A Technical and Regulatory Overview', Keller and Heckman LLP, November 2004, pp.1–6.

[48] From Wikipedia 'IEEE 802.20', viewed on 19 March 2005, <http://en.wikipedia.org/wiki/802.20>.

[49] Engles. M, 'Wireless OFDM Systems How to make them work?', Kluwer Academic Publishers, London 2002.

[50] 'Third Generation (3G) Wireless White Paper', *Trillium Digital Systems Inc*, USA 2000.

[51] Miller, L. E., 'Generic Network Model', Wireless Communication Technologies Group, NIST, 2 April 2001.

[52] Gibson, J.D., "Speech Coding Methods, Standards, and Applications", IEEE Circuits and Systems Magazine, Fourth Quarter 2005, pp. 30 – 49.

[53] *3GPP TS 26.090 v10.0.0; Mandatory Speech Codec speech processing functions Adaptive Multi-Rate (AMR speech codec); Transcoding functions, (Release 10); 3GPP, 2011.*

Chapter 2

WiMAX: Worldwide Interoperability for Microwave Access

There is a silent battle being fought within the broadband communication industry between the proponents of wireless broadband communications and fixed broadband communications. The battle is for supremacy of access techniques between the haves of fiber in the ground and the have-nots who also have wireless infrastructure. Within the wireless communication arena as well, the usual debate of what is broadband and which technology is broadband or 4G also rages. For the consumer, both battles are a welcome sign of a better future for broadband access. Every battle could lead to a better future and newer technologies at reduced costs.

What Is Broadband Communication?

Over the years, the term "wireless broadband" has been attached to network speeds of varying capacities, including 3G (UMTS) at 2 Mbps, HSDPA at more than 8 Mbps, and IEEE 802.11a, b, and g at 11 and 54 Mbps. Fixed broadband access normally refers to DSL (256 kbps) and its variants such as ADSL, VDSL2 and VDSL2+, and SONET. In 2007, the FCC in the United States redefined what should be termed broadband to uplink/downlink speeds of 738 kbps. This definition is an update of an existing definition that saw a speed of 400 kbps being called broadband. Hence in today's terms, for a communication service to be broadband, the speed available to a subscriber should be at least up to 738 kbps.

Each broadband technology, however, has its niche application areas so they can coexist. DSL is a fixed technology, usually a buried cable (underground), and hence requires the support of above-ground technologies such as 3G, which is not only wireless, but also provides for mobility access that DSL cannot offer. WiMAX (IEEE 802.16 and 802.16e), or the so-called Worldwide Interoperability for Microwave Access, defines a point-to-multipoint wireless network operating in the frequency range of 2 to 66 GHz. It also complements 3G (FDD) to provide long-range wireless broadband access at TDD and hence also offers long-range mobility at broadband access. It was ratified by ITU-R in 2007 to operate in the 2.5 to 2.69 GHz frequency band. In building solutions are provided by WLAN and also femtocells.

In a classic paper by Shannon in 1948, he established that a non-fading channel can support at least received energy per bit per noise of value –1.6 dB:

$$\left({E_b}\Big/{N_0} = -1.6 \text{ dB} \right)$$

This value depends not only on the noise spectral power, but also on the transmitted power, size (or number) of antennae, path loss, interference, and receiver sensitivity. The objectives of modern communication systems have been to reach this limit using many bit packing techniques including quadrature modulation techniques, smart antennae, and MIMO (multiple-input and multiple-output). This limitation means smaller cells if more bits per second per Hertz (broadband) are to be achieved without using large antennae and large transmit power. Large transmit power is economically not good and also health-wise dangerous. The challenge to communication engineers is to achieve broadband communications so that bits per second per Hertz is maximized such that Shannon's channel capacity equation is obeyed or so that we have:

$$\eta_{max} = \log_2\left(1 + S/N\right)$$

In this equation, S is signal power and N is noise power. Shannon, through this equation, has clearly indicated that there is a limit to communication capacity and that limit is set by all the factors that include the signal and noise powers. Hence, the challenge is to raise signal power and diminish the noise power. These two objectives are not complementary and cannot be achieved without sacrifices. In today's terms, this means using MIMO and also having smaller cells—here lies one of the driving factors toward femtocells. Hence, to increase capacity, we need to shrink the cells and reduce the signal-to-interference ratio. WiMAX says increase the cells. Therefore, something smart needs to be done to achieve that.

Wireless Last Mile

The dream of the wireless community is therefore to provide wireless connectivity to homes at speeds comparable to fixed Internet solutions that use cable, T1 (1.544 Mbps), and DSL. The connection from a service provider network into the house is often referred to as the *last mile* solution. WiMAX has been touted as a possible solution and this can be offered through either the support of picocells or femtocells, or both. WiMAX has the potential to deliver broadband services at ranges close to 70 km and at speeds potentially up to 350 Mbps [1]. The aim is for WiMAX to provide fixed, nomadic, portable, and mobile wireless broadband connectivity without direct line-of-sight with a base station. The WiMAX Forum earmarks a typical cell radius of 3 to 10 km, and WiMAX Forum Certified systems are expected to deliver capacity of up to 40 Mbps per channel portable and fixed access.

The objective is for WiMAX to also be used as a last mile broadband wireless access technology. In this situation, the broadband connectivity to homes and premises allows an operator to support businesses and homes with wireless broadband services. The method of delivery is a point-to-multipoint single-hop communication between a base station and multiple receiving points (subscriber stations).

WiMAX is also to be used as a wireless backhaul. In this mode it operates in a multihop scenario. Therefore, the intention is also to support broadband wireless mesh networks. Although not conceived at the initial stage to service mobile hosts, this multihop provision established a foundation for it to support mobility. One of the most viable deployment situations for this form of WiMAX is for it to be used by operators as a backhaul for cellular networks. This allows them to transport voice and broadband services (e.g., IPTV and Mobile TV) from the edge of the cell to the core of the network using multi-subscriber-station mesh-networking.

Furthermore, if the dream of a cell radius of 10 km is achieved, WIMAX base stations can provide broadband access to networks in motion deployed in commercial aircraft (Figure 2.1). For this case, direct LOS is possible provided the mitigating factors of the channel (such as rain and cloud cover) are addressed.

A typical WiMAX to the air situation is illustrated in Figure 2.2, where a commercial aircraft carries a WiMAX mobile router and communicates through the Internet using a land-based WiMAX network.

At lower heights where emergency aircraft such as helicopters used by the police, fire brigades, and emergency response systems operate, WiMAX can also provide suitable broadband coverage with less rapid handover between base stations compared with commercial and military air craft.

WiMAX can also support IPTV (Internet Protocol television) (Figure 2.3) and it allows a service provider to offer the same programming wirelessly to both residential and mobile customers with significant improvement in QoS compared with existing IPTV offerings.

Video on Demand (VOD) can also be offered by the service provider fairly easily to also residential customers and individuals on the move.

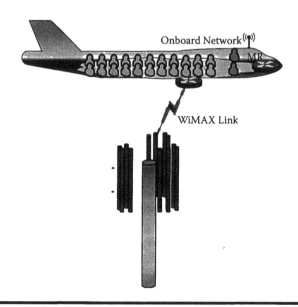

Figure 2.1   WiMAX broadband communications with a commercial aircraft.

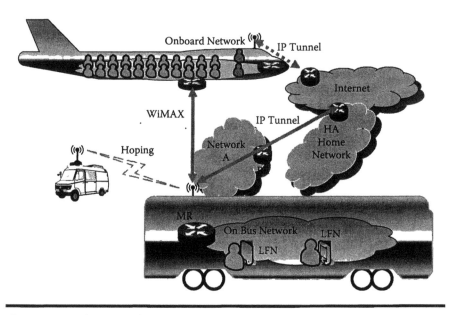

Figure 2.2   WiMAX-to-the-Air.

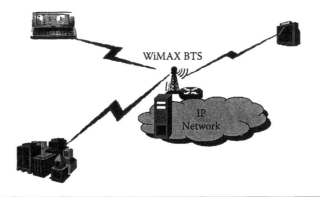

Figure 2.3   WiMAX IPTV.

Using OFDM, whereby the orthogonal channels are spaced very close to each other through the application of inverse FFT and avoiding inter-symbol interference, large data capacity increases can be achieved. WiMAX also applies other broadband technologies such as adaptive modulation and coding (AMC) and adaptive forward error correction (AFEC). Hence, a typical WiMAX base station can provide enough bandwidth "to cater to the demands of more than fifty businesses with T1" level services and "hundreds of homes with high-speed Internet access" [2]. The WiMAX base station provides a one-to-many or point-to-multipoint single-hop access for many homes or businesses. WiMAX also provides wireless cellular backhaul in a mesh or multihop architecture. Thus, it has the capability to mesh network many subscriber stations (SSs) onto a WiMAX base station (BS).

Physical Layer

At the physical layer of the IEEE 802.16 air interface, two frequency bands of operation are specified either for:

- IEEE 802.16 at 10 to 66 GHz or
- IEEE 802.16a at 2 to 11 GHz

Depending on the operation bandwidth (e.g., 20, 25, or 28 MHz) and the modulation and coding techniques adopted, WiMAX can support data rates in the range of 32 to 130 Mbps. The IEEE 802.16 standard specifies various air interfaces:

- 802.16e Mobile WiMAX
- 802.16j Mobile Multihop WiMAX

The Mobile Multihop Relay Task Group (IEEE 802.16j) intends to develop the mobile multihop relay technique standard for increasing the coverage range and achieving high transmission rates. As such, this task group has objectives for the future to evolve WiMAX to fulfill the requirements of future 4G mobile communication networks. Through multihop relay, the coverage range can be increased significantly at low cost. Coverage range often shrinks due to the shadow fading effect experienced by base stations. Relay communication mitigates this by ensuring a smooth and continuous large rate over a large distance. Thus, coverage is increased at low cost. Furthermore, by adopting selective power amplification, it amplifies the desired signal per situation.

WiMAX Spectrum Overview

In most parts of the work, both unlicensed bands in 2.4 GHz and 5 GHz and various licensed frequency bands are specified, including licensed bands in 2.3 GHz, 2.5 GHz, and 3.5 GHz (Table 2.1). The ITU-R ratified 3G WiMAX in the band 2.5 to 2.69 GHz. A 256 FFT OFDM modulation is popular in the unlicensed 5.8 GHz band using a flexible channel plan from 1.5 MHz to 20 MHz per channel.

IEEE 802.16 and 802.16a Air Interfaces

Typically, line-of-sight (LOS) propagation is used in the 10 to 66 GHz band between a BS and an SS. A single carrier (SC) air interface used for this band is wireless single carrier (SC).

The 802.16 standard defines four air interfaces for different frequency bands and in these four regimes; it can support data rates from 32 to 130 Mbps, depending on the type of modulation and channel bandwidth (e.g., 20, 25, 28 MHz). The first air interface is defined for LOS communications. In the 6 to 66 GHz band, signal propagation should use direct line-of-sight-communications. This band is suitable for rural situations and flat terrains where there are few or no obstacles in the path taken by the signal from source to destination. The remaining three air interfaces are defined for the 2 to 11 GHz band and are for non-line-of-sight situations that typically occur in urban regions and also in difficult terrain environments such as

Table 2.1   WiMAX Spectrum

Region	Spectrum (Unlicensed) (GHz)	Spectrum (Licensed) (GHz)
China	2.4; 5	3.4–3.43; 3.5–3.53
Europe		3.4–3.6
Latin America		3.4–3.6
USA		2.5–2.7
ITU		2.5–2.69

in hilly and mountainous regions. These three air interfaces are described in the next section.

IEEE 802.16a Air Interfaces

Three different air interfaces using non-line-of-sight (NLOS) communication are used in the 2 to 11 GHz band. They are:

1. *WirelessMAN-SCa* for single-carrier modulation.
2. *WirelessMAN-OFDM* using OFDM-based transmission with 256 subcarriers. The MAC scheme for the subscriber stations for this air interface is based on time-division multiple access (TDMA).
3. *WirelessMAN-OFDMA* uses OFDM-based transmission and 2048 subcarriers. The MAC scheme for the subscriber stations for this air interface is based on orthogonal frequency division multiple access (OFDMA). In this scheme, different groups of subcarriers are assigned to different SSs.

The quality of a wireless link between a base station and a subscriber station is a function of the channel fading and interference. The effective channel capacity is modified accordingly. Hence, adaptive modulation and coding is used to enhance the data transmission rate. Using AMC, the radio transceiver adjusts the transmission rate according to channel quality or the signal-to-noise ratio (SNR) at the receiver (see Shannon's channel capacity equation). Using error, the Reed-Solomon (RS) codes concatenated with an inner convolution code (or turbo codes and space-time block codes (STBCs)), error correction is achieved.

IEEE 802.16d and IEEE 802.16e Air Interfaces

Both 802.16d and 802.16e evolved from 802.16a and "use advanced physical layer techniques to support NLOS communications" [3]. The IEEE 802.16e standard is designed to support user mobility. A single base station is used in this standard to serve several mobile subscriber stations in the service area. Both the MAC and physical layers are reengineered to support mobility requirements at the physical layer. "The IEEE 802.16g standard (under development) aims to support mobility at higher layers (transport and application) and across the backhaul network for multi network operation" [3].

Medium Access Control (MAC)

A connection-oriented MAC protocol is used in WiMAX MAC. Through this means, subscriber stations are provided with the mechanism to request bandwidth from the BS. Each subscriber station has a 48 bit MAC address so that in IPv6 and beyond, the stations can be identified uniquely by extending this host MAC identity to the EUI-64 host suffix. Therefore, the roaming hardware identification is possible. Each connection to the base station is therefore identified using a 16 bit connection identifier (CID). The BS broadcasts data to all SSs in the same network in the downlink. "Each SS processes only the MAC protocol data units (PDUs) containing its own CID and discards the other PDUs" [3].

The IEEE 802.16 MAC has an innovative approach for allocating transmission permissions to the subscriber stations through a combination of FDM and TDM. The standard supports the so-called grant per SS (GPSS) mode of bandwidth allocation. In GPSS, a portion of the available bandwidth is granted (allocated) to each of the SSs. Each SS is responsible for allocating its share of the bandwidth among the corresponding connections. Hence, the IEEE 802.16/WiMAX supports both frequency division duplex (FDD) and time division duplex (TDD) transmission modes.

In TDD, "a MAC frame is divided into uplink and downlink subframes. The lengths of these subframes are determined dynamically by the BS and broadcast to the SSs through downlink and uplink map messages (UL-MAP and DL-MAP) at the beginning of each frame. Therefore, each SS knows when and how long to receive and transmit data to the BS. In the uplink direction, a subframe also contains ranging information to identify an SS, information on the requested bandwidth, and data PDUs for each SS. The MAC protocol in the standard supports dynamic bandwidth allocation. In this case, each SS can request bandwidth from the BS using a BW-request PDU. There are two modes to transmit BW-request PDUs: contention mode and contention-free mode (e.g., polling). In contention mode, an SS transmits BW-request PDUs during the contention period in a frame, and a backoff mechanism is used to resolve the contention among the BW-request PDUs from multiple SSs. In contention-free mode, each SS is polled by the BS, and after receiving the polling signal from the BS, an SS responds by sending the BW-request PDU. Due to predictable delay, contention-free mode is suitable for QoS-sensitive applications" [3].

"To provide access control and confidentiality in data transmission, IEEE 802.16 provides a full set of security features [4], which are implemented as a MAC sublayer functionalities. In addition to the single-hop point-to-multipoint operation scenario, the IEEE 802.16a standard also defines signaling flows and message formats for multihop mesh networking among the SS (i.e., client mesh). In this scenario several BSs can communicate with each other. Data traffic from an SS is transmitted through several BSs along the route in the mesh network to the destination BS or an Internet gateway" [3].

QoS Framework and Service Types in WiMAX

The IEEE 802.16 WiMAX standard defines a QoS framework for different classes of services. The major types of services, each having different QoS requirements, are supported.

- *Unsolicited grant service (UGS).* Constant-bit-rate (CBR) traffic is supported by this service type. For this, the BS allocates a fixed amount of bandwidth to each of the connections in a static manner. This helps in minimizing both delay and jitter. UGS is suitable for traffic with very strict QoS constraints for which delay and loss need to be minimized (e.g., commercial IPTV).
- *Polling service (PS).* This service supports traffic for which some level of QoS guarantee is required. There are two types of PS: real-time polling service (rtPS) and non-real-time polling service (nrtPS). The difference between the two subtypes lies in the tightness of the QoS specifications (i.e., rtPS is more delay sensitive than nrtPS). In addition to delay-sensitive traffic, "non-real-time Internet traffic can use polling service to achieve a certain throughput guarantee. The amount of bandwidth required for this type of service is determined dynamically based on the required QoS performance and the dynamic traffic arrivals for the corresponding connections" [3].
- *Best-effort (BE) service.* This service is defined for traffic with no QoS guarantee (e.g., Web and e-mail traffic). The bandwidth allocated to BE service depends on the bandwidth allocation policies for the other two types of service. In particular, any bandwidth left after serving UGS and PS traffic is given to BE service.

Advantages and Disadvantages of IEEE 802.16e

There is a common misconception that is prevalent about WiMAX that it will provide up to 70 Mbps over a distance of up to 70 miles (or 113 km). The obtainable data rate from every air interface decreases with speed and range and with moving stations. Therefore, practically, these system performance values are not achievable with today's versions of WiMAX. In ideal situations with a single subscriber and direct line-of-sight, the expected rate should be about 10 Mbps and at about 10 km. These values will decrease with the mobility of the host. Mobile WiMAX, however, can meet the data rate requirements of DSL and cable-like services.

In urban environments where line-of-sight communication is usually impossible, speeds will be lower than 10 Mbps and the achievable ranges will be close to about 3 km cell range. These values will decrease with the speed of the terminals. The most likely achievable speeds in a multi-user environment could range from 2 to 10 Mbps. Even so, these rates are much better than UMTS. Like the situation in 3G, spectrum purchases will take place and new equipment and installations will increase the OPEX. Therefore, the initial capital investment for WiMAX could still be significant. In a nutshell, therefore, the major advantages of WiMAX boil down to:

1. The ability for a single base station to serve many subscribers at broadband speeds; a single station can serve hundreds of users

2. Easier and faster deployment of new networks compare to fixed networks
3. Broadband speeds at line-of-sight ranges and over larger cell ranges; speed of 10 Mbps at 10 km with line-of-sight
4. Interoperability due to a common standard being used

The disadvantages also are significant and include:

1. Non-line-of-sight connections in urban areas where large populations and subscribers exist
2. Interruptions from atmospheric effects such as rain (weather conditions such as rain degrade the signal)
3. Interference from other wireless infrastructure operating in the same frequency range used by WiMAX
4. The high cost of spectrum
5. The use of multiple wideband frequencies
6. High power consumption by WiMAX, which will limit rural installations to where power can easily and cheaply be obtained as solar energy; use could also be found to be expensive
7. Initial high operational costs for equipment and installation
8. At higher speeds (long range), delay increases and therefore QoS also degrades

WiMAX Compared with Other Technologies

WiBRO

A TDD mode version of WiMAX (WiBRO) was proposed by Korea Telecom. The broadband wireless Internet technology called WiBRO operates in the licensed 2.3 GHz frequency band and can support both fixed and mobile users at a channel bandwidth of 9 MHz. WiBRO uses a MAC frame size of 5 ms and AMC to achieve enhanced transmission rates. The MAC scheme is based on OFDMA. Its QoS framework supports four service types, as in the IEEE 802.16 standard.

HiperMAN

HiperMAN, the high-performance radio MAN, is a creation in Europe of the Broadband Radio Access Networks (BRAN) group of the ETSI. HiperMAN is designed to operate in the 2 to 10 GHz (mainly in the 3.5 GHz) band. HiperMAN also has a defined QoS framework and uses AMC and dynamic power allocation for NLOS communications.

The standard also supports a mesh configuration. Both the WiBRO technology and HiperMAN standards are compatible with IEEE 802.16a and 802.16.

WiMAX and 3G

WiMAX, ratified in October 2007 by the ITU-R as a 3G standard (and included as part of the IMT-2000 family of standards), is a complementary rather than replacement technology for 3G wireless systems. Traditional 3G systems primarily target mobile voice and data users but WiMAX systems are optimized to provide large-range, high-rate wireless connectivity for a large set of services and applications that require QoS guarantees.

Table 2.2 compares 3G, IEEE 802.16e, and IEEE 802.20 MBWA and MobileFi [3]. Using WiMAX as wireless backhaul, a cellular network can provide high bandwidth with large coverage area. Using topology management to reduce the network deployment costs, the number of WiMAX links in a backhaul network can be reduced significantly compared to a ring topology.

Table 2.2   Comparison between IEEE 802.16e, 802.20, 3G, and MBWA and MobileFi

	3G	*IEEE 802.1/WiMAX*	*IEEE 802.20/ MobileFi*
Objective	To provide voice and data services to mobile users	To provide BWA to fixed and mobile users	To provide mobile broadband connections to mobile users
Frequency	2 GHz	2–10 GHz	3.5 GHz
Channel bandwidth	<5 MHz	>5 MHz	<20 MHz
Transmission rate	Up to 10 Mbps (HSDPA from 3GPP)	10–50 Mbps	>16 Mbps
Cell radius	Up to 20 km	Up to 50 km	—
Mobility	Full mobility functions (IP mobility, roaming, handoff, paging)	IP mobility	Full mobility functions and inter-technology handoff
Mobile speed	Up to 120 km/h	60 km/h	Up to 250 km/h
Multiple access	CDMA	TDMA or OFDMA	—
MAC frame size	10 ms	<10 ms	<10 ms

WiMAX and 4G

There are several evolution paths to the fourth-generation networks, including the long-term evolution (LTE). Compared with WLANs that are more suitable for stationary/quasi-stationary users with high throughput connection requirements, "cellular networks are more efficient for voice-oriented and limited throughput mobile data services" [3]. WiMAX networks can provide very-high-speed wireless connectivity in the presence of mobility.

WiMAX and IEEE 802.20/MobileFi

"IEEE 802.20 (also called MobileFi) is being designed specifically for MBWA services. This standard will be optimized for transport of IP services for fixed and mobile users. IEEE 802.20 will operate in the licensed bands below 3.5 GHz and provide data transmission speeds over 10 Mbps for user speeds up to 250 km/h" [3].

Mobility Management

In IEEE 802.16e, mobility management is handled by a dedicated mobility agent (MA) sublayer that is located on top of the MAC layer. Therefore, an MSS can move from one serving BS to a target BS. Several parameters are used for triggering this handover. Handover can be triggered based on signal fading or changing the attached access point.

Mobile WiMAX versus 3G-1xEVDO and HSPA

Many features of Mobile WiMAX that were developed to support throughput performance are shared by 3G networks (HSDA and 1xEVDO). The common features include fast scheduling, adaptive modulation and coding, hybrid ARQ, and bandwidth-efficient handoff framework. These common features are given in Table 2.3. The Mobile WiMAX physical layer is based on OFDMA technology. It is scalable and the channel bandwidth is variable. The application of new technologies such as IFFT employed for Mobile WiMAX results in low equipment complexity. With an all-IP core network, it has simpler management and endows Mobile WiMAX systems with significant CDMA-based 3G techniques, including:

- Quality-of-service (QoS)
- Scalable channel bandwidth
- Orthogonal multiple access in the uplink
- Tolerance to multipath and self-interference
- Fractional frequency reuse

Table 2.3 Attributes of Mobile WiMAX and HSPA and 1xEVDO

Features	Mobile WiMAX	1xEVDO	HSPA
Standard	802.16e	CDMA2000/IS-95	UMTS
Bandwidth	5, 7, 8.75, 10	1.25 MHz	5 MHz
DL peak rate	46 Mbps DL/UL-3 32 Mbps DL/UL-1 (10 MHz BW)	3.1 Mbps	14 Mbps
UL peak rate	7 Mbps DL/UL-3 4 Mbps DL/UL-1 (10 MHz BW)	1.8 Mbps	5.8 Mbps
Duplexing	TDD	FDD	FDD
Multiple Access (UL)	OFDMA	CDMA	CDMA
Downlink (DL)	OFDMA	TDM	FDD - TDM
Frame size	5 ms	1.67 ms (DL) 6.67 ms (UL)	2 ms (DL) 2, 10 ms (UL)
Modulation (DL)	QPSK/16QAM, 64QAM	QPSK/8PSK, 16QAM	QPSK/16QAM
Modulation (UL)	QPSK/16QAM	BPSK, QPSK/8PSK	BPSK/QPSK
Scheduling	Fast scheduling in DL and UL	Fast scheduling in DL	Fast scheduling in DL
H-ARQ	Multichannel asynchronous CC	Fast 4-channel synchronous	Fast 6-channel asynchronous CC
Handoff	Network optimized hard handoff	Virtual soft handoff	Network-initiated hard handoff
TX diversity and MIMO	STBC, SM	Simple open-loop diversity	Simple open- and closed-loop diversity
Beamforming	Yes	No	Yes (dedicated pilots)
Coding	CC Turbo	Turbo	CC Turbo

- Spectrally efficient TDD
- Advanced antenna technology
- Frequency-selective scheduling

Mobile WiMAX is based on OFDM and OFDMA technology, and is more suited to broadband wireless data communications.

An OFDM/OFDMA-based system can support a large range of advanced antenna technologies. With TDD support, Mobile WiMAX can also dynamically adjust the downlink/uplink ratio. This provides greater flexibility and spectral efficiency advantages for carrying varied types of broadband traffic. Mobile WiMAX is endowed with excellent QoS and implements several service classes and level agreements so that the needs of varied customers can be met.

Both the EVDO and HSPA are based on fixed FDD channel allocations with fixed asymmetric downlink and uplink ratios determined by the difference in downlink/uplink spectral efficiency.

Some of the typical characteristics of some broadband technologies are provided in Table 2.4.

Packet Header

This section describes the format of the WiMAX packet called the MAC protocol data unit (MPDU). There are two forms of the MPDU: the generic MAC header (GMH) and the bandwidth request header (BRH). The GMH is a 6-byte field and is followed by the payload (the MAC service data unit or MSDU) and then the 4-byte CRC field. The GMH carries transport the connection ID.

In the bandwidth request format, the GMH is replaced with the BRH and the MSDU field is replaced with the MAC message management part. The transport connections carry the MAC service data units.

Network Entry

Network entry is the process through which an SS gains access to the WiMAX network. It consists of five generic steps (Figure 2.4):

1. The subscriber station scans or searches for an appropriate downlink signal from a base station. With the signal, the SS establishes its channel parameters.
2. The information on ranging allows the SS to set its PHY parameters fairly well and then to establish its primary management channel through that base station. Through this channel capability negotiation, authorization and key management are done.
3. Using the PKM Protocol, authorizes the SS to the base station.

Table 2.4 Characteristics of Broadband Technologies (SONET not included)

Broadband Technology	Standard	Range (km)	Frequency (GHz)	Speed (Mbps) Downlink	Speed (kbps) Uplink	Mobility
ADSL	xDSL			8	0.144	Fixed
ADSL2+	XDSL			25	0.144	Fixed
Cable		NA		1–2	0.128–0.384	Fixed
WiMAX	802.16	10	2–11	70		Fixed
WiMAX	802.16e	<3.5	2–6	30		Mobile
WiFi	802.11a	0.10	5	54		Fixed
WiFi	802.11b	0.10	2.4	11		Fixed
WiFi	802.11g	0.10	2.4	54		Fixed
UMTS	WCDMA	6	1.8, 1.9, 2.1	0.144–2		Mobile/Fixed
EDGE	2.5G	6	1.9	0.348		Mobile/Fixed
HSPA	3G+		1.8, 1.9, 2.1	8		Mobile/Fixed
Femtocell	3G+		0.8, 1.9, 2.3			Fixed indoor

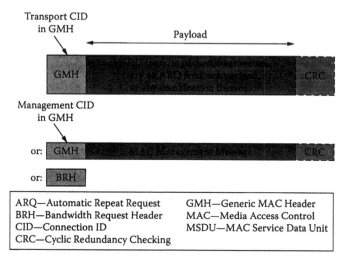

Figure 2.4   MAC protocol data unit.

4. The SS then registers with the base station by sending a request message to the base station. In the reply from the base station, the SS is assigned a connection ID for a secondary management connection.
5. The SS and BS then create transport connection using a MAC_create_connection request. A request to create a dynamic transport connection carries information on whether an encryption is required at the MAC level.

Security Architecture

WiMAX security is implemented as a privacy sublayer at the bottom of its MAC protocol's internal layer [4] with the main goal of providing access security and confidentiality of the data link.

The WiMAX security architecture contains five components:

1. Security associations
2. X.509 certificate profile
3. Privacy key management (PKM) authorization
4. Privacy
5. Key management and encryption

Each one of these components is discussed below.

Security Association

The objective of security associations (SAs) is to maintain the security state pertinent to an association. Two types of security associations are defined: ·

1. Data security association
2. Authorization security association

Data Security Association

Only the data SA is explicitly defined by the IEEE. The data SA is used to protect the transport connections between one or more SSs and a base station. There are three types of SA: *static, dynamic,* and *primary.* Primary SAs are established during link initialization, static SAs are configured on the BS, and dynamic SAs are constructed per need for dynamic transport connections. The data SA consists of [4]:

■ A 16-bit SA identifier, or SAID
■ A cipher using DES in a cipher block chaining (CBC) mode is used to protect the data exchanged over the connection; other cryptographic algorithms may be used
■ Two traffic encryption keys (TEKs) to encrypt data: the current operational key and a TEK for when the current key expires
■ Two 2-bit key identifiers, one for each TEK
■ A TEK lifetime set to a default value of half a day but with a minimum value of 30 minutes and a maximum value of 7 days
■ A 64-bit initialization vector for each TEK
■ An indication of the type of data SA

The operation of the data security association is as follows. To secure a transport connection, a subscriber station must make a request for the association by first initiating a data SA using a *create_connection* request. In a multicast situation, the standard allows many connection IDs to share the same SA. "On network entry, IEEE 802.16 automatically creates an SA for the secondary management channel" [4]. A fixed SS may have two or three SAs—one for the secondary management channel, a separate SA each for uplink and downlink connections, or one for both uplink and downlink transport connections. A multicast group also requires an SA that is shared among group members [4].

Authorization Security Association

The authorization security association is not explicitly defined by the IEEE. It, however, consists of:

■ An X.509 certificate identifying the SS
■ A 160-bit authorization key (AK); correct use of this key demonstrates authorization to use IEEE 802.16 transport connections
■ A 4-bit identifier for the AK

- A time to live (lifetime) for AK, ranging from one to 70 days; the default lifetime is set to seven days
- A key encryption key (KEK) (a 112-bit Triple-DES key) for distributing TEKs

Let:

a || b represent the concatenation of two numbers a and b;

$a \oplus b$ represents the exclusive OR of a and b as "a XOR b";

a'' is the octet a repeated n times

Truncate-N(.) be to throw away all but the first N bits of the argument

Then the construction of the KEK is as follows:

$$KEK = Truncate - 128\left(SHA1\left(\left(\left(AK \mid \mid O^{44}\right) \oplus 53^{64}\right)\right)\right)$$

where SHA1 is the Secure Hash standard.

- A downlink hash function-based message authentication code (HMAC) key providing data authenticity of key distribution messages from the BS to the SS; the authentication key is constructed as:
  The uplink HMAC key is constructed as:

$$Downlink\ HMAC\ Key = SHA1\left(\left(AK \mid \mid O^{44}\right) \oplus 3A^{64}\right)$$

- An uplink HMAC key providing data authenticity of key distribution messages from the SS to the BS:

$$Uplink\ HMAC\ Key = SHA1\left(\left(AK \mid \mid O^{44}\right) \oplus 5C^{64}\right)$$

- A list of authorized data SAs: an authorization SA is state shared between particular BSs and SSs [4]. It is assumed by design that these two stations will maintain the AK as a secret. "Base stations use authorization SAs to configure data SAs on the SS" [4].

X.509 Certificate Profile

Communicating parties in WiMAX are authenticated using X.509 certificates. Details of the X.509 certificate profile are as follows:

- Version of X.509 certificate format
- Certificate serial number
- The signature algorithm Public Key of the Certificate issuer—RSA encryption with SHA1 hashing
- Certificate issuer
- The validity period of the Certificate
- Identity of the Certificate holder; if the subject is the SS, it includes the station's MAC address
- The signature of the issuer, which is the digital signature of the Abstract Syntax Notation 1 Distinguished Encoding Rules (ASN.1 DER), encoding of the rest of the certificate
- Certificate holder's public key, which also identifies how the public key is used, and is limited to RSA encryption
- Signature algorithm, identical to the certificate issuer's signature algorithm

Two certificate types are defined by the standard: manufacturer certificates and SS certificates. No certificates are defined for the base station.

The SS certificate identifies a particular SS and it includes its MAC address in the subject field. The manufacturer certificate identifies the manufacturer of the WiMAX device. The certificate may be a self-signed certificate or signed by a certifying third party. The manufacturer certificate's public key is used by the BS to verify the identity of the SS (verification of its certificate). Positive verification is proof that the device is genuine. The assumption is that the SS maintains the private key corresponding to its public key in some sort of secrecy to prevent attackers from easily compromising it. In practice, if an attacker has control of the SS for long enough and has some power over it to force revelation, this measure can be compromised.

PKM Authorization Protocol

A privacy key management authorization protocol is used to distribute tokens to authorized SSs. It is used for message exchange between the SS and a BS. The protocol consists of three messages between them. The objectives for the messages are for verification of the SS and for the BS to decide if the SS will be authorized to access the WMAN channel. The sequence of messages is as follows:

Message 1: SS → BS Cert (Manufacturer (SS)). This message is sent by SS to the BS. Through it sends its Certificate to the BS. The BS uses it to decide if the SS is a trusted device.

Message 2: SS → BS Cert(SS)||Capabilities||SAID. SS sends a second message after the first containing the X.509 certificate of SS, security capability, and identity SAID.

Message 3: BS → SS RSA − Encrypt(PubKey(SS), AK||Lifetime||SeqNo||SAID List)

Table 2.5 Terms Used for PKM Authorization Protocol

Term	Description
AK	Authorization key
Capabilities	Supported authentication and data cipher algorithms of SS
SAID	The secure link (connection ID) between SS and BS
SAIDList	List of SA descriptors, each with an SAID, SA type, and the SA cipher suite
Lifetime	A 32-bit unsigned number, the number of seconds before AK expires
SeqNo	A 4-bit value for the AK
$A \rightarrow B:M$	A sends B a message with content M
Cert(Manufacturer(SS))	X.509 certificate identifying the SS manufacturer
Cert (SS)	X.509 certificate with the SS public key
RSA-Encrypt(k, a)	Instruction to RSA-OAEP encrypt its second argument **a** under key k
PubKey (SS)	The public key of SS as given in Cert (SS)

The base station uses Cert (SS) public key in Message 2 to construct Message 3, which is a reply to SS. Provided the SS is a valid device or that its identity is verified (i.e., its certificate is verified to be authentic) and if it is authorized to access the WMAN channel, Message 3 is therefore issued and instantiates an authorization security association (SA) between the SS and BS.

Because it is assumed that only base stations and SS possess the AK, correct use of the association key is a demonstration of the authorization to access the channel. The terms in Table 2.5 [4] are used for the description of the PKM protocol.

Privacy and Key Management

The PKM protocol consists of two or three messages for privacy and key management between a BS and SS. The objective of Message 1 and Message 2 are for the SS and BS, respectively, for authentication and also to detect forgeries in the network. Partial authentication of SS to BS is supported but authentication of BS to SS is not supported. Message 3 is used by the BS to configure a required SA. In this section, parameters in Table 2.6 are used [4].

The PKM process is as follows. Whenever the BS wants to force rekeying of data security association or to create a new security association, it issues the Message 1. This message is optional and is used for rekeying.

Table 2.6   Terms Used in PKM

Term	Description
[…]	Optional message
SAID	The ID of the data SA being relayed or created
SeqNo	The AK used for the exchange
NewTEK	The initialization vector of the TEKs of next generation, remaining lifetime, sequence number for the data SA specified by SAID
	The sequence number of the TEK is 1 greater, modulo 4, than the OldTEK sequence number
OldTEK	The initialization vector of the TEKs of previous generation, remaining lifetime, sequence number for the data SA specified by SAID
	The sequence number of the TEK is 2 bits long
HMAC (1)	The HMAC-SHA1 digest of SeqNo ‖ SAID under AK's downlink HMAC key
HMAC (2)	The HMAC-SHA1 digest of SeqNo ‖ SAID under AK's uplink HMAC key
HMAC (3)	The HMAC-SHA1 digest of SeqNo ‖ SAID ‖ OldTEK ‖ NewTEK under AK's downlink HMAK key

Message 1: BS → SS:SeqNo‖SAID‖HMAC(1)

Computation of HMAC (1) is used by the SS to detect forgeries in the network.
Message 2 is issued by the SS to request SA parameters:

Message 2: SS → BS:SeqNo‖SAID‖HMAC(2)

For this message, the SS takes its SAID from the authorization protocol SAIDList. Separate Message 2 is used for different SAs. By computing HMAC (2), the BS can detect forgeries in the network. A valid HMAC (2) authenticates SS to BS.

If HMAC (2) is valid and SAID identifies one of the SS's SA, then the BS uses Message 3 to create and configure an SA:Message 3: BS → SS:SeqNo‖SAID ‖OldTEK‖NewTEK‖HMAC(3)

The current and active SA is OldTEK and the new SA is NewTEK. NewTEK prescribes the parameters to be used for the new SA when OldTEK expires. The base station uses Triple DES to encrypt the old and new TEKs with the authorization SA KEK using electronic code book (ECB) mode. By computing HMAC (3), the SS can

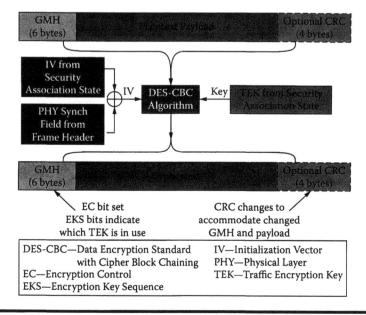

Figure 2.5   802.16 TEK encryption. (From: *FIPS PUB 180-1, Secure Hash Standard,* National Institute of Standards and Technology, April 1995; http://csrc.nist. gov/ CryptoToolkit/tkhash.html.)

detect forgeries. Because only the SS can unscramble the AK sent in Message 3 of the authorization protocol and because the AK is not predictable, a valid HMAC (2) authenticates the SS to BS. No reverse authentication of BS to the SS is required by the protocol. This is one of the security flaws of current versions of WiMAX. Because BSs are not required to authenticate themselves, a masquerading BS could fool some SSs.

Encryption

The MAC protocol data unit (MPDU) is encrypted using DES-CBC. The 6-byte GMH and the optional 4-byte CRC parts of the packet are not encrypted. The TEK method used is indicated by 2 bits in the GMH (Figure 2.5). The MPDU initialization vector (IV) is computed by performing an exclusive OR of the security association SA initialization vector from the most recent GMH with the contents of the PHY synchronization field. The MPDU IV is predictable, and no data authentication is provided by the IEEE 802.16.

Summary

This chapter provided a detailed summary of WiMAX as a wireless broadband access technology with potential to deliver speeds comparable with fixed networks and at large link ranges up to 70 km. WiMAX also defines within the standard several QoS specifications, which is a major improvement over the traditional mobile communication standards such as GSM and UMTS. By including definite security specifications and mobile WiMAX, it is a technology that can support services across various public transport regimes such as buses, trains, aircraft, and in maritime applications including the armed forces. WiMAX competes with LTE for the same market and can be used for wireless backhaul, last mile, and in-building solutions.

Practice Set: Review Questions

1. The WiMAX standard is really the IEEE:
   a. 802.11a
   b. 802.11b
   c. 802.16e
   d. 802.11g
   e. 802.11n
2. WiMAX channel frequencies can be in the frequency bands around:
   a. 2 to 11 GHz
   b. 10 to 66 GHz
   c. 20 to 40 GHz
   d. (a) and (b) above
   e. None of the above
3. WiMAX is a:
   a. 1G network
   b. 2G network
   c. 3G network
   d. 4G network
   e. (c) and (d) above
4. Mobile WiMAX is the IEEE standard:
   a. 802.16n
   b. 802.16e
   c. 802.16j
   d. (b) and (c)
   e. (a) and (b)

Exercises

1. Distinguish between the WiMAX standards 802.16, 802.16a, and 802.16e.
2. What do you think is the main reason for not requiring the authentication of a WiMAX base station to a subscriber station?
3. What are the three fields in a WiMAX MAC packet?

4. What is the difference between a GMH and a BRH?
5. List the air interfaces for WiMAX.
6. What is the difference between the air interfaces in the 2 to 11 GHz range and the 6 to 66 GHz range?
7. What is a security association?
8. How many types of security associations are defined in the WiMAX standards?
9. How is an SS authenticated to the base station that it is attached to?
10. Describe how communicating parties are authenticated.

References

[1] S.J. Vaughan-Nichols, Achieving Wireless Broadband with WiMax, *IEEE Comp.*, 37(6), June 2004, 10–13.
[2] M. Chatterjee and S. Sengupta, Feedback-Based Real-Time Streaming over WiMAX, *IEEE Communications Magazine*, February 2007, pp. 64–71.
[3] D. Niyato, E. Hossain, and J. Diamond, IEEE 802.16/WiMAX-Based Broadband Wireless Access and Its Application for Telemedicine / E-Health Services, *IEEE Wireless Communications*, February 2007, pp. 72–82
[4] D. Johnston and J. Walker, Overview of IEEE 802.16 Security, *IEEE Security and Privacy*, 2(3), May–June 2004, 40–48.
[5] FIPS PUB 180-1, Secure Hash Standard, National Institute of Standards and Technology, April 1995; <http://csrc.nist. gov/CryptoToolkit/tkhash.html>.

Chapter 3

SCTP and Vertical Handoff

Seamless vertical handoff is the ability of a mobile terminal to successfully or simultaneously attach to different access points in an integrated wireless network infrastructure in a way that makes the physical movement two types of wireless networks transparent and preserves application-level connectivity. To achieve seamless handoff, handoff decision algorithms, handoff metrics, and mobility handling to maintain ongoing user connections must be addressed. Traditional handoff decision metrics based on the received signal strength (RSS) and other physical layer parameters are insufficient for the challenges of a heterogeneous wireless system because the upper-layer applications are really interested in metrics related to network conditions (such as available bandwidth and delay) and user preferences. To make an intelligent and better decision on which wireless network should be chosen to deliver each service via the network that is the most suitable for it, the following context-aware metrics have been proposed for use in addition to the RSS measurements: perceived QoS and QoS requirements of sessions; user preferences such as a preferred wireless access network; terminal capabilities such as supported access networks, protocols, and available resources; status of the networks; location of the mobile host; and monetary costs involved in changing wireless networks.

Vertical Handoff

Vertical handoff involves changing the access interface across disparate networks, typically resulting in changing the mobile node's IP address and administrative realm.

Given the complementary characteristics of WLAN (faster, short-distance access) and WWAN (slower, long-range always-connected access), it is compelling to combine them to provide ubiquitous wireless access [1, 2]. Mobile devices are increasingly being equipped with multiple network interfaces that enable the mobile user to utilize WLANs in hot spots and switch to WWANs when the coverage of a WLAN is not available or the network condition in WLAN is not good enough. We refer to such a switch procedure as cross-access technology handoff or *vertical handoff.* Mobile users should access the Internet via an integrated heterogeneous network when they are free to move. An efficient mobility management scheme is crucial in this integrated network access situation. Mobility management consists of support for location management, which tracks and locates the mobile terminal (MT) for the delivery of incoming calls; support for handoff, which provides continuity of an ongoing connection in spite of movements between and across WWANs and WLANs; and support for personal mobility, which is the ability of the user to access his/her personal services, independent of terminal type or point of attachment.

Horizontal handoff, on the other hand, only deals with the switch between base stations (BSs) or access points (APs) in the same wireless network and is supported by all terrestrial wireless technologies in homogeneous environments. The next-generation all-IP wireless networks will need to support seamless vertical handoff in heterogeneous network architectures that encompass multiple access technologies. Seamless vertical handoff involves the ability of the mobile terminal to successfully or simultaneously attach to different APs in the integrated wireless network infrastructure in a way that makes the physical movement transparent and preserves application-level connectivity.

Mobility management solutions exist for all major layers of the Internet protocol stack. Link-layer mobility protocols avoid IP address changes while network-layer protocols hide them from the layers above. Transport-layer mobility protocols maintain a continuous connection between two endpoints over address changes. Session- and application-layer solutions reestablish transport-layer connections after an address change. All these solutions have their advantages and disadvantages.

The Mobile IP (MIP) Protocol is a natural candidate to support smooth vertical handoff. MIP provides network-layer mobility by enabling a mobile node (MN) that migrates from its home network to be addressable by the same home IP address across a foreign network the MN is visiting. MIP [3, 4] is a widely studied network-layer mobility management scheme in which a home agent (HA) and a foreign agent (FA) are used to bind the home address of a mobile node (MN) to the care-of-address at the visited network and provide packet forwarding when the MN is moving between IP subnets. Triangular routing of all incoming packets to the MN via the home network can cause additional delays and waste of bandwidth capacity. To optimize the routing performance, improve the handoff latency and packet loss, and resolve scalability problems associated with MIP, several new schemes such as Fast Handover Protocol [5] and Handoff Aware Wireless Access Internet Infrastructure (HAWAII) [6] have been proposed. The solutions, including MIP, significantly rely on newly introduced network infrastructures such as the HA and FA.

Session Initiation Protocol (SIP) extensions were proposed to extend the protocol to support host mobility [7, 8] from end-to-end, alleviating some of the shortcomings associated with MIP and its route optimization variants. However, this solution applies only to applications that use SIP.

A novel transport-layer scheme to support WWAN-WLAN vertical handoffs uses the multi-homing feature of mobile Stream Control Transmission Protocol (mSCTP), without support of network routers or special agents. This scheme is designed to solve many of the drawbacks of MIP. Because the transport layer is the lowest end-to-end layer in the Internet protocol stack and most of the applications in the Internet are end-to-end, a transport-layer mobility solution is a natural candidate for an alternative mobility scheme. Because the transport layer controls data flows, a transport-layer approach to vertical handoff enables the end nodes to adapt the flow and congestion control parameters quickly, thereby offering the potential for significant performance enhancements. Furthermore, in the transport-layer approach, no third party other than the end-point nodes participate in vertical handoff leading to fast handoff, and no addition or modification of network components is required, which makes this approach universally applicable to all current and future wired and wireless network architectures. Implementing a transport-layer approach to vertical handoff requires changes to existing transport-layer protocols. When considering the potential that a mobile terminal could be in contact with multiple APs at the same time, then other protocols might offer simpler starting points. A good candidate is the Stream Control Transmission Protocol (SCTP) because an SCTP association (that is, a relationship) can use multiple addresses simultaneously.

The Internet Engineering Task Force (IETF) has standardized the SCTP [11] as a reliable transport protocol to expand the scope beyond the Transmission Control Protocol (TCP) and User Datagram Protocol (UDP). The design of SCTP absorbed many of the strengths of TCP and incorporated several new features that are not available in TCP (such as multi-homing and multi-streaming). Multi-homing features and dynamic address reconfiguration (DAR) extension [12] of SCTP, referred to as the mobile SCTP (mSCTP) [13], have been developed to support vertical handoff between heterogeneous wireless networks.

The mobile SCTP, in the present form, targets seamless handoff of mobile nodes or sessions initiated by mobile nodes. To support a communication session initiated by a fixed correspondent node toward a mobile node, the mSCTP must be used along with an additional location management scheme such as MIP, SIP, or Reliable Server Pooling (RSerPool) [21].

Transport-layer mobility provides persistent connections as long as only one node of a connection changes its point of attachment to the network at a time. Transport-layer mobility may fail and the connection may break only in the case where both nodes of a connection move simultaneously and thereby change their network addresses simultaneously. In such a situation, a mobility solution based on the RSerPool protocol suite at the session layer arranges handoffs that are transparent for applications, and provides for efficient network-wide registration and lookup of peers.

WWAN-WLAN Vertical Handoff

There are two generic ways of interworking between WLANs and cellular networks—namely, *loose coupling* and *tight coupling*.

Tight Coupling Architecture

With a tight coupling architecture, the WLAN is connected to the cellular data core network in the same manner as any other radio access network (RAN), such as GPRS RAN and UMTS terrestrial RAN (UTRAN). The rationale behind the tight coupling is to make the WLAN appear to the cellular core network as another cellular access network. The WLAN emulates functions that are natively available in 3G/GPRS RANs. In this case, the WLAN data traffic goes through the 3G/GPRS core network before reaching the external PDNs (public data networks). In this way, especially the mechanisms for mobility, the QoS and security of the 3G/GPRS core network can be reused.

The existing 3G/GPRS protocols in the mobile, in particular, the logical link control (LLC), Subnetwork Dependent Convergence Protocol (SNDCP), GPRS mobility management (GMM), and session management (SM), are used in both a standard GPRS cell and a WLAN area. The different networks share the same authentication, signaling, transport, and billing infrastructures, independent of the protocols used at the physical layer on the radio interface.

Tight coupling architecture features have many benefits, including:

- Seamless service continuation across WLAN and 3G/GPRS
- Reuse of 3G/GPRS AAA (authentication, authorization, and accounting)
- Reuse of 3G/GPRS infrastructure and protection of cellular operator's investment
- Support of lawful interception for WLAN subscribers
- Increased security, because 3G/GPRS authentication and ciphering can be applied on top of WLAN ciphering
- Common provisioning and customer care
- Access to core 3G/GPRS services such as short messaging service (SMS), multimedia messaging service (MMS), and location-based services

However, tight coupling also has some disadvantages:

- Tight coupling is primarily tailored for WLANs owned by cellular operators and does not easily support third-party WLANs. The same operator must own both the WLAN and the 3G/GPRS parts of the network because the 3G/GPRS core network directly exposes its interfaces to the WLAN.
- There are some cost and capacity concerns associated with the connection of a WLAN to an SGSN (serving GPRS support node). For example, the

throughput capacity of an SGSN is sufficient for supporting thousands of low-bit-rate GPRS terminals but is not sufficient for supporting hundreds of high-bit-rate WLAN terminals.

■ Tight coupling cannot support legacy WLAN terminals that do not implement the 3G/GPRS protocols.

Loose Coupling Architecture

Loose coupling is defined as the utilization of WLAN as an access network complementary to the 3G/GPRS network, utilizing the subscriber databases in the 3G/GPRS network but featuring no data interfaces to the 3G/GPRS core network. The intent of WLAN-3G/GPRS interworking is to extend 3G/GPRS services and functionality to the WLAN UE and user, respectively. Thus, the WLAN effectively becomes a complementary radio access technology to the 3G/GPRS system. The WLAN bypasses the 3G/GPRS network and provides direct data access to the external PDNs.

Loose coupling is mainly based on standard IETF protocols for AAA and mobility. It is therefore not necessary to introduce cellular technology into the WLAN. Roaming can be enabled across all types of WLAN implementations, regardless of who owns the WLAN, solely via roaming agreements.

There are several advantages to the loose coupling architecture:

■ It allows independent deployment and traffic engineering of WLAN and 3G/GPRS networks.
■ While roaming agreements with many partners can result in widespread coverage, including key hotspot areas, subscribers benefit from having just one service provider for all network access.
■ It allows a WISP (wireless Internet service provider) to provide its own public WLAN hotspot, interoperate through roaming agreements with public WLAN and 3G/GPRS service providers, or manage a privately installed enterprise WLAN.

The loose coupling architecture offers several advantages over the tight coupling architecture. Therefore, it has emerged as a preferred architecture for the integration of WLAN with 3G/GPRS networks.

Main Features of SCTP and Mobile SCTP

SCTP: Stream Control Transmission Protocol

SCTP is an IP-based transport protocol that was defined by the IETF Signaling Transport Working Group for the transport of signaling data over IP networks. Recognizing that other applications could use the capabilities of SCTP, the IETF

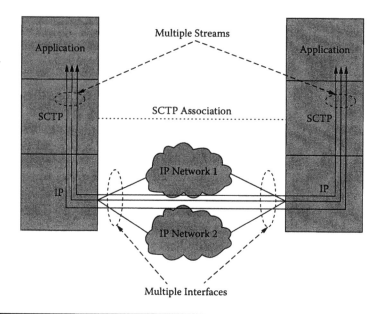

Figure 3.1 A schematic view of an SCTP association. (From S. Fu, and M. Atiquzzaman, *IEEE Communications Magazine*, April 2004, 64–76. With permission.)

has embraced SCTP as a general-purpose transport layer protocol above the IP layer of the Internet protocol stack. It is designed to expand the scope beyond TCP and UDP. An SCTP connection, called an *association*, provides novel services such as *multi-homing*, which allows the end points of a single association to have multiple IP addresses, and *multi-streaming*, which allows for independent delivery among data streams. Figure 3.1 shows the position of SCTP in the Internet protocol stack, and also illustrates an SCTP association using multi-homing and multi-streaming. The main interesting features of SCTP include multi-homing and multi-streaming.

Multi-Homing

An *endpoint* is the logical sender/receiver of SCTP packets. It is represented by a set of transport addresses; a *transport address* consists of a network layer address and a port number. In SCTP, all transport addresses of an endpoint must share the same port number. Thus, in practice, an SCTP endpoint is identified with a non-empty set of IP addresses and a single port number. Each transport address can belong to only one endpoint at a time. An SCTP association is a protocol relationship between two SCTP endpoints. Specifically, an SCTP association between two hosts *A* and *B* is defined as:

$$\{[a\ set\ of\ IP\ addresses\ at\ A] + [Port\text{-}A]\} + \{[a\ set\ of\ IP\ addresses\ at\ B] + [Port\text{-}B]\}$$

Figure 3.2 Multi-homed hosts. A₁ and A₂ represent two IP addresses for the endpoint host A. B₁ and B₂ represent two IP addresses for host B. (From A.L. Caro et al., *IEEE Computer*, 36, 11, 2003. With permission.)

Multi-homing allows an association between two endpoints to span across multiple IP addresses or network interface cards. As Figure 3.2 shows, a multi-homed host is accessible through multiple IP addresses. Currently, SCTP uses multi-homing only for redundancy, not for load sharing. Any of the IP addresses on either host can be used as a source or destination address in the IP packet. If one of its IP addresses fails, the destination host can still receive data through an alternate source interface. SCTP normally sends packets to a destination IP address designated the primary address, but can retransmit lost packets over the secondary address. The built-in support for multi-homed endpoints by SCTP is especially useful for achieving fast recovery from fault conditions in environments that require high availability of the applications. SCTP keeps track of each destination address's reachability through two mechanisms: ACKs of data chunks and *heartbeat* chunks—control chunks that periodically probe the status of a destination.

Multi-Streaming

Multi-streaming allows data from the upper layer application to be multiplexed onto one channel (called association in SCTP). Sequencing of data is done within a stream; if a segment belonging to a certain stream is lost, segments from that stream following the lost one will be stored in the receivers' stream buffer until the lost segment is retransmitted from the source. However, data from other streams can still be passed to the upper layer application.

Within streams, SCTP uses *stream sequence numbers* (SSNs) to preserve the data order and reliability for each data chunk. Between streams, however, no data order is preserved. This approach avoids TCP's head-of-line (HOL) blocking problem, in which successfully transmitted segments must wait in the receiver's queue until a TCP sending endpoint retransmits any previously lost segments.

Partial Reliability

Unlike TCP, which provides reliable deliveries, and UDP, which provides unreliable deliveries, SCTP has a partial reliability mechanism by which a user can specify a reliability level on a per-message basis. This reliability level defines how persistent an SCTP sender should be in attempting to communicate a message to the receiver (for example, never retransmit, retransmit up to a certain number of times, or retransmit until lifetime expires). The partial reliability mechanism benefits applications such as real-time multimedia traffic that are transferred during periods of poor QoS due to network congestion or path failures.

Mobile SCTP

In the standard SCTP, the endpoints exchange all the IP addresses before the SCTP association is established, and these IP addresses remain static during the session. The DAR extension [12] for SCTP defines two new chunk types (ASCONF and ASCONF-ACK) and several parameter types (Add IP address, Delete IP address, Set Primary IP address, etc.). It enables each endpoint to add or delete an IP address to or from an existing association, and to change the primary IP address for an active SCTP association using address configuration (ASCONF) messages. The extension allows an end host to signal to the peer endpoint which IP address is preferable as the primary address. Therefore, in a handoff situation where connectivity to two networks may be given, a mobile device can signal to its peer which network is the preferred destination to send data to and thus improve data throughput. An SCTP implementation with its DAR extension defines mobile SCTP (mSCTP) [13].

Reliable Server Pooling

While SCTP provides improved network-level fault tolerance, it does not improve node reliability. If a node (server) fails, the service it provides is interrupted. A solution to this problem is to have multiple servers providing the same service so that if a server fails, then its clients can arrange an application layer fail-over to another server to continue the service.

The Reliable Server Pooling (RSerPool) protocol suite defined by the Reliable Server Pooling Working Group of the IETF [21] focuses on providing server redundancy using server pools. It provides for efficient networkwide registration and lookup of peers. It is possible to build systems without single points of failure using a combination of SCTP and RSerPool. The RSerPool architecture uses three classes of elements:

- *Pool element (PE):* a server entity that has registered to a pool.
- *Pool user (PU):* a client being served by one PE.
- *Name server (NS):* provides a translation service for and supervises the PEs.

A pool is a set of servers providing the same service. A pool handle, which is a byte vector of arbitrary length, is a logical pointer to a pool. To become a PE for a specific pool, a server must register itself with the pool handle of the pool at its home NS. This NS will supervise the PE to make sure that it is working and informs the other NSs about the new PE. The Aggregate Server Access Protocol (ASAP) [20] is used between the PEs and NSs. It provides registration, re-registration, and de-registration of PEs, supervision of PEs using session keep-alive messages, and detection of NSs using server announcements. Pool handles are only valid within an operational scope. The protocol used by NSs within an operational scope to synchronize their name spaces is called the Endpoint Name Resolution Protocol (ENRP) [22].

If a client wants to be served by a PE belonging to a pool identified by a specific pool handle, it sends a name resolution request to an NS. The NS will respond with a subset of all transport addresses, which can be used to access the PEs. This communication also uses ASAP. The selection of the PE is realized in two steps:

1. The NS can select a subset of all PEs and their transport addresses in the pool. This selection can be based on the requested transport capabilities and/or the pool policy such as round-robin or least used.
2. The PU must select one of the PEs in the given subset. This can also be based on the pool policy.

Using mSCTP for WWAN-WLAN Vertical Handoff

A scheme to support WWAN/WLAN vertical handoff using mSCTP's multi-homing feature, where the handoff is accomplished at the transport layer without requiring any modification to the IP infrastructure, is discussed further in this section. Due to the multi-homing feature of mSCTP, an endpoint's network interface can be added into the current association if it is possible for the interface to establish a connection to the Internet via an IP address. The capabilities of mSCTP to add, change, and delete the IP addresses dynamically during an active SCTP association provides an end-to-end WWAN/WLAN vertical handoff solution. The scheme is a simpler network architecture than that required by network layer and application layer solutions because no addition or modification of network components is required. The main idea of the handoff procedure is to exploit multi-homing to keep the old data path alive until the new data path is ready to take over the data path transfer, and thus achieve a low-latency, low-loss handoff between adjacent subnets. The MN that initiates an SCTP association with a correspondent node (CN) may be in IPv4/IPv6 networks.

Association Type Considered in Mobile SCTP

Associations considered in mobile communications can be classified according the following two types:

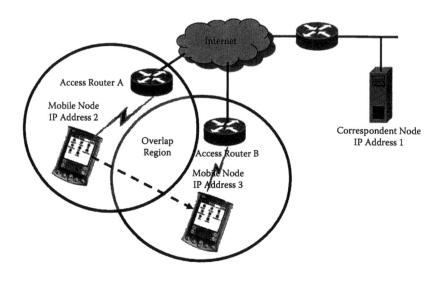

Figure 3.3   mSCTP vertical handoff. (From S. J. Koo, M. J. Chang, and M. Lee, *IEEE Communications Letters,* **March 2004. With permission.)**

1. Association initiated by the MN toward the fixed CN
2. Association initiated by the fixed CN toward the MN

The mobile associations in type (1) seem to be a natural extension of the client/server model, in which the mobile host originating the session can be viewed as a client, while the counter endpoint functions as a server.

On the other hand, type (2) requires the additional location management functionality for the association originator to find the current location of the mobile node and to keep track of the location changes.

The mSCTP, in its present form, is targeted for seamless handoff of mobile session type of type (1). To support association type (2), the mSCTP must be used along with an additional location management scheme such as MIP, SIP, or RSerPool [21].

Association Initiated by the MN toward the Fixed CN

Single-Homing CN

In this case, the CN is configured with only one IP address, say, CN_IP. An MN initiates an association with a CN. After initiation of an SCTP association, the MN moves from access router (AR) A in a WWAN cell to AR B in a WLAN cell, as indicated in Figure 3.3. The resulting SCTP association consists of IP address 1 (CN_IP) for the CN and IP address 2 (WWAN_IP) for the MN. Then the handoff procedural steps (i.e., obtain an IP address, add new IP address, change primary

IP address, and delete IP address) described below are repeated whenever the MN moves to a new location until the SCTP association is released.

Step 1. *Obtain an IP address for a new location*: The handoff preparation procedure begins when the MN moves into the overlapping region of a WWAN cell and a WLAN cell. Once the MN receives router advertisements from the new access router (AR B), it initiates the procedure of obtaining a new IP address, WLAN_IP. This can be accomplished using IPv6 stateless address autoconfiguration (SAA).

Step 2. *Add the new IP address to the SCTP association*: After obtaining a new IP address, WLAN_IP, the MN's SCTP informs the CN's SCTP that it will use a new IP address by sending an SCTP ASCONF chunk to the CN with parameters set to "add IP address" and WLAN_IP. The MN receives the responding ASCONF acknowledgment (ASCONF-ACK) chunk from the CN.

Step 3. *Change the primary IP address*: While the MN continues to move toward AR B, it needs to change the new IP address into its primary IP address according to an appropriate rule. The WWAN-to-WLAN vertical handoff is triggered by the MN (based on mobility information such as movement of the MN, and signal strength from the old and new ARs) sending an ASCONF chunk with the parameters set to "set primary address" and WLAN_IP, which results in the CN setting the MN's primary address to WLAN_IP. After the MN receives an ACK from the CN, the WLAN_IP address becomes the primary choice, and the network traffic between the MN and the CN is routed through the WLAN. This eliminates the infamous triangular routing problem inherent in MIP. Similarly, the WLAN-to-WWAN handoff is triggered by the MN sending an ASCONF chunk with parameters set to "set primary address" and WWAN_IP. In this case, after the MN receives an ACK from the CN, the WWAN_IP becomes the primary choice, and the network traffic between the MN and the CN is routed through the WWAN.

Step 4. *Update the location manager*: The handoff scheme supports location management by employing a location manager that maintains a database that records the correspondence between the MN's identity and the current primary IP address. Therefore, the MN should update the location manager's relevant entry with the new IP address once the "set primary address" procedure is completed successfully.

Step 5. *Delete the old IP address from the SCTP association*: When the MN moves out of the coverage of the WLAN cell, no new or retransmitted data packets should be directed to WLAN_IP. The MN sends an ASCONF message with parameters set to "delete IP address" and WLAN_IP to request that the CN release the IP address WLAN_IP from its routing table. After the MN receives an ACK from the CN, it deletes WLAN_IP from its address list, and WLAN_IP is released from the SCTP association.

Dual-Homing CN

In this case, the CN is configured with two IP addresses, say, CN_IP1 and CN_IP2. At the beginning of the vertical handoff procedure, WWAN_IP and CN_IP1 are the primary IP addresses of the MN and CN, respectively. The procedure adopted in this case is as follows:

Step 1. *Obtain an IP address for a new location*: The MN initiates the procedure of obtaining a new IP address, WLAN_IP, when it moves into the overlapping region of a WWAN cell and a WLAN cell.

Step 2. *Add the new IP address to the SCTP association*: After obtaining a new IP address, WLAN_IP, the MN's SCTP informs the CN's SCTP that it will use a new IP address by sending an SCTP ASCONF chunk to the CN with parameters set to "add IP address" and WLAN_IP. The MN receives the responding ASCONF-ACK chunk from the dual-homed CN bundled with an ASCONF to request the MN to add the CN's secondary IP address CN_IP1 into the association. The MN then sends an ASCONF-ACK to confirm the completion of the add IP address process.

Step 3. *Change the primary IP address*: The vertical handoff triggering process is different from that of the single-homed CN. In this case, because both the MN and CN are dual-homed, the MN can directly set the CN's secondary IP address as the primary address in its host routing table and start to send data on the new link.

Step 4. *Delete the old IP address from the SCTP association*: When the MN moves out of the coverage of the WLAN cell, no new or retransmitted data packets are directed to WLAN_IP. The MN sends an ASCONF message with parameters set to "delete IP address" and WLAN_IP to request that the CN release the IP address WLAN_IP from its routing table. When the dual-homed CN responds to the MN's request, the CN bundles an ASCONF to request the MN to delete the CN's secondary IP address from the association. The MN then sends an ACK to the CN to confirm completion of the delete IP process.

Association Initiated by the Fixed CN toward the MN

The handoff scheme requires a location manager for the fixed CN to locate the current position of the MN when a new association is to be set up by the CN. To support this type of association, the mSCTP is used along with an additional location management scheme such as MIP, SIP, or RSerPool [21].

When a location manager is used, the location management and vertical handoff procedures are performed in the following sequence:

1. The MN updates the location manager with the current primary IP address.
2. When the CN wants to set up a new association with the MN, the CN first sends a query to the location manager with the MN's identity (home address, domain name, etc.).
3. The location manager replies to the CN with the current primary IP address of the MN.
4. The CN sends an SCTP INIT chunk to the MN's new primary IP address to set up the association.

Location Management for mSCTP

As stated above, to support an association initiated by the fixed CN, the mSCTP must be used along with an additional location management scheme such as MIP, SIP, or RSerPool [21].

Use of mSCTP with Mobile IP

In this scenario, MIP is used to locate a mobile host and then for the HA to forward the data packet (SCTP INIT chunk) to the MN. The succeeding process for SCTP association initiation, including SCTP INIT-ACK, COOKIE-ECHO, and COOKIE-ACK, are done directly between the MN and the peering host (the CN), not via the HA. After an SCTP association is successfully set up, the mSCTP is used for providing seamless handoff for the MN.

Use of SCTP with SIP

In this scenario, each host uses SCTP instead of TCP/UDP as the transport protocol. After the call setup by SIP signaling, the SCTP is used for data transport and seamless handoff.

The SIP provides location management functionality using SIP REGISTER messages. When the MN moves into a visiting network, it updates its current location (e.g., IP address or SIP URL) by sending a SIP Register (with a Contact Header) to the (home) SIP Registrar server. The Registrar server then updates the location database as indicated by the REGISTER message.

When a call setup is requested with the MN, the (home) SIP proxy server interrogates the location database to locate the MN and then relays the SIP INVITE message to the (visiting) SIP Proxy server up to the MN. Once the SCTP association is established via SIP signaling, the data transport between the two concerned hosts is done according to the mSCTP handoff mechanisms.

Use of SCTP with RSerPool

RSerPool can be used for location management. A mobile server registers a pool handle such that it becomes part of a pool. It is allowed that a pool consists of one pool element only. A client (mobile or not) must know the pool handle of the mobile server it wants to talk to. It sends a name resolution request to one of the ENRP name servers and gets back the current IP addresses. Because the ENRP NSs within an operational scope share its state, it is not important which ENRP NS is contacted. If the mobile server (MS) changes its IP address, it re-registers at the home ENRP NS. So the pool handles can be used to address a server with changing IP addresses. If either the MN or the MS change their addresses due to handoffs, the mSCTP can be used to handle this, except for the case where the MC (mobile client) and the MS change their addresses simultaneously. The NS functionality can be compared to a location register in mobile networks because the ENRP ensures that all NSs of the operational scope get the updated pool data.

Simultaneous Handoffs

If both communicating nodes are mobile, the situation may occur that both nodes move to new networks simultaneously, and therefore change their addresses at the same time. Then each node is unable to inform the other about its address change because all known addresses of the peer node have become invalid. Therefore, mSCTP fails, that is, the SCTP association breaks. To cope with this problem, there are some possible solutions.

mSCTP and MIP

Each mobile peer is always reachable via its home address using MIP. Packets sent to this address are tunneled via the HA and FA to the mobile device. However, it is more efficient to use mobile SCTP and MIP because all data reaches the SCTP endpoints directly instead of through the HA. Two IP addresses can be assigned to the SCTP association at the mobile host: a permanent home address (HoA) and a care-of address (CoA). The HoA is kept unchanged throughout the whole life of the association, while the CoA will be assigned by the current network interface. In situations where both peers move simultaneously, the respective ASCONF requests should be sent to a peer's HoA, and will therefore reach the peer. All user data can then be sent on a direct path to the last known peer CoA. Any new ASCONF request should be sent to the HoA to guarantee its arrival, even in the case of simultaneous address changes.

mSCTP and RSerPool

RSerPool inserts a layer between the transport and application layers that relieves the application layer from managing communication sessions. When the transport connection breaks due to simultaneous movement of the two endpoints, the RSerPool layer ensures establishment of a new transport association and triggers an application-specific fail-over procedure.

In the RSerPool framework, at least one node registers as a PE of a server pool having a unique handle. Consider a communication session between two nodes based on the RSerPool. The receiver node registers under a unique pool handle, after which the sender node can establish the association with the help of an NS and then connect to the resolved transport address of the receiver node.

The mSCTP fails, that is, the SCTP association breaks, in the case of simultaneous mobility of the two nodes. The RSerPool session concept can be used to reestablish a new SCTP association using the new addresses and continuing the RSerPool session. Depending on the application, the impact of this session failover for the application can be very small. The receiver node re-registers its new transport address but under the same previous pool handle with its NS. The ENRP ensures that all NSs of the operational scope get the updated pool data. Using an appropriate pool policy, the sender node is now able to let the NS resolve the pool handle to the new transport address. Then, it can establish a new association and execute an application-specific fail-over procedure. After that, the application can continue the communication session.

Summary

This chapter discussed mobility from both the terminal and network points of view. It showed how mobility is supported in each layer of the OSI through different protocols such as Mobile IP, which supports mobility in the network layer. Tight and loose couplings were discussed within the context of vertical mobility. SCTP is well explained as a means of signaling in the transport layer, including how a single endpoint can be multi-homed or used in multiple transport addresses. This address consists of multiple network layer addresses linked to a port number. In the standard SCTP, the endpoints exchange all the IP addresses before the SCTP association is established, and these IP addresses remain static during the session. A scheme to support WWAN/WLAN vertical handoff uses mSCTP's multi-homing feature, wherein the handoff is accomplished at the transport layer without requiring any modification to the IP infrastructure. Due to the multi-homing feature of mSCTP, an endpoint's network interface can be added to the current association if it is possible for the interface to establish a connection to the Internet via an IP address. The capabilities of mSCTP to add, change, and delete the IP addresses dynamically during an active SCTP association provide an end-to-end WWAN/WLAN vertical handoff solution. The scheme is a simpler network architecture than those required by network layer and application layer solutions because no addition or modification of network components is required.

Practice Set: Review Questions

1. Vertical and horizontal handoff are the same.
   a. True
   b. False
2. Vertical handoff is handoff within the same network.
   a. True
   b. False
3. Horizontal handoff helps maintain connectivity within the same network.
   a. True
   b. False
4. Multi-homing means:
   a. An Internet node having multiple IP addresses
   b. One Internet IP address is associated with many IP nodes
   c. One IP address for one Internet node
   d. All of the above
   e. None of the above
5. Multi-streaming allows:
   a. Data from the upper layers to be distributed into many streams
   b. Data from the upper layers to be multiplexed together onto one channel
   c. Sending many multimedia streams to many ports
   d. None of the above
   e. All of the above
6. SCTP provides reliable server pooling because:
   a. It facilitates the use of many application servers
   b. It allows multiple servers to provide the same service to negate service failures
   c. If a node fails in SCTP, the service is not interrupted
   d. All of the above
   e. None of the above

Exercises

1. List context-aware handoff metrics.
2. Distinguish between mobility management at different layers of the OSI.
3. What are the major limitations of Mobile IP?
4. How do you couple WLAN to work with cellular networks?
5. Distinguish between tight and loose coupling of WLAN to cellular networks.
6. What are the major disadvantages of tight coupled WLAN to cellular networks?
7. What are the advantages of loose coupling of WLAN to cellular networks?

References

[1] A.K. Salkintzis, C. Fors, and R. Pazhyannur, WLAN-GPRS Integration for Next-Generation Mobile Data Networks, *IEEE Wireless Communications*, October 2002, pp. 112–124.

[2] M.M. Buddhikot et al., Design and Implementation of a WLAN/CDMA2000 Interworking Architecture, *IEEE Communications Magazine*, November 2003, pp. 90–100.

[3] C. Perkins, IP Mobility Support for IPv4, *IETF RFC 3220*, January 2002.

[4] C. Pekins et al., Mobility Support in IPv6, <http://www.ietf.org/html.charters/mobil-eip-charter.html>.

[5] R. Koodli (Ed.), Fast Handovers for Mobile IPv6, IETF draft-ietf-mobileip-fast-mipv6-08.txt (work in progress).

[6] R. Ramjee et al., HAWAII: A Domain-Based Approach for Supporting Mobility in Wide-Area Wireless Networks, *IEEE/ACM Transactions on Networking*, 10(3), June 2002, 396–410.

[7] F.Vakil et al., Supporting Mobility for TCP with SIP, IETF Draft, work in progress, July 2001.

[8] T. Kwon et al., Mobility Management for VoIP Service: Mobile IP vs. SIP, *IEEE Communications Magazine*, October 2002, pp. 66–76.

[9] Q. Zhan, C. Guo, Z. Guo, and W. Zhu, Efficient Mobility Management for Vertical Handoff between WWAN and WLAN, *IEEE Communications Magazine*, November 2003, pp. 102–108.

[10] C. Guo, Z. Guo, Q. Zhan, and W. Zhu, A Seamless and Proactive End-to-End Mobility Solution for Roaming Across Heterogeneous Wireless Networks, *IEEE Journal on Selected Areas in Communication*, June 2004, pp. 834–848.

[11] R. Stewart et al., Stream Control Transmission Protocol, IETF RFC 2960, October 2000.

[12] R. Stewart et al., Stream Control Transmission Protocol (SCTP) Dynamic Address Reconfiguration, draft-ietf-tsvwgaddip-sctp-08.txt, Sept. 2003, work in progress.

[13] M. Riegel and M. Tuexen, Mobile SCTP, draft-riegel-tuexen-mobile-sctp-03.txt, Aug. 2003, work in progress.

[14] S. Fu and M. Atiquzzaman, SCTP: State of the Art in Research, Products, and Technical Challenges, *IEEE Communications Magazine*, April 2004, pp. 64–76.

[15] A.L. Caro et al., SCTP: A Proposed Standard for Robust Internet Data Transport, *IEEE Computer*, 36(11), November 2003, pp. 56–63.

[16] S. Fu et al., TraSH: A Transport Layer Seamless Handover Scheme, Tech. Rep., Comp. Science, University of Oklahoma, www.cs.ou.edu/~atiq, November 2003.

[17] L. Ma et al., A New Method to Support UMTS/WLAN Vertical Handover Using SCTP, *IEEE Wireless Communications*, August 2004, pp. 44–51.

[18] S.J. Koo, M.J. Chang, and M. Lee, mSCTP for Soft Handover in Transport Layer, *IEEE Communications Letters*, March 2004, pp. 189–191.

[19] S.J. Koh et al., Mobile SCTP for Transport Layer Mobility, draft-sjkoh-sctp-mobil-ity-04.txt, June 2004, work in progress.

[20] R. Stewart, Q. Xie, M. Stillman, and M. Tuxen. Aggregate Server Access Protocol (ASAP), IETF Reliable Server Pooling Working Group, May 2003, draft-ietf-rserpool-asap-07.txt, work in progress.

[21] M. Tuxen, Q. Xie, R. Stewart, M. Shore, L. Ong, J. Loughney, and M. Stillman, Requirements for Reliable Server Pooling, IETF RFC 3237, January 2002.

[22] Q. Xie, R. Stewart, and M. Stillman, Endpoint Name Resolution Protocol (ENRP), IETF Reliable Server Pooling Working Group, May 2003, draft-ietf-rserpool-enrp-06.txt, work in progress.

[23] T. Dreibholz, A. Jungmaier, and M. Tuxen, A New Scheme for IP-based Internet Mobility, *28th Annual EIII Conference on Local Computer Networks*, <http://tdrwww.exp-math.uni-essen.de/dreibholz/rserpool/rserpool-publications/LCN2003.dpf>, 2003.

Chapter 4

Mobility

One major limitation of the current Internet is its inability to provide and support mobility. This chapter discusses the solutions on mobility for IP communications. Mobility solutions are engineered differently in different layers of the OSI, including the physical, data-link layers, network, session, transport, and the application layers. We discuss a solution for mobility in relation to SIP in the application layer.

The mobility of nodes is handled in the application layer. The problem is that a node that changes from one link to another is not able to maintain communication at the new location without changing or acquiring a new IP address. This is undesirable, as a node should be allowed to keep its IP address in the new or foreign network. In IPv4 and IPv6, Mobile IP is one of the solutions to this problem and is discussed in subsequent sections of this chapter.

A key objective in next-generation IP networks is seamless mobility across heterogeneous networks that ensure continuity of communications. Typically, it is essential not to break communications while moving within the coverage ranges of a 3G network into a WLAN coverage and into the coverage area of another type of network when they exist in a location. This should happen under low, medium, and high speeds as well.

The migration of communication from one network to another at the same point and maintaining sessions or communication is a serious requirement in modern communications. For example, there should be no break in communication when a mobile node moves from within the domain of one wireless LAN (WLAN) to the domain of another WLAN, or from one ad-hoc network to another. The solution to this problem deals with the so-called *vertical handover* between networks and is discussed.

Mobility can be described in terms of the terminal, the entity moving, and in terms of the speed of the entity. The speeds of the entity define nomadic, cellular, micro, and macro mobility. Nomadic mobility is the ability of a user or terminal to change the network point of attachment or access as he/it moves with the service session completely stopped and restarted again when a new access is gained. In nomadic mobility, the device moves and the continuity of service across the two attachment points is limited or nonexistent. Hence, the user's current session could be completely stopped and reactivated when a new attachment is gained.

In cellular mobility, the user maintains a point of attachment to the network and this attachment is handed over from base station to base station with service continuity. This is made possible by implementing seamless handover between either neighboring networks and/or neighboring BSCs. What is important is that the user should not realize when this handover has taken place, as the handset does not show any break in communication. As such, there is no need to restart the access to the network.

Micro mobility occurs when there is frequent and fast handoff within a small geographical coverage area. This could result due to small cell sizes or fast-moving terminals or users as in fast vehicles. Macro mobility involves fewer handoffs and occurs in larger geographical coverage areas. This too occurs within the coverage areas of large cells and also at very low speeds of the terminals and users. For example, large cells are used in rural areas. That also means that even when communication is established under fast-moving vehicles, the rate of handoff is still low because of the large sizes of the cells.

In this chapter, the term "IP mobility" is used to encompass the various types of mobility in IP networks. The applications of IP mobility include monitoring and surveillance systems, security, tele-health, telematics, vehicular networks, personal area networks, mobile entertainment, tele-education, remote control, and emergency response systems.

Continuously, the issue of IP network mobility has emerged as areas previously uncovered with communication networks are being covered, such as commercial and fleets of military aircraft, buses, personal automobiles, trains, taxi cabs, and other forms of mobile transport. IP network mobility therefore is first geared toward remaining connected to provide communication, entertainment, security, and also health/emergency support systems. It is essential to distinguish this type of mobility from that of single nodes. In IP network mobility, there are multiple nodes in the IP network subnet, all moving — including the router that supports them. Hence, the issue here is IP-layer mobility. IP-layer mobility is facilitated by IPv6. For the communication to be maintained, a node in such a mobility scenario must be able to change the point of attachment or IP subnet without breaking communication. However, changing IP subnet means changing the IP address and also the routing directives! It is therefore essential to find a solution to this problem because, traditionally, changing the IP address leads to a break in communication and also changing the IP subnet breaks routing. It is essential therefore not to break routing or communication in IP network mobility. These mobility solutions are explained in this chapter.

Distinction between Mobility Types

Mobility can be grouped broadly in terms of host mobility, network mobility, and ad-hoc network mobility. Host mobility is covered generally by Mobile IP and is discussed in another section. The second type of mobility is network mobility. The whole network is mobile and this mobility is usually due to the fact that the platform on which the router is deployed is mobile. This mobility scenario includes the case when hosts are mobile as well as when the hosts are fixed, as in Figure 4.1. Network mobility also includes personal area networks when the network is carried by a moving person.

Mobility in the ad-hoc networking sense is desirable in most uncovered areas where it is necessary for hosts to form networks without the support of an existing fixed infrastructure. Mobility, in general, affects system performance. It reduces the system capacity (throughput) and also can introduce significant signal-to-noise degradation. There are various mobility solutions that have been proffered. We have discussed mobility in the application layer (SIP), and the next section discusses mobility in the network layer in terms of Mobile IPv6.

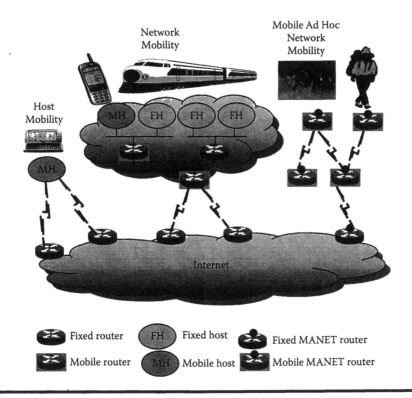

Figure 4.1   Distinction between mobility types.

Mobility in the Network Layer

In the current Internet, all IP addresses are location dependent and act as both an identifier and a locator. However, when a node moves, we lose the ability to use the same address to locate it wherever it is. This problem in the mobility of the node can be solved. Therefore, for the Internet to support mobility, we need a location-independent identifier for mobile hosts and a location-dependent address. The location-independent identifiers enable any user who wants to communicate with the mobile host to do so with the unique identifier and not be bothered about its current location. A location-dependent address is also required for a mobile host so that packets for it can be addressed and routed to it wherever its current location is. These two requirements cannot be satisfied with a single IP address. Therefore, a mechanism that provides identity and also permits locating the host/routing with packet delivery is required.

Mobile host mobility solution is provided as part of IPv6 (RFC 3775) mobility framework. In this case, the mobile host must obtain a topologically correct IP address at its new location. Similarly, IP network mobility support is provided in the so-called NEMO basic support in RFC 3963. In this case, only the mobile router needs to obtain a topologically correct IP address, and nodes attached to the router do not need to change their points of attachment.

Mobile IPv6 Host Mobility Solution

IPv6 provides mobility support through its mobile IP framework. This section is a summary of this framework or solution. We use the following acronyms:

AR: access router (foreign network router)
CN: correspondent node
HA: home agent (home network router)
MN: mobile node (also called mobile host [MH] and mobile router [MR])
MNP: mobile network prefix
HoA: home address of mobile node
CoA: care-of address of mobile node

Being able to locate a mobile host in its current location is essential not only for the support of mobility, but also for roaming between networks owned by various operators and also for handoff between different access types such as UMTS, GPRS, or WLAN. Furthermore, enhancing a node's ability to choose between different access and network types provides it with the means to choose the network with the best reception and quality of service based on its location. Not only does it gain the best coverage in doing so, but also could lead to better quality of service, cheaper costs, and speed.

Using Mobile IPv6, each host has two addresses. One is used for its identification and the second is used for routing the data to it. A permanent HoA is used for

Figure 4.2 Home address using network prefix_h.

identifying the mobile host and for direct communication from all correspondent nodes (Figure 4.2). This is issued by the home network through the HA. The second address, the CoA, which an MN obtains using auto-configuration, allows the HA to tunnel packets to the MN as it receives them from the CNs. Thus, to the CN, it appears as if the MN is always at its home network.

The HA keeps information on the binding or association between the two addresses. At the foreign network through agent advertisement, the MN acquires the prefixes of the visited networks and attaches them to its host suffix to create the CoA. By retaining the host suffix in its new location and changing its network prefix, an MH gains a CoA. Hence, it can still be recognized and also its current location can be identified. In this manner, Mobile IPv6 helps the node separate its identity from its location (Figure 4.3).

The home agent helps the mobile host manage its mobility and access to the Internet, and connections/sessions established with the mobile host can therefore survive mobility. In Figure 4.4, the mobile node is visiting network A and therefore is given the prefix for the network A, which is used to complete its 128 bit address.

Figure 4.3 Care-of-address using network prefix_f.

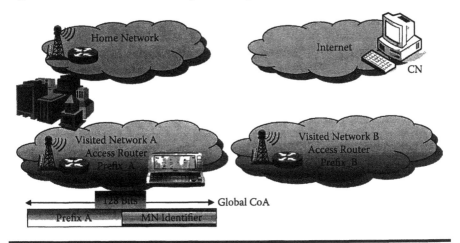

Figure 4.4 Mobile node visiting network A with IPv6 mobility.

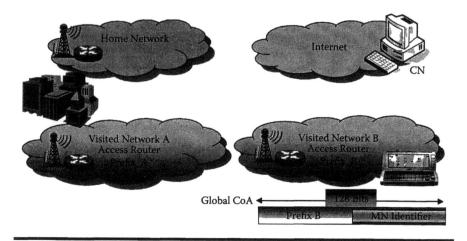

Figure 4.5   Mobile node visiting network B with IPv6 mobility.

Hence, its location can be determined on the Internet. When the mobile node moves to network B with Prefix_B (Figure 4.5), it relinquishes the Prefix_A and takes on the Prefix_B and uses it to complete its IPv6 address.

The HA keeps track of the CoA address of the MH and uses tunneling to send packets from the CN to an MH. It uses IP-IP tunneling and creates a virtual connection between the CoA of the MH and the HA (Figure 4.6). In this manner, packets sent to the home address are sent to the CoA and the MH appears to the HA as if it is at home.

Using a proxy neighbor discovery, the home agent is able to redirect packets on its home links that were destined for the home address of the MN to its link layer address.

One of the main features of Mobile IPv6 mobility is the elimination of the role of a foreign agent (FA). It is not used. The objective is to offer direct communication between the mobile host and its correspondents. A mobile node is always reachable through its home address. Each correspondent node is able to obtain a binding update from the binding cache and, through it, learn the location of the mobile node by processing the binding update options. The home agent tracks and collects packets meant for the mobile node, encapsulates and tunnels them, and sends them directly to the mobile node. It also does the reverse tunneling from the mobile node to the correspondent node.

Detection of movement is done using nearest neighbor discovery. Because of mobility, there is need for binding update cache management. Each time the

IPv6 Tunnel Header (CoA-HA)	IPv6 Header (HA-CN)	Data

Figure 4.6   Mobility in IPv6.

mobile node is attached to a foreign network, it sends a binding update to its home agent. Each binding update has a time to live, after which the binding is assumed stale. The mobile node is also required to keep a list of all the correspondent nodes to which it sends its binding update. Therefore, a mobile node may have several care-of-addresses and the primary care-of-address is the one it reports to the home agent. The mobility of a host increases its exposure to security violations. Hence, it is necessary to protect the binding update messages and a need for authentication with visited networks and IPSec can be used to achieve these.

A mobile node communicates with correspondent nodes through the HA. The home agent tunnels the packets to the CN.

Route Optimization

IPv6 mobility includes a feature called route optimization. At the inception of communications, the correspondent node reaches the mobile node through its HA, which tunnels the packets to the mobile node where it is. This sequence of communication is therefore: CN → HA → MN. This communication allows the MN to know about the correspondent node and hence can communicate with it directly: MN → CN using its HoA. Alternatively, the mobile node can communicate with the correspondent node through the HA that tunnels packets to the CN. Once communication is established between the MN and the CN, the CN can communicate with the mobile host directly: CN → MN. Therefore, the HA is not a single point of failure anymore and, furthermore, direct communications reduces delays. When the communication distance between the correspondent node and the mobile node CN → MN is shorter than the distance CN → HA → MN, authentication is required between the CN and the MN.

Mobile IP Network Mobility

The NEMO (NEtwork in MOtion) framework is used to address the issue of the mobility of a network when the whole network is moving. A network that is mobile allows always-on IP connectivity for all the terminals attached to it. It uses an MR, which is a mobile IP client that is configured as a stub network. Hence, the whole subnet is mobile and the terminals that are attached to it are oblivious to this. Traffic to the MR is routed through its HA, and traffic from the terminals attached to the MR goes through the MR. Figure 4.7 depicts a NEMO deployed in moving vehicles.

The aircraft, bus, and ambulance in the figure all carry MRs and the ambulance can access the Internet by hopping through the bus or by a roadside gateway (not shown in the figure). Communication from and to the bus (aircraft) uses an IP tunnel to the HA.

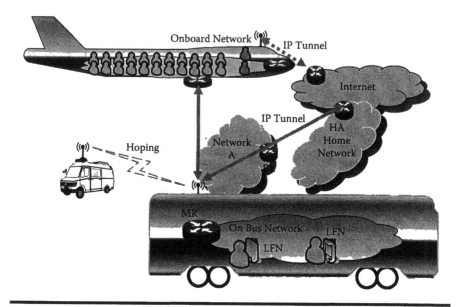

Figure 4.7 NEMO in moving vehicles.

Summary

After reading this chapter, you should be able to explain how mobility is engineered in an IP network. Mobility is grouped broadly in terms of host mobility, network mobility, and ad-hoc network mobility. Host mobility is covered generally by Mobile IP. The mobility of nodes is handled in the application layer. The migration of communication from one network to another at the same point and maintaining sessions or communication is a serious requirement in modern communications. Mobility is described in terms of the terminal or the entity moving and in terms of the speed of the entity. The speeds of the entity define nomadic, cellular, micro, and macro mobility. Nomadic mobility is the ability of a user or terminal to change the network point of attachment or access as he/it moves with the service session completely stopped and restarted again when a new access is gained. In cellular mobility, the user maintains a point of attachment to the network, and this attachment is handed over from base station to base station with service continuity. In the current Internet, all IPv4 addresses are *location dependent* and act as both an identifier and a locator. However, when a node moves, we lose the ability to use the same address to locate it wherever it is. IPv6 solves this problem to a point using Mobile IPv6, wherein each host has two addresses. One is used for its identification and the second is used for routing of data to it. A permanent home address (HoA) is used for identifying the mobile host and for direct communication from all correspondent nodes (CNs). This is issued by the home network through the home agent (HA). The second address, the care-of address, which a mobile node obtains using auto-configuration, allows the HA to tunnel packets to the mobile node as it receives them from the CNs. This chapter also explained network mobility—that is, the mobility of a whole network. A network that is mobile allows always-on IP connectivity for all the terminals attached to it. It uses a mobile router (MR), a mobile IP client that is configured as a stub network. Hence, the entire subnet is mobile and the terminals that are attached to it are oblivious of this.

Practice Set: Review Questions

1. Mobility has many definitions, including:
   a. Network mobility
   b. Host mobility
   c. Service mobility
   d. Session mobility
   e. All of the above
2. IPv6 supports host mobility with the aid of Mobile IPv6 and:
   a. Two host addresses
   b. One host address
   c. Just the host MAC address
   d. None of the above
   e. All of the above
3. The care-of address created by IPv6 in a visited network is formed by:
   a. The host prefix and the prefix of the visited network
   b. The host prefix and an additional host prefix
   c. Retaining its old host address
   d. Swapping the last 16 bits of the host prefix
   e. Exchanging its host prefix for a new host prefix
4. In IPv6 when a node moves away from its home network:
   a. Its location is lost completely
   b. It retains the location of its home network
   c. It gains a new location through the prefix of the visited network
   d. It cannot be reached at all
   e. It cannot access the Internet in any way again
5. The home agent of a mobile node:
   a. Uses labels to send packets to the mobile host
   b. Uses a new IP address to reach the mobile host
   c. Has no way of sending packets to the mobile host
   d. Uses tunneling to send packets to the mobile host
   e. All of the above

Exercises

1. List all the mobility types.
2. Distinguish all types of mobility.
3. Compare and contrast Mobile IPv4 and Mobile IPv6.
4. What is route optimization in Mobile IPv6?
5. When is re-authentication necessary in Mobile IPv6?

Chapter 5

IPv6

Throughout its history, the Internet has seen repeated improvements and enhancements to patch the inherent flaws in its initial design—and they are many. The Internet is traditionally not plug-and-play because nodes and networks need elaborate set-up and configuration processes. Other prominent flaws include its ability to support only best-effort QoS, lack of security, inability to support mobility in the real sense, and the limited address space provisioned in IPv4.

The fast-dwindling IPv4 address space led to the creation of IPv6 to replace it. IPv4 address space was limited to 32 bits or about 4 billion addresses. To provide for a large number of hosts without wasting too much space in the IP packet on overhead, the designers of TCP/IP settled on an address size of 32 bits. In 1969, this looked like a lot of addresses! In 2009, this address space is definitely far too small if every human being on Earth were to request one for a network node. Although the introduction of NAS, PPP, DHCP, and CIDR helped to reclaim some of the addresses on offer at any point in time, it soon became obvious that something other than patching IPv4 was required to solve the problem. The consumption of IP addresses so far can be modeled using approximately a fifth-order polynomial. The model shows that by the year 2013, it is likely that all the free IPv4 addresses will have been distributed. This knowledge may have informed network equipment (routers, hubs, and bridges) manufacturers to adopt enabling new equipment to support both IPv4 and IPv6.

IPv6 provides a reasonable but also timely means to reengineer the IPv4 in terms of adding features that support quality of service (QoS), multicast, and also mobility. It also provides the chance to include other new features such as binding updates and to simplify the IPv4 packet header structure.

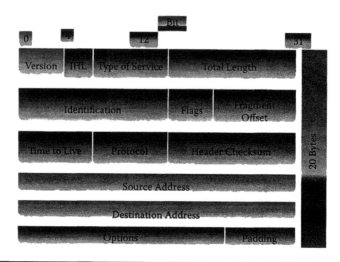

Figure 5.1   IPv4 packet header fields.

Quick Refresher in IPv4

Although this chapter is not about IPv4, it is however essential to introduce it to gain familiarity with where we are coming from and migrating to. IPv4 addresses are grouped into five classes (A, B, C, D, and E). Class E is reserved and has not been distributed. IPv4 address ranges include:

> Class A: 1.0.0.1 to 126.255.255.254
> Class B: 128.1.0.1 to 191.255.255.254
> Class C: 192.0.1.1 to 223.225.254.254
> Class D: 224.0.0.0 to 239.255.255.255—reserved for multicast groups
> Class E: 240.0.0.0 to 254.255.255.254—reserved

Class D addresses are reserved for multicast communications. The greater majority of IPv4 addresses in use worldwide fall within the classes A, B, and C. In addition to the normal payload (data) that is carried by an IP packet, the IP packet consists of a header section that is used to aid network protocols to decide the destination of the packet, the source of the packet, and what protocols within the network should do with the packet as it is routed from the source to the destination.

The IPv4 header is 20 bytes long and consists of header fields as shown in Figure 5.1.

- *Version:* currently 0100 (4) for IPv4 but will be 0110 (6) in IPv6.
- *Internet Header Length (IHL):* length of header in 4-byte words; Min. = 5.
- *Type of Service (TOS):* the type of service is one of:
  - Maximize reliability

- Maximize precedence
- Minimize delay
- Maximize throughput
■ Total Length: the total length (size) of the IP datagram in bytes.
■ Identifier: a unique number to identify a datagram.
■ Flags: only two flags are defined:
  - "More" bit: used in segmentation and reassembly
  - "Don't fragment" bit: used if fragmentation is not allowed
■ Fragment Offset: identifies fragment position in the non-fragmented PDU.
■ Time-To-Live: every time the packet is routed, TTL is reduced by 1. If TTL becomes zero, the packet is discarded.
■ Protocol: identifies which upper layer protocol to deliver the packet to.
■ Header Checksum: detects errors in the header only and must be recalculated by each router.
■ Source Address: IPv4 address where the data originated.
■ Destination Address: address where the datagram is going.
■ Options and Padding: some options are:
  - Security (very minimal!)
  - Route recording
  - Source routing
  - Stream identification
  - Time stamping
  - Padding may be required to make the total header length an integer multiple of 4 bytes
■ *Data:* this is the most important part, the actual data (payload) in the packet!

IPv6

Internet Protocol version six (IPv6) is still a relatively new protocol that was engineered to account for some of the deficiencies in IPv4. It specifically addressed the limited address space, increasing each address to 128-bit length instead of the old 32 bits. Notably, if this new address space is shared for each square inch on the Earth's surface, there will be about 100 addresses per square inch. The future is, however, unknown to man. Extensions to living on other planets and space settlements may mean that this address space could be shared not only with man-machine identifications on Earth, but also in space and other planets. Hence, it is too early to conclude anything about the adequacy of this address space for future IP communications.

IPv6 is a complete rebuild of IPv4 and is designed to cope with the Internet of today and in the future. Both forward and backward compatibility are built into IPv6. A lot of people have worked on its design and hence it is a better and well-engineered protocol than its parent IPv4. One of its biggest drawing cards is the inherent auto-configuration to support plug-and-play. This feature reduces the financial burden

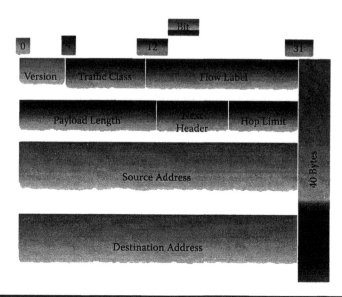

Figure 5.2   IPv6 packet header fields.

required for human interventions to configure network nodes. A server-less auto-configuration is essentially useful for nontechnical users of the Internet to quickly get going and also for those familiar with the Internet working to spend minimum time in setting up network nodes and terminals that are used to access the Internet.

The IPv6 header is also highly simplified compared with IPv4. This simplification is vital for more efficient performance of routers and network elements because they spend less time unraveling the IP header fields to make decisions about routing and where to send packets and which applications should be used to recover them. It also facilitates the embedded security mechanisms and the often extra patches that we are so used to in IPv4.

Although the size of the IPv6 header is twice that of the IPv4, there are fewer fields to consider. Figure 5.2 shows the IPv6 header. It has eight objects as opposed to thirteen objects in IPv4.

The packet header provides a description of the IP packet to other protocols in the communication infrastructure, so that they understand it and know how to handle it. It consists of a source address (128 bits) and a destination address (also 128 bits long). It contains a version field (6), a traffic class that aids in the definitions and handling of quality-of-service issues, a flow label, the length of the payload, and the hop limit (which is similar to the TTL I IPv4). Each field is described further in this section.

- *Version (4 bit):* same as before.
- *Traffic Class (8 bit):* similar in functionality to "Type of Service" field in IPv4. Mechanism to allow differentiated services to be implemented. As yet undefined values but commonly used by proprietary or *de facto* "standards."

- *Flow Label (20 bit):* a unique label assigned to a source/destination pair for a particular traffic flow that allows routers to forward packets based solely on that flow label value (i.e., a MPLS-like mechanism).
- *Payload Length (16 bit):* length of datagram payload in bytes. Option fields are included in this.
- *Hop Limit (8 bit):* same as TTL in IPv4.
- *Source Address:* 128-bit IPv6 address of originating entity.
- *Destination Address:* 128-bit IPv6 address of destination (not necessarily the final destination if a routing header is in use).
- *Next Header (8 bit):* identifies type of header immediately following the IPv6 header (if any).

Apart from the extra length of the addresses and the header being twice the size of the IPv4 header, there are other differences worth noting. The existing TOS field is expanded and renamed Traffic Class. No details of implementation are standardized as yet—some proprietary systems use it. IPv6 does not provide full QoS but what is provided is better than what we have now in IPv4. In the real sense, IPv6 cannot support QoS any more than IPv4. Other protocols and infrastructure built on top of or beside IP have to take care of that.

The Flow Label field is a facility for a tag switching or label switching concept (as in MPLS). It simplifies routing and forwarding for traffic streams. It also provides less switching delay and higher throughput. It has the ability to treat streams individually.

IPv6 has a vastly improved security. Its inherent "IPSec" refers to the combination of authentication header and encapsulating security payload. This is not entirely unique to it because the same features can also be implemented in IPv4 if desired. Together, the features provide confidentiality, data origin authentication, connectionless integrity, anti-replay service (a form of partial sequence integrity), and limited traffic flow confidentiality.

Next Header Options

The Next Header field points to the next header, if one is present; if not, then the value in this field is essentially "TCP"—that is, the IP payload. Extra headers are used only when the features specified are necessary. The next header options include:

- Hop-by-Hop Options
- Destination Options: for some destinations along the route
- Routing header
- Fragment header: this applies to the source only!
- Authentication header (see RFC2402)

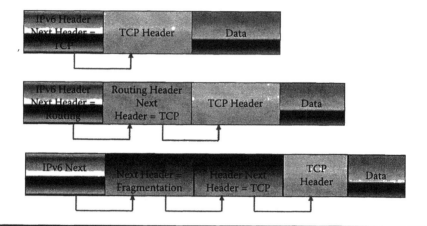

Figure 5.3   Format of next header field.

- Encapsulating Security Payload header (see RFC2406)
- Destination Options header: applies to the final destination only
- Upper-Layer header

The structure of the IPv6 Next Header field is therefore as in Figure 5.3.

The extension headers are processed by only the destination node, with the exception of the hop-by-hop options header. The extension headers that are currently defined include the Hop-by-Hop Options; Routing; fragmentation; security, which includes authentication (RFC 2402 and next header=51); and the Encapsulating Security Payload (ESP, RFC 2406, and next header=50), and the destination options.

From the control plane point of view, the following distinctions can be made between IPv4 and IPv6 (Figure 5.4).

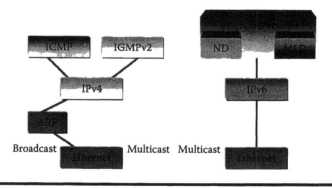

Figure 5.4   Control plane views of IPv4 and IPv6.

Figure 5.5   Structure of typical IPv6 address.

Structure of IPv6 Addresses

One of the design goals for the 128-bit addresses is to allow more hierarchical address allocation and simplification of configuration and routing. IPv6 addresses consist of two parts: a 64-bit network prefix that is used for locating the network where it is and also a 64-bit host suffix that is used for identifying the host within the network (Figure 5.5). The network prefix part identifies the network to which the host is connected, and this depends on the location of the network. When a host moves from one network to another, it only needs to change its network prefix part of the address. This provides a means for reaching it, locating it, and maintaining sessions.

The long IPv6 addresses mean that representation in decimals is hard to remember. For example, an address such as 194.153.11.222.128.17.135.44.240.36.97.66. 205.221.52.4 is more difficult to remember. This can, however, be simplified using hexadecimal numbering to C299:0BDE:8011:872C:F024:6142:CDDD:3404. Although shorter, it is still difficult to remember. Hexadecimal numbering is the standard approach for presenting IPv6 addresses. There are, however, other differences that have been introduced to simplify the addresses. IPv6 uses the colon ":" as address separators, as opposed to the dot "." in use in IPv4.

IPv6 addresses are simplified if many digits are zeros. For example,

        DEAD:BEEF:0000:0000:0073:FEED:F00D

becomes

        DEAD:BEEF::73:FEED:F00D

and "::" can replace a number of adjacent zero groups. This is permitted once in an address. Only one string of zeros can be compressed; otherwise, string lengths are ambiguous. Therefore, we can compress the following addresses:

        FF80:0:0:0:0:0:0:100 = FF80::100

and

        0:0:0:0:0:0:0:1 = ::1

For a long time to come, IPv4 and IPv6 networks will coexist. Hence, there is a need to find a way of converting IPv4 addresses to IPv6 addresses. This is done fairly

Table 5.1   Address Prefixes

Allocation	Prefix	Fraction of Address Space
Reserved for compatibility with IPv4	**00000000**	
Provider-based Unicast address	**01000000**	
Multicast address	**11111111**	**FF00::/8**
Link local	**1111 1110 10**	**FE80::/10**
Site local	**1111 1110 11**	**Deprecated**
Place holder	0:0:0:0:0:0:0:0	No address is available
Loop back address	0:0:0:0:0:0:0:1	For sending packets to self

Figure 5.6   Structure of IPv4 address as an IPv6 address.

easily using the double colons. For example, 138.25.40.1 becomes ::138.25.40.1 and the last 32 bits are still in decimal with "." separators. Hence, for this case, all the leading 96 bits are zeros.

There are four different types of IPv6 addresses for unicast, multicast, anycast, and link-local addresses. The first 8 bits of the addresses are reserved for use for address allocation purposes as shown in Table 5.1. About 15 percent of the address space is reserved.

For example, a provider-based unicast address is shown in Table 5.1. The remaining 5 bits after the 010 prefix are reserved for the definition of the registry specifications for each continent. IPv4 addresses in IPv6 address space are of the form in Figure 5.6.

Now let's take a closer look at the different types of IPv6 addresses and explain their structures. In the discussions that follow, it should be understood that the Interface ID retains its 64-host suffix length and the modifications discussed happen in the Network prefix.

Unicast Addresses

Three forms of unicast addresses are described: the site local unicast addresses, link local unicast addresses, and the global unicast addresses. The provider-based site local unicast addresses have the form in Figure 5.7.

The address describes or contains the specific identifications for the registry, provider, subscriber, subnet, and the interface. The multicast address is clearly

Figure 5.7   Local unicast address format.

different. The concept of scope is introduced in IPv6 addresses. The scope is equal to the local link, or Scope = local link (or virtual LAN or subnet). The scope can only be used between the nodes of the same local link and cannot be routed.

Unicast addresses have a five-digit registry identification that describes the agency that is responsible for allocating network addresses in that geographical region. The following agencies are responsible for the following geographical regions:

- Africa: AfriNIC
- Asia Pacific countries: APNIC
- Europe, Middle-East, and Central Asia: RIPE NCC
- Latin America and Caribbean: LACNIC
- North America: ARIN

The site-local unicast address format has been deprecated.

The ID of the provider in the IPv6 address is a 16-bit ID of the Internet service provider obtained from the registry. The identity of a subscriber is a 24-bit number and is obtained from the provider. The identity of the subnetwork is a 16-bit number. The identity of the interface is a 48-bit unique number within a subnetwork. Therefore, a typical unicast address is of the form:

```
FEC0::<subnet id>:<interface id>
```

A site in IPv6 is defined as a network of links. The site local addresses are also scoped as Scope = site. The scope is applicable to only the nodes of the same site. This is similar to how private IPv4 addresses are defined. The format of the site local addresses is:

```
FEC0:0:0:<subnet id>:<interface id>
```

This consists of 16-bit subnet addresses or a total of 64K subnets. A site should be numbered before connecting to the Internet.

Unicast local addresses (ULAs) are assigned with the intention of possessing a degree of uniqueness. As such, they have globally unique prefixes that are assigned with intentions that they will be used for local communications inside a site and not outside the site. Therefore, they are not expected to be routable in the open or global Internet. They are addresses for local consumption within a site and are routable within the site. Hence, they have limited regional scope. Although they may be routable within a limited set of known sites, they are local to the sites.

Figure 5.8   Unicast local address format.

There are two ways to assign the ULAs, either using locally assigned local addresses or centrally assigned local addresses. There are several purposes for making addresses local. They are ISP independent and therefore can be used for internal communications within a site without intermittent or permanent Internet connectivity. They also have well-known prefixes and therefore facilitate filtering at site boundaries. However, being site limited, if they are leaked to the open Internet through DNS, routing, or some other means, conflict with other addresses is avoided.

ULAs have a format (Figure 5.8) with the prefix FC00::/7. The prefix identifies the local unicast addresses. The address prefix is 7 bits long and bit L has two definitions for when the bit is set or not set:

L=0 will be defined in future
L=1 if the prefix is assigned locally

The global ID field is 40 bits long and the subnet ID is again 16 bits long. Hence, the interface ID retains its 64 EUI-64 format. Generally, the ULAs are allocated using a pseudo-randomly allocated format. This eliminates any relationships between the allocated addresses and clarifies the fact that the addresses are local and not to be routed globally.

Centrally versus Locally Assigned Local Unicast Addresses

A central approach for allocating ULAs was announced by the IETF in February 2005 in RFC 4193. One of the intentions is to avoid address assignment conflicts. Furthermore, using a centrally assigned method, addresses can be "escrowed," thereby resolving the possibility of duplicate address assignments [4].

When creating centrally assigned local addresses, the L field is set to 0. Hence, this clearly identifies the prefix as distinct from when L=1 in the traditional approach for allocating ULAs.

The global unicast addresses (RFC 3587) are of the form with 45-bit prefixes and are normally assigned to a zone, site, or a combination of many subnetworks/links (Figure 5.9). The subnet ID in this case describes a subnet within a site.

Figure 5.9   Global unicast address.

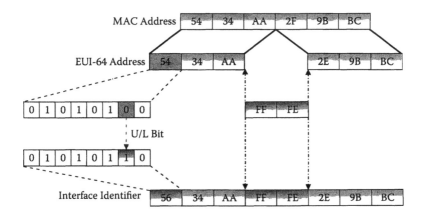

Figure 5.10   EUI-64 conversion of MAC address to interface identifier.

The global routing prefix is 45 bits long, and the subnet ID is 16 bits long. The remaining 64 bits are allocated to the interface ID and have the form of the EUI-64 format.

Interface Identity

In all the ULAs, the lowest 64 bits of the IPv6 address field have been allocated as the interface identity. This identity address field is allocated in several ways.

Using the 48-bit MAC, the interface ID can be auto-configured and expanded to the required EUI-64 format. This ID can also be assigned using DHCPv6, or manually configured or automatically generated using a pseudo-random format. This helps in reducing the ease of interface ID piracy. The IEEE proposed method of expanding the 48-bit MAC ID to a 64-bit interface ID is as in Figure 5.10.

In this conversion scheme, the lower byte of the MAC address is modified and the two bytes FF and FE are inserted in the middle of the MAC address. Bit 7 (U/L bit) of the lower byte of the MAC address is set to 1. Thus, as in Figure 5.10, we convert the lower byte from hex 54 to hex 56 and insert FFFE in the middle of the MAC address. This completes the EUI-64 conversion. The MAC address 54 34 AA 2E 9B BC becomes the interface identifier 56 34 AA FF FE 2E 9B BC.

Multicast Addresses

Multicast addresses in IPv6 in many ways resemble unicast addresses. Each one of them has an 8-bit prefix, 4-bit flags, and a 4-bit scope. The remaining 112 bits are used for specifying the group ID. A multicast address looks like that in Figure 5.11. All IPv6 multicast addresses are therefore of the form:

```
FF<flags><scope>::<multicast group>
```

Figure 5.11 Multicast address.

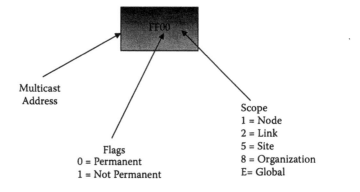

Figure 5.12 IPv6 multicast address flags.

The pictorial form is given in Figure 5.12.

The low-order flag is used to represent permanent/transient group. The remaining 3 bits in the flags are reserved. The scope fields are as shown in the diagram. The remaining are reserved.

For example, the addresses for nodes and routers on a local network could be:

```
FF02::1 refers to all nodes on the local network
FF02::2 refers to all routers on the local network
```

In a solicited-node multicasting, the address of a node is:

```
FF02:0:0:0:0:1:FF00::/104
```

The address is formed by appending the lower 24 bits of the IPv6 address. When a node is required to join for every unicast and anycast address, it is assigned the address (global unicast address)

```
3FFE:0B00:0C18:0001:0290:27FF:FE17:FC0F
```

Solicited multicast address:

```
FF02:0000:0000:0000:0000:0001:FF17:FC0F
```

Table 5.2 gives some of the reserved multicast addresses [1]. Note that the addresses with "X" in the scope field mean all legal scope and they are valid over all scope ranges.

Table 5.2 Reserved Multicast Addresses

Address	Scope	Use
FF01::1	Interface local	All nodes
FF01::2	Interface local	All routers
FF02::1	Link local	All nodes
FF02::2	Link local	All routers
FF02::3	Link local	Unassigned
FF02::4	Link local	DVMRP routers
FF02::5	Link local	OSPFIGP
FF02::6	Link local	OSPFIGP designated routers
FF02::7	Link Local	ST routers
FF02::8	Link Local	ST hosts
FF02::9	Link local	RIP routers
FF02::A	Link local	EIGRP routers
FF02::B	Link local	Mobile-agents
FF02::D	Link local	All PIM routers
FF02::E	Link local	RSVP-encapsulation
FF02::1:1	Link local	Link name
FF02::1:2	Link local	All-DHCP-agents
FF02::1:FFXX:XXXX	Link local	Solicited-node address
FF05::2	Site local	All routers
FF05::1:3	Site local	All-DHCP-servers
FF05::1:4	Site local	All-DHCP-relays
FF05::1:1000	Site local	Service location
FF05::1:13FF	Site local	

Figure 5.13   Anycast address.

Anycast Address

The IPv6 addressing architecture defines an "anycast" address. An anycast address in IPv6 is assigned to one or more network interfaces. These interfaces typically belong to different nodes and they have the property that a packet that is sent to an anycast address is routed to the nearest interface having that address [2]. The format of the anycast address is as shown in Figure 5.13. The anycast address definition allows the anycast address to be "used in a source route to force routing through a specific Internet service provider, without limiting routing to a single specific router providing access to that ISP" [2].

Reserved Subnet Anycast Addresses

According to Johnson and Deering, "within each subnet, the highest 128 interface identifier values are reserved for assignment as subnet anycast addresses" [2]. The type of IPv6 addresses used in the subnet determines the reserved subnet anycast addresses. This is indicated in the prefix of the address.

When the IPv6 address is required to have 64-bit interface identifiers in EUI-64 format, the universal/local bit must be set to 0 (local) in all reserved subnet anycast addresses. This is to indicate the non-globally unique interface identifier address.

This type of IPv6 address currently consists of those having format prefixes 001 through 111, except for multicast addresses (1111 1111) [3]. When the interface identifier is a 64-bit EUI-64 format, the structure of the reserved anycast addresses is given in Figure 5.14. Hence, the interface identifier field is also a 64-bit number formed by the concatenation of the bits from the anycast ID field and 57 bits.

Mobility

IPv6 addresses mobility much better than IPv4. Increasingly, mobile routers are being deployed to service a range of new services enabled by IPv6. As more and more subscribers use personal digital assistants (PDAs), laptops, and mobile

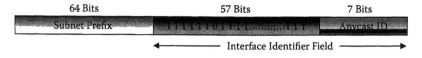

Figure 5.14   Anycast address in EUI-64 format.

devices, and new applications based on sensors are deployed, mobility issues will also become increasingly important. New cases of vehicular networks deployed in aircraft, buses, and personal automobiles mean that the domain of fixed access networks is being gradually extended to mobile networks with new mobility requirements. Therefore, it is not unusual to expect a mobile vehicle containing several or many mobile nodes or mobile subnets and each one bearing its own IPv6 address. Mobility solutions are provided differently for different layers of the OSI. This includes application layer solutions (SIP), network layer solutions (Mobile IP), transport layer solutions (mSCTP), and the data-link layer solutions. These solutions address terminal, application, service, session, personal, nodes, and network mobility. This section addresses only the mobility of network nodes. Other sections of the book address different types of mobility, including the chapters on SIP. Mobility in IPv6 is supported by Mobile IPv6. Chapter 6 discusses mobility based on IPv6.

Summary

This chapter explained IPv6 and by reading it, you should gain knowledge of why it is preferred over IPv4 and how IPv6 differs from IPv4. It specifically addressed the limited address space increasing each address to 128-bit length instead of the IPv4 32 bits. IPv6 provides a reasonable but also timely means to reengineer IPv4 in terms of adding features that support quality-of-service (QoS), multicast, and also mobility. It also provides the ability to include other new features such as binding updates and to simplify the IPv4 packet header structure. The IPv6 header is also highly simplified compared with IPv4 but has twice the size of the IPv4 header. One of the design goals for the 128-bit addresses is to allow more hierarchical address allocation and simplification of configuration and routing. IPv6 addresses consists of two parts, a 64-bit network prefix that is used for locating the network and also a 64-bit host suffix that is used for identifying the host within the network. There are four different types of IPv6 addresses for unicast, multicast, anycast, and link-local addresses. IPv6 addresses mobility much better than IPv4. Mobility in IPv6 is supported.

Practice Set: Review Questions

1. Is IPv6 just a new version of IPv4?
   a. Of course, with just a new version name.
   b. No, it is a completely a new Internet protocol.
   c. It is IPv4 with only increased address space.
   d. It is IPv4 with only new security features.
   e. It is IPv4 with only QoS added to it.

2. What is new in IPv6?
   a. It has a larger address space compared with IPv4.
   b. It has a more compact header.
   c. It uses twice the number of bytes in the header.
   d. None of the above.
   e. All of the above.
3. The care-of address created by IPv6 in a visited network is formed by:
   a. The host prefix and the prefix of the visited network.
   b. The host prefix and an additional host prefix.
   c. Retaining its old host address.
   d. Swapping the last 16 bits of the host prefix.
   e. Exchanging its host prefix for a new host prefix.
4. In IPv6 when a node moves away from its home network,
   a. Its location is lost completely.
   b. It retains the location of its home network.
   c. It gains a new location through the prefix of the visited network.
   d. It cannot be reached at all.
   e. No node on the Internet can reach it again.
5. The structure of address representation in IPv6:
   a. Is the same as in IPv4.
   b. Uses octets but is longer than in IPv4.
   c. Uses 64 bits for the network prefix and 64 bits for the host suffix.
   d. Only (b) is correct.
   e. Both (b) and (c) are correct.

Exercises

1. Draw and correctly label the IPv6 address structure.
2. Discuss how IPv6 addresses can be simplified, and why.
3. How does an IPv6 router recognize IPv4 addresses? Discuss fully.
4. How can you distinguish between unicast, multicast, link local, and site local IPv6 addresses? Discuss.
5. Discuss a method of converting an EUI-64 MAC address to an interface identifier in IPv6.

References

[1] IETF, IPv6 Multicast Address Assignments, RFC 2375, 1998.
[2] D.B. Johnson and S.E. Deering, Reserved IPv6 Subnet Anycast Addresses, RFC 2526, March 1999.

[3] R. Hinden and S. Deering, IP Version 6 Addressing Architecture, RFC 2373, July 1998.
[4] J. Palet, IPv6 Basics, APNIC Training, Taiwan, 2006.

Chapter 6

RSVP: Resource Reservation Protocol

Traditional broadcasting of data assumes that there is a path from the source to the destination and that there is no need to establish or provision bandwidth, links, or buffers. Hence, whatever is received at the destination is subject to the limitations of the path separating the transmitter and the receiver. This simple model, which has worked for telecommunication engineers over many years, fails miserably when the data to be transported is digital and in packet form. It fails at various points along the route to the destination because routers and links may have varying bandwidths and speeds and, because they were not originally configured for any particular set of multimedia rate, their efficiencies are limited. In some cases, they work perfectly because they meet the requirements of the communication and in others they fail because the requirements for resources are higher. In traditional Internet, the resources between the source and destinations include buffers, links, and CPU speeds.

Resource Reservation Protocol (RSVP)

RSVP was conceived as a signaling protocol that permits requesting for resources along the path that data must take to its destination. Resource reservation messages are used to obtain permissions from admission control software with respect to the availability of desired resources to efficiently transport voice and video over IP-routed networks.

Three communication situations under which resources need to be reserved are apparent. They are during:

1. Point-to-point communications
2. One-to-many multicast communications
3. Many-to-many multicast communications

In point-to-point communications, the objective is to establish a virtual path between the source and destination. Therefore, resources must be reserved at the intermediate points or routers. At the routers it is essential to reserve

1. Buffers to hold packets in transit
2. Links at required bandwidth and data rate or capacity
3. CPU speeds and capacity to ensure that packets are transported from the source to destination in real-time

Issues to Consider in Resource Reservation

There are definite issues that must be considered in the course of making reservations. Let us first consider resource reservation from the sender's point of view.

Change of Path in Point-to-Point Communication

Consider Figure 6.1, consisting of three routers between the two end nodes. The path from node H1 to node H2 consists of three routers. There is just a direct path and there is no ambiguity in the route.

The goal of the RSVP is to reserve resources at routers R1, R2, and R3 and including the links $(L_k; 1 \leq k \leq 4)$ bridging them to form a virtual path between the hosts H1 to H2. In Figure 6.1, the *connect* request message is sent by H1 to H2. The message contains the flow specifications (flowspec), which are used for reserving resources at the routers. When R2 *accepts* the request, each router reserves the

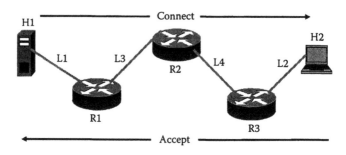

Figure 6.1 Point-to-point communication.

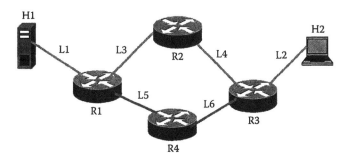

Figure 6.2   Path changes in resource reservation.

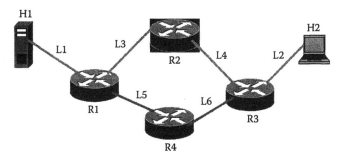

Figure 6.3   Router R2 is either congested or has failed.

required links/capacity, buffers, and router capacities for the flowspec. The reservations are made only when the routers in the path receive the accept message or reply. In practice, depending on the conditions at the routers and links, the paths taken by packets from H1 to H2 may differ. This is where ambiguity can result. Suppose that resources have been reserved as in Figure 6.1 and after this, conditions in the route change dramatically (e.g., congestion takes place in one or more routers), forcing decisions to be taken for packets to be sent through alternate routes when they exist, as in Figure 6.2. What happens to the resources reserved? When might the route through L5-R4-L6-R3 be taken to reach H2? There are six obvious situations shown in Figures 6.3 through 6.5.

There are six situations when it might be necessary to consider abandoning the route on which resources were reserved. These are:

1. Router R2 fails completely and is not functioning.
2. Router R2 is congested and too busy to serve the real-time application (Figure 6.3).
3. Link L3 has failed.
4. Link L3 is too noisy so that its capacity is well below that required in the RSVP (Figure 6.4).

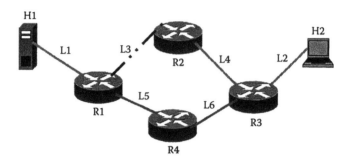

Figure 6.4   Link L3 has failed or is experiencing high interference.

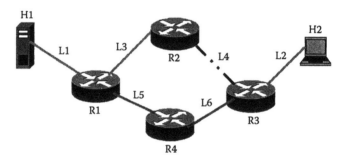

Figure 6.5   Link L4 has failed or is experiencing high interference.

5. Link L4 has failed.
6. Link L4 is too noisy, causing its capacity to fall well below that required in the RSVP (Figure 6.5).

In these six situations, how does the application know about the problems? How do the hosts H1 and H2 reestablish connection, reserve resources at R4, and maintain QoS requirements?

Differing Link and Router Capacities in Point-to-Multicast Communications

In the current Internet and across many networks and countries, link capacities are not unique but differ depending on the goals and budget of the owners of the links and intermediate networks. Therefore, point-to-point reservation is applied recursively in point-to-multicast communications so that each path to each end host is reserved as in point-to-point. Hence, the problems emanating from such reservations are inherited. However, an additional problem emerges. Links are not always of the same capacity, and routers in the networks also may be of differing

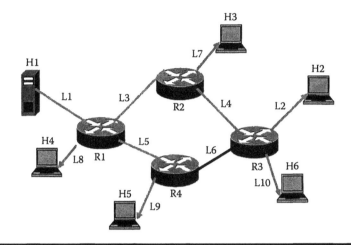

Figure 6.6 Point-to-multicast communications.

capacities. In such situations, the link and router with the smallest capacity determine the reservations.

In Figure 6.6, link L6 has the smallest capacity among all the links and defines the reservation capacity for the paths to hosts H2 and H6.

The second problem in point-to-multicast reservations is how to add and delete new hosts to the multicast communications. In this case, the source reinitializes the reservations afresh with all the receiving hosts.

Many-to-Many Multicast

In many-to-many multicast reservations, several sources participate independently and reserve resources to destinations (Figure 6.7). Two problems emerge. How does one reserve resources and optimize for the length of the path to ensure that packets take the shortest routes to their destinations? How can one guard against over-reserving of resources due to multiple requests from many sources?

In Figure 6.7, both H4 and H1 are reserving resources independently to other hosts and, as shown at R3, there is a double request by both sources.

It has been observed that when sources initiate reservations, the process does not scale. Low utilization of resources or links result, and it is also difficult to accommodate heterogeneous receivers. Furthermore, overheads at the sender points are high. It is therefore preferable to merge the processes of resource reservation and routing. Next we consider resource reservation from the receiver point of view.

Receiver-Initiated Resource Reservation

Receiver-initiated resource reservation (Figure 6.8) is more sensible for several reasons to correct some of the problems from source-initiated resource reservations:

1. Receivers know the amount of bandwidth or capacity they need or can handle for a particular service and therefore should reserve them for themselves.
2. The burden of joining and leaving the multicast process is left to the joining end-point and the source need not worry about it.
3. Optimizing for path and capacity is enhanced in the routers.

RSVP Design Philosophy

The role of the RSVP is to deliver QoS requests to routers or switches along the chosen path, and the path can be set up using either routers or ATM. The RSVP is

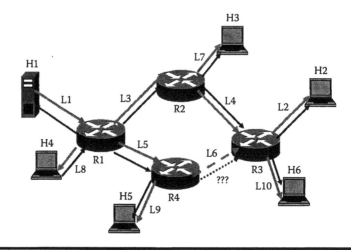

Figure 6.7　Many-to-many multicast reservation.

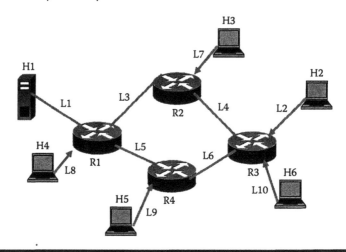

Figure 6.8　Receiver-initiated resource reservation.

therefore a signaling protocol and not a routing protocol. The RSVP assumes the availability of routing protocols such as IGRP (Interior Gateway Routing Protocol), IS-IS (Intermediate System to Intermediate System), and BGP (Border Gateway Protocol). The reservation is distinct from the routing protocol. Soft state reservations are made so that reservations are not permanent states in the devices. This is often referred to as "soft" state because the handling of dynamic route changes should not be an exception but a norm. By maintaining soft states, routers are constantly updating their reservations in response to new RSVP requests. Therefore, the protocol should expect route changes independent of the Internet routing protocols.

The system is constantly and periodically refreshed with the desired state. The refresh states or messages are identical or repetitions of the original resource requests. The refresh messages are useful not only to keep the reservations alive, but also to create new paths if routing has changed or there is a link failure. In such circumstances, the refresh messages will create a new resource reservation state on the new segment of the path. If, however, a new path cannot be created by the refresh message, an error message is returned to indicate that the requested resources are not available.

What happens to the QoS, or how can it be maintained due to dynamically changing routes? When the path from source to destination fails, the new route created by the refresh message may not have the same QoS as the original one and hence the QoS required to support the service is lost. This loss could be temporal until a new path segment is established with a refresh statement. The new path segment may have a lower than expected QoS, which poses a problem. If a "hard" state protocol such as ATM (Asynchronous Transfer Mode) is used, it will drop the connection if the route fails or if QoS cannot be guaranteed.

The RSVP is designed to allow heterogeneous bandwidth requirements, particularly in cases when multiple receivers are involved in multicast sessions. Each endpoint can get or request a different QoS. This is achieved by merging requests or using different QoS layers.

The RSVP has the ability to scale to include a large number of participants because the number of messages traveling upstream is reduced through merging. Therefore, fewer states are required at the routers. It will only ask for states to be kept but cannot provide them by itself.

RSVP reservations do not specifically identify which packets can use the reservations. Different reservation styles are also supported through the use of filters. These solutions are inherent in the seven design principles of the RSVP [1]:

1. Accommodate heterogeneous receivers.
2. Adapt to changing multicast group membership.
3. Exploit the resource needs of different applications in order to use network resources efficiently.
4. Allow receivers to switch channels.
5. Adapt to changes in the underlying unicast and multicast routes.

6. Control protocol overhead so that it does not grow linearly (or worse) with the number of participants.
7. Make the design modular to accommodate heterogeneous underlying technologies.

Therefore, for the design of RSVP, the following principles are followed:

- Receiver-initiated reservation
- Separating reservation from packet filtering
- Providing different reservation styles
- Maintaining "soft state" in the network
- Protocol overhead control
- Modularity

RSVP Reservation Messages

RSVP depends strongly on the use of reservation messages. The resource requirements of applications vary widely. For example, both video-conferencing and audio have very high probability of requiring high quality to meet the viewing and listening satisfaction of users. An HDTV stream requires very high bandwidth and a VoIP stream requires very small delay and low delay variability between the packets. Therefore, the RSVP model provides an avenue for the applications to request different reservation styles using messages to match the type of service being requested or for economic (SLA) requirements. RSVP supports three reservation styles, as given in Table 6.1.

The *wildcard filter reservation style* provides for sessions in which all the sources need similar service guarantees. This also implies that the sources have some capability to limit their output, as in audio-conferencing. The wildcard filter is useful

Table 6.1   RSVP Reservation Styles

Reservation Style	Definition	Style
Wildcard filter	One reservation is used by all senders. The reservation propagates toward all sender hosts and extends to any new senders that join the session	WF({*Q})
Fixed filter	One reservation per source	FF(S{Q}), FF(S1{Q1}), FF(S2{Q2})...
Shared explicit	One reservation is shared by listed upstream senders. Receiver can explicitly specify the senders	SE((S1, S2, ...){Q})

Table 6.2   RSVP Message Types

Message Type	Function
PATH	This is sent by the source to announce the existence of resource and also to specify which parameters should be used for transmitting
RESV	This is a speculative message transmitted in hope of reserving resources
CONFIRMATION	This is a receiver's message to signal successful resource reservation
TEARDOWN	This message is used to delete or relinquish reservation
ERROR	An error message to signify abnormal performance or condition such as a reservation failure (e.g., PATH error, RESV error)

in multicast sessions. The *fixed filter reservation style* requires one reservation per source and targets sources such as video distributions. The *shared explicit reservation style* is provided for when one reservation is shared by several sources, particularly to support concurrent viewing as in video sources, and each requires a different QoS.

Filters provide support for heterogeneity within the network. Some receivers are attached to slow links and others to fast links. Therefore, receivers attached to slow links can still participate in multicasting and flows by using a filter to limit the portion of a flow that it can handle. By making some of the filters dynamic, receivers can therefore modify flow properties. This is also useful when a receiver is listening to multiple flows. Filters also reduce loading of network elements and improve bandwidth management.

The RSVP operates in a client/server manner. It also defines five data objects (Table 6.2), which carry resource reservation information that is not critical to its RSVP. Each of the five messages in Table 6.2 carries several subfields. The messages are used at some points in the establishment of RSVP state.

RSVP Packet Header

The RSVP packet header fields are given in Figure 6.9. The version signifies the RSVP protocol number, the Flags are undefined, and the Message Type is typically the RSVP Signal Types, which are:

- PATH
- RESV
- Path Error
- RESV Error

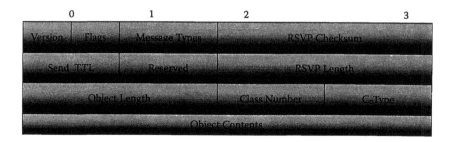

Figure 6.9 RSVP packet header fields.

- PATH Tear
- RESV Tear
- RESV Confirmation

The checksum is calculated over the entire message and the Send_TTL (send time-to-live) is identical to the IP time-to-live (TTL). The Object Length is an indicator of the individual object variable. The Class Number specifies the object class and C-Type signifies the class type of the object.

RESV Messages

Whenever a receiver decides to communicate with a sender, it sends a reservation message upstream using the same route as the one for the PATH message. The RESV message contains the following fields:

- Session ID
- The previous hop address (RSVP neighbor downstream)
- Reservation style
- A flow descriptor (depending on reservation style, different combinations of flow and Flowspec are used)
- Options (integrity, policy data)

If the reservation request is successful, then the router or host installs packet filtering in its forwarding database and this database is interrogated when the node or router has a packet to send upstream. It is used to classify packets into different classes. Next, the flow parameters for the QoS-enabled data are also passed to the packet scheduler. The scheduler forwards packets at a rate that is compliant with the description of the flows using the parameters signaled.

If the RESV request fails at any of the intermediate nodes, an error message containing the cause of failure is generated and transmitted to the requestor.

Merging of reservations can take place if two or more RESV messages for the same source pass through the same router. The device attempts to merge them. The

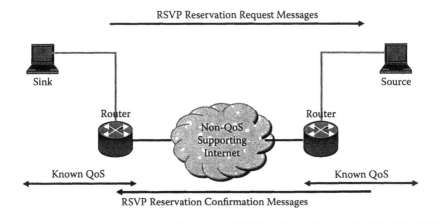

Figure 6.10   Non-RSVP networks in the middle.

merged reservation request is forwarded as a single request to the next upstream node. This helps increase the network's efficiency.

Handling Non-RSVP Intermediate Networks

Naturally, it is possible to encounter somewhere in the wide area network a non-QoS supporting network. How does the RSVP handle this? If there is an intermediate network with routers that do not understand the RSVP, both the PATH/RESV messages are forwarded through the non-RSVP network because to it, they are just like any other IP packets. Consider the network in Figure 6.10 with a non-QoS supporting Internet in the middle. The two routers at the edges of the network communicate with their neighbors through the cloud as if they were side by side. The performance of the intermediate network is limited to as just best-effort Internet and mitigates the quality of the reservations. Therefore, the performances across the spots must be estimated and communicated to the receivers using ADSPEC messages.

TEARDOWN Message

A TEARDOWN message is used when either of the two communicating parties wishes to terminate the session. There are two specific forms:

1. PathTear
2. ResvTear

The *PathTear message,* when invoked by the source, propagates across the network and releases all the reserved resources for the communication, thereby making them available to other users. The message removes all the source's states in addition to all the reservations made for the data flow from the source.

The *ResvTear message,* when invoked by the receiver, propagates upstream and removes only the Reservation State of the receiver. In principle, the TEARDOWN message is optional to the protocol because of the soft state. With time, all reservations expire automatically if they are not refreshed manually. However, if the time set for the states to be refreshed is long, the reserved resources could be held for a long time if they are not released gracefully. This wastes resources and prevents access to them by other users.

ResVConf Message

The RSVP allows a receiver to query the network to determine the state that has been reserved for its session. The explicit RESV-CONFIRM object is required in a RESV message because the protocol's "single pass" model does not provide the receiver with any information about the success of its request. To check on the state that has been reserved, a RESV message with a RESV-CONFIRM object containing the IP address of the receiver is propagated upstream. At each router, the flowspec associated with the reservation is compared with the reservation from the downstream router. When the reservation in the current router is equal to or greater than the receiver's reservation, a halt is put to the propagation of the RESV-CONFIRM. The receiver is then returned a ResVConf message. The message contains the flowspec values of the current router.

There are two issues associated with this approach. First, all messages are transmitted unreliably using UDP/IP and could be lost. Second, when messages are merged during multicast reservations, before the messages reach the source they may encounter a router with a reserved state from a previous RESV of another receiver. As a result, the merged RESV will fail at some point upstream [2]. The router will still send a ResVConf downstream, indicating that the RESV has been successful when in the real sense it failed.

Details of RSVP Operation

In the operation of RSVP, both senders and receivers join a multicast group and this is normally done outside the domain of RSVP. There are three groups of message types:

1. Senders advertise their presence by issuing Sender-to-Network messages. The messages are of two types:
   - Path request, in which the sender makes its presence known to the network elements
   - Path teardown, in which the sender requests the paths set up at each router to be torn down
2. Receivers subscribe to the sender's data streams. This too is done through the use of two message types:

- Reservation requests, in which the receiver specifies its request to reserve resources between it and senders
- Reservation teardown message, which asks for the reserved resources to be released back to the network
3. Network-to-end-elements messages. There are typically two types of this message:
  - Reservation error
  - Path error

Sender-to-Network Messages

Senders use path messages to identify themselves. The path messages contain five elements:

1. *session id:* This is a destination or multicast group address and port. A session in this case is a group of parameters that describe the reservation and includes unique information to differentiate the traffic flow associated with the session. The identity of a session therefore is: *destination address + protocol + port.*
2. *flowspec:* This is a requirement descriptor including a required stream bandwidth. The flowspec in a reservation request consists of the following parameters:
   - A service class
   - An "Rspec" (R for 'reserve'): it defines the desired QoS.
   - A "Tspec" (T for 'traffic'): it defines and describes the type of data flow.
   These parameters are determined by the integrated service models and are opaque to RSVP.
3. *filtering spec:* This describes how the packets from this source should be filtered. It is the header pattern to use for this packet classification.
4. *previous hop:* This is the address of the terminal forwarding the message.
5. *refresh time:* This is the time duration for the path message to time out.

The path messages are therefore used by the network elements to label or encode the paths between the senders and receivers. For example, when a router processes a PATH message, it establishes a "PATH state" whose content is taken out of the fields in the path messages (Figure 6.11). They therefore form the states that routers keep as their states. The PATH state records information about the IP address of the sender, its policy, and a description of its QoS class. Routers also record the name of the source and the previous router that forwarded the message in the form: *Element[source ⇒ forwarding router]*.

For example, in R2 we will have: $L2[H1 \Rightarrow R1]$. In this case, the source of the message is host H1 and it was forwarded by router R1. The PATH message consists of the following four fields:

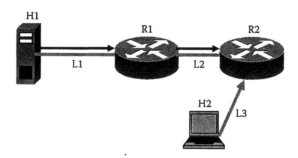

Figure 6.11 Path state.

1. Session ID (session identity)
2. Address of the previous hop of the upstream RSVP neighbor
3. A descriptor of the sender (filter + Tspec)
4. Options (integrity object, policy data, ADspec)

The ADspec (advertised specification) is a set of modifiable network parameters that are used to describe the QoS capability of the path between the source and destination (e.g., service classes supported).

Reservation messages originate at receivers and carry reservation requests that are forwarded upstream toward the sender. At every intermediate node, a reservation request triggers two general actions:

1. *Make a reservation on the link.* This requires passing the request to both the admission control and policy control. If any of the two tests fails, the reservation request is rejected. The RSVP process then returns an error to the receiver that initiated the request. If both tests are successful, then the node sets its packet classifier to select the data packets defined by the filter spec. It also communicates with the data-link layer to retrieve the desired QoS defined in the flowspec.
2. *Send the reservation request upstream.* The request is propagated forward to the next network nodes. The propagated request that is sent does not necessarily have the same flowspec as the one received from downstream.

Development of Path States

Using the above example, we develop the state of the paths using the path method:

$$\textbf{path}(\text{TSpec, Element, filter spec, refresh time})$$

The state of each path is formed by a combination of the variations defining the "Transmission specification," the network element that is either a router (Rk) or

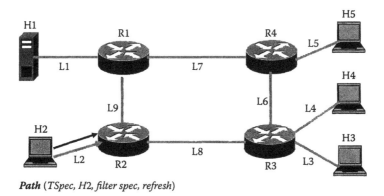

Path (*TSpec, H2, filter spec, refresh*)

Figure 6.12   Path state for H2 (stage 1).

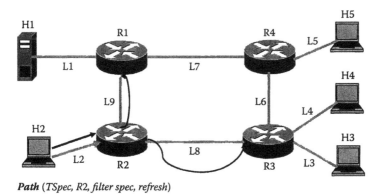

Path (*TSpec, R2, filter spec, refresh*)

Figure 6.13   Stage 2 of path state definition.

host (*Hj*). The transmission specification characterizes the data flow from the physical appearance of the packet stream (e.g., headers, packets per second, etc.). It identifies and differentiates the QoS requests. In the following figures, H1 and H2 are senders, and H3, H4, and H5 are the receivers (see Figure 6.12).

The states of the routers as in Figure 6.13 are shown in Table 6.3. H1 is both the sender and the forwarder of packets.

Table 6.4 and Figure 6.14 show that router R2 forwards packets sourced from host H2 to routers R1 and R3.

Table 6.3   State of Routers in Figure 6.11

Router State	R1	R2	R3	R4
Inbound Link	L2[H2 ⇒ R2]			

Table 6.4 Router R2 Forwards Packets Sourced from Host H2 to Routers R1 and R3

Router State	R1	R2	R3	R4
Inbound Link	L9[H2 ⟹ R2]	L2[H2 ⟹ H2]	L8[H2 ⟹ R2]	

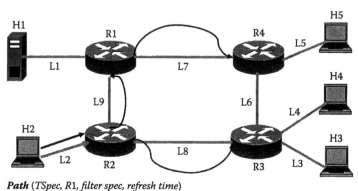

Path (*TSpec, R1, filter spec, refresh time*)

Figure 6.14 Stage 3 of path state definition.

The complete path state for H2 is therefore as shown in Table 6.5 with the accompanying path diagram in Figure 6.15.

Receiver-to-Network Messages

The second type of message is sent by the receiver to the network. The objective in the receiver reservation messages is to specify its requirements to be able to handle the service and the messages contain:

1. Desired bandwidth
2. Packet filter
3. Filter type

Three types of filters can be specified:

1. *No filter:* implies that any packet destined for the receiver can use the reservation without modifications.
2. *Fixed filter:* packets from a specified set of senders only can use the reservation; packets from other sources are excluded.
3. *Dynamic filter:* specifies that the set of senders whose packets can use the reservation is variable with time; this option provides for the possibility of new senders that might require filters.

Table 6.5 Complete Path State for H2

Router State	R1	R2	R3	R4
Inbound Link	L9[H2 ⇒ R2]	L2[H2 ⇒ H2]	L8[H2 ⇒ R2]	L7[H2 ⇒ R1]

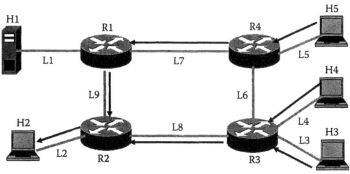

Path (*TSpec, H2, filter spec, refresh*)

Figure 6.15   Complete path state for H2.

In this case, the pair (**reservation, source**) becomes part of the state of the router and this state is maintained at all the routers in the path between the source and receiver. The following are examples taken from [1, 3] with some modifications and explanations.

When no filters are specified by the end-points or hosts, several problems emerge during reservation. First, which host does the receiver actually listen to per timeout of the multitudes of the terminals that are permitted to use the reservation? Suppose an endpoint consumes more bandwidth than the one reserved by the reserving node. Suppose two nodes transmit simultaneously at the reserved bandwidth adding up to twice the reserved bandwidth. How does the network handle this situation? Suppose all the hosts in the network individually reserve some bandwidth b. How does the network handle this over reservation of resources? Suppose individual hosts reserve multiple reservations. Is the network overcommitted in terms of bandwidth because of that? These questions emerge from no filter reservation.

Reserving Link Bandwidth and Forwarding Reservation Messages

In this example [1, 3], each of the five hosts is able to send and receive packets and they belong to the same multicast group.

No Filter Reservations

This section illustrates what happens when a router makes reservations and the filter flag is turned off. For each of the following steps, we use a figure and a table to

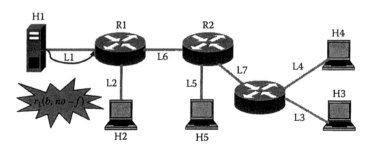

Figure 6.16 No filter resource reservation.

Table 6.6 Router State Table Contains a Definition and Label of the Links and Its Matching Network Element

Router State	R1	R2	R3
Inbound	L1[H1], L2[H2], L6[R2]	L6[R1], L5[H5], L7[R3]	L3[H3], L4[H4], L7[R2]
Outbound			

show what happens at the network nodes and the states of the routers. We consider the case when H1 is making reservations for one stream only, as in the starting Figure 6.16. The router state table contains a definition and label of the links and its matching network element (Table 6.6).

As shown in Table 6.6, no reservation has been made and each router receives path messages from all the sources with the f-flag off. Suppose H1 wants to make a reservation for bandwidth b. It wants to receive from any sender but one stream at a time. What does the reservation look like? Let us designate the reservation by the expression: $r_1(b, no - f)$. This request is signaled to the router R1 by host H1 as in Figure 6.16. Table 6.7 also shows the state of the router R1 in column R1. As a result of the request for bandwidth, the following reservation is made in R1. The router performs two functions when it reserves the requested resource:

1. It reserves the bandwidth b in the link L1 to H1.
2. It forwards the reservation request to all the network elements connected to it in its PATH database (that is, in this case to H2 and R2; Figure 6.17).

Table 6.7 State of Router 1

Router State	R1	R2	R3
Inbound	L1[H1], L2[H2], L6[R2]	L6[R1], L5[H5], L7[R3]	L3[H3], L4[H4], L7[R2]
Outbound	L1(b)		

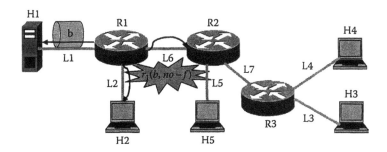

Figure 6.17 Bandwidth b reserved and request propagated to R2 and H2.

As a result of this function, the router state changes to include the reservation it has made and is given as Table 6.7.

The reservation functions performed by R1 are repeated by the next sets of network routers and in this case only R2 has a link forward to R3. Therefore, R2:

1. Reserves b bandwidth between it and R1.
2. Forwards the reservation request to H5 and R3, as shown in Figure 6.18.

The router state is also modified as shown in Table 6.8.

Bandwidth needs to be reserved on link L7 and this is done by router R3 as shown in Figure 6.18.

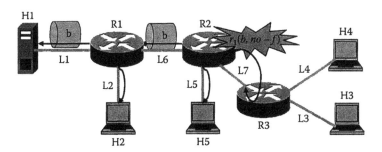

Figure 6.18 Bandwidth reserved on link L7 by router R3.

Table 6.8 States of Routers R1 and R2

Router State	R1	R2	R3
Inbound	L1[H1], L2[H2], L6[R2]	L6[R1], L5[H5], L7[R3]	L3[H3], L4[H4], L7[R2]
Outbound	L1(b)	L6(b)	

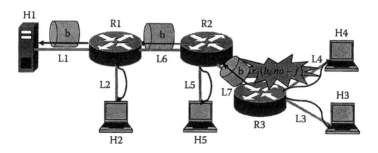

Figure 6.19   Bandwidth reservation on link L7.

1. R3 reserves b bandwidth to R2.
2. R3 forwards the reservation message to hosts H3 and H4.

Finally, reservations are made in the remaining links as well so that the overall reservation is shown in Figure 6.20. Reservations at the hosts are made also to match the service requests.

With these complete reservations made, H1 is able to receive packets from any host in the network. The time required to make the reservations at the routers and hosts and the time it takes to receive a reply that all the reservations have been made are not discussed in this sections but they do add to the overall quality of service offered by the network as seen by H1. If routers and hosts are not of the same type and depending on how busy they are, the time each one takes to make the reservations will differ.

Table 6.9   States of Routers R1, R2, and R3

Router State	R1	R2	R3
Inbound	L1[H1], L2[H2], L6[R2]	L6[R1], L5[H5], L7[R3]	L3[H3], L4[H4], L7[R2]
Outbound	L1(b)	L6(b)	L7(b)

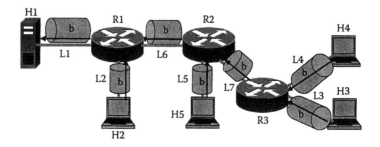

Figure 6.20   Complete reservation state for H1.

Table 6.10 Complete Reservation States at Routers R1, R2, and R3

Router State	R1	R2	R3
Inbound	L1[H1], L2[H2], L6[R2]	L6[R1], L5[H5], L7[R3]	L3[H3], L4[H4], L7[R2]
Outbound	L1(b), L2(b), L6(b)	L5(b), L6(b), L7(b)	L3(b), L4(b), L7(b)

In making the reservations, hosts are not concerned with which one of them H1 might choose to listen to. H1 also does not know if all the hosts might choose to send messages at the same time. For example, at router R3, if H3 and H4 transmit simultaneously, the reserved bandwidth at link L7 will be inadequate. The router in such circumstances might choose to interleave their packets before sending them.

Additional reservation requests from H2, H3, H4, and H5 in a multicast situation are handled in the same form as in this example. A complete router path in that case is bi-directional with b bandwidth in either direction. That ensures that all the hosts can receive at least a stream. The router state in that case is shown in Table 6.10.

Filtered Reservations

In the RSVP, the entity reserving the resources does not decide which packets can use the reserved resources. It only specifies for "whom" the reservation is made, and this "whom" is not the packets but the entity requesting the resources. The decision on which packets can use the reservation is taken by a separate function, called a packet filter. This filter can be fixed or dynamic. In the dynamic case, the receiver can change the filter during the course of the reservation. The filter selects packets that can use the reserved resources and the setting is done by the reserving node. The settings of the filter can be changed without changing the amount of reserved resources [1].

"A "fixed-filter" reservation allows a receiver to receive data only from the sources listed in the original reservation request, for the duration of the reservation. A "dynamic-filter" reservation allows a receiver to change its filter to different sources over time [1].

In Figure 6.21, consider that hosts H1, H4, and H5 are senders and hosts H2, H3, H4, and H5 are members of a multicast group. The routers R1, R2, and R3 received PATH messages with the filter flag f-set from all the sources. This forces each of the routers to undertake two actions:

1. Maintain a list of sources that can send on each output link.
2. Keep a record of the previous hop that forwarded the sent data from the source.

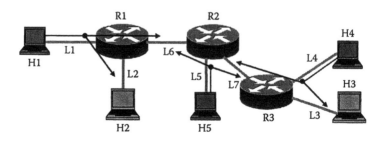

Figure 6.21 Filtered reservation.

Table 6.11 Complete Reservation States at Routers R1, R2, and R3

Router State (Output Links)	R1	R3	R3
Source/List of Previous Hop	L2[H1 ⇒ H1, H4 ⇒ R2, H5 ⇒ R2] L6[H1 ⇒ H1]		

Thus, in Figure 6.21 with H1 sending the record kept at R1 for output links L2 and L6 are as shown in Table 6.11.

Fixed-Filter Reservation

A "fixed-filter" reservation allows a receiver to receive data only from the sources listed in the original reservation request, for the duration of the reservation [1]. That is, the receiver, when making the reservation, explicitly specifies the senders it wants to receive data from. Consider in this example that H4 is the sender and H2 the receiver. H2 is the originator of the reservation and makes the reservation to receive one stream from H4.

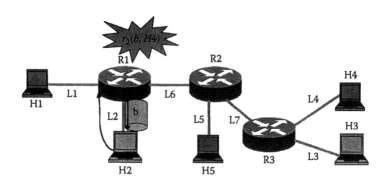

Figure 6.22 H2 originates reservation to receive data from H4.

Table 6.12   Reservation in Router R1

Router State (Output Links)	R1	R3	R3
Source/List of Previous Hop	L2[H1 ⇒ H1, H4 ⇒ R2(b), H5 ⇒ R2] L6[H1 ⇒ H1]		

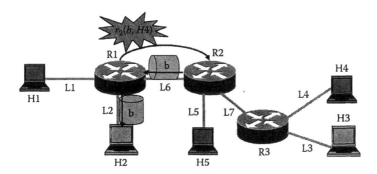

Figure 6.23   R1 forwards reservation to R2.

Table 6.13   Updated List of Previous Hops

Router State (Output Links)	R1	R3	R3
Source/List of Previous Hop	L2[H1 ⇒ H1, H4 ⇒ R2(b), H5 ⇒ R2] L6[H1 ⇒ H1]	L6[H5 ⇒ H5, H4 ⇒ R3(b)] L5[...], L7[...]	

The reservation request causes router R1 to reserve bandwidth b on link L2. It then looks up H4 before forwarding the reservation request to R2. R2 therefore reserves bandwidth b between it and router R1. Suppose there are two routes to the host H4. Which route will the request be forwarded to?

After this sequence, R2 forwards the reservation to router R3, where the reservation process continues with the overall reservations as in Table 6.14 and Figure 6.24.

Table 6.14   Complete Fixed-Filtered Reservation States

Router State (Output Links)	R1	R3	R3
Source/List of Previous Hop	L2[H1 ⇒ H1, H4 ⇒ R2(b), H5 ⇒ R2] L6[H1 ⇒ H1]	L6[H5 ⇒ H5, H4 ⇒ R3(b)] L5[...], L7[...]	L7[H4 ⇒ H4(b)] L4[...], L3[...]

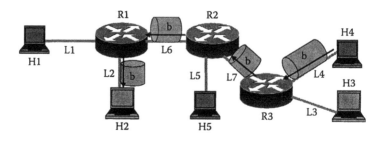

Figure 6.24   Complete fixed-filtered reservations.

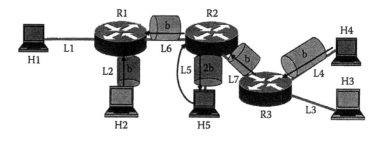

Figure 6.25   Dynamic filtered reservations.

Dynamic Filter Reservation

In dynamic filter reservation, the receiver that made the reservation can change its filter to accommodate different sources at different times so that the use of resources is determined by the changed filter.

Let us assume that host H5 can accept two streams and makes such a reservation request for 2b bandwidth and from any two sources. Therefore, router R2 reserves 2b bandwidth on L5 and forwards the reservation request. The reservations so far are as in Figure 6.25.

As a result of the forwarded reservation request to routers R1 and R3, only one of them needs to reserve bandwidth reservation on a link to ensure that the limit set is 2b. Let router R1 reserve the bandwidth on link L6 toward router R2 (Figure 6.26). Therefore, on link L5, host H5 can receive 2b bandwidth stream from host H1 and host H4 (Table 6.15).

Summary

By reading this chapter, you should gain an understanding of why RSVP was conceived as a signaling protocol, a protocol that permits requesting for resources along the path that the data must take to its destination. The goal of the RSVP is to

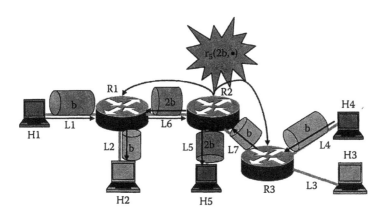

Figure 6.26   Complete dynamic-filtered reservations.

Table 6.15   Complete Dynamic-Filtered Reservation States

Router State (Output Links)	R1	R3	R3
Source/List of Previous Hop	L6[H1 ⇒ H1](b) L1[...], L2[..., (b)]	L5[H1 ⇒ R1, H4 ⇒ R3(2b)] L6[...](b), L7[...]	L7[H4 ⇒ H4(b)] L4[...], L3[...]

reserve resources at the routers along the path from source to destination. This allows for maintaining QoS requirements. In current Internet and across many networks and countries, link capacities are not unique but differ depending on the goals and budgets of the owners of the links and intermediate networks. Therefore, point-to-point reservation is applied recursively in point-to-multicast communications so that each path to each end host is reserved as in point-to-point. Hence, the problems emanating from such reservations are inherited. However, an additional problem emerges; links are not always of the same capacity, so routers in the networks may also be of differing capacities. In such situations, the link and router with the smallest capacity determine the reservations. Receiver-initiated resource reservation (Figure 6.8) is more sensible for several reasons to correct some of the problems from source-initiated resource reservations. A receiver knows the amount of bandwidth or capacity it needs or can handle for a particular service and therefore should reserve them for themselves. The burden of joining and leaving the multicast process is left to the joining endpoint and the source need not worry about it. Optimizing for path and capacity is enhanced in the routers. RSVP reservations do not specifically identify which packets can use the reservations. Different reservation styles are supported also through the use of filters. The RSVP depends strongly on the use of reservation messages. The RSVP model provides an avenue for the applications to request different reservation styles using messages to match the type of service being requested. Reservation messages originate at receivers and carry reservation requests that are forwarded upstream toward the sender. At every intermediate node, a reservation request triggers two general actions to *make a reservation on the link* and to *send the reservation request upstream*.

Practice Set: Review Questions

1. Which of the following represent a situation under which reservations should be made?
   a. Point-to-point
   b. One-to-many multicast
   c. Many-to-many multicast communications
   d. All of the above
   e. None of the above
2. In point-to-multicast situations, the reservation is for:
   a. Only one path for all the end-points
   b. Point-to-point reservation for each end-point
   c. End-points may share groups of multiple reservations
   d. No reservation is required
   e. Only the nearest end-point needs reservations
3. The path from a source to an end-point consists of three links of capacities 100 kbits, 1 Mbit, and 64 kbits. What is the capacity of the path?
   a. 1.164 Mbits
   b. 100 kbits
   c. 1 Mbits
   d. 64 kbits
   e. All of the above

Exercises

1. Draw the router state figure for the example in Table 6.8 showing the reservations.
2. Draw the router state figure when only H1 and H2 make reservations. What does the router state table look like? Show the message propagation steps.
3. Draw the router state figure when only H1 and H3 make reservations. What does the router state table look like? Show the message propagation steps.
4. Draw the router state figure when only H2 and H4 make reservations. What does the router state table look like? Show the message propagation steps.

5. Draw the router state figure when only H2 and H4 make reservations for two streams. What does the router state table look like? Show the message propagation steps.

6. How many streams can flow from R1 to R2 if all routers make reservations for two streams each?

References

[1] L. Zhang, S. Deering, D. Estrin, S. Shenker, and D. Zappala, RSVP: a new resource reservation protocol, *IEEE Communications Magazine, 50th Anniversary Commemorative Issue*, May 2002, pp. 116–127.

[2] Vodafone Australia, Future Technology Group, Introduction to IP with Carrier Extension, IIR Training Notes 2000.

[3] K. Jeffay, The Resource Reservation Protocol RSVP, 1999, <http://www.cs.unc.edu/~jeffay/courses/comp249f99>.

Chapter 7

MPLS: MultiProtocol Label Switching

Initially, the objective for developing the Internet was purely to transport data from one terminal to the other without considerations for QoS management (best effort was the aim), speed, scalability, and traffic engineering. As time progressed, the need for QoS increased, leading to protocols such as MPLS (MultiProtocol Label Switching). The original driver for introducing the MPLS was to make routers faster. It was observed that ATM switches, which used virtual path switching, were faster than IP routing. Trying to match a long IP address in a router is slower than using purely short-length labels lookup in routers. Achieving this allows devices to match the performance of ATM switches while doing the same job of data routing. Furthermore, MPLS enables integration of IP and ATM, so that ATM can be mapped to IP. In conventional IP packet routing, the destination IP address in the packet header is used to forward a packet to the destination. MPLS helps to decouple this IP packet-forwarding dependence to the information carried in the IP packet header. By not first inspecting the header of the packet before it decides to send a packet, MPLS proves to work independent of the source of data. It leads to the emergence of new routing functions and forwarding paradigms not supported by IP routing. For example, MPLS enables multiprotocol lambda switching, virtual private networks, layer 2 transport, traffic engineering, and guaranteed bandwidth services.

MPLS has so far been found to meet the bandwidth management and service requirements for the proposed all-IP backbone next-generation networks. Transforming networks to all-IP means new requirements must be addressed to ensure service quality comparable to circuit-switched networks. MPLS addresses scalability and QoS-based routing as well as service quality measures in IP networks.

It has been found to coexist with Frame Relay and Asynchronous Transfer Mode (ATM) networks. In both the existing Internet and emerging all-IP networks, new bandwidth-hungry and delay-sensitive services have been developed—for example, IPTV and VoIP. The performances of these emerging services depend on guaranteed bandwidth with low delays at the network backbone. These resource demands put a severe strain on current networks that must ensure that the customer receives uninterrupted streams of data services at reasonable bandwidths and in most cases real-time. MPLS is positioned to help meet the demands of real-time services by playing the important roles of packet forwarding, switching, and routing in the next-generation networks.

MPLS improves packet-forwarding performance in the network by simplifying packet forwarding through routers with layer 2 switching fabric. Because the mechanism for forwarding packets is easy to implement, MPLS can increase network performance and QoS offerings. MPLS therefore also provides the means for classifying and differentiating between services and thereby provides the means for service level agreements and guarantees. Furthermore, MPLS integrates ATM to IP networks, providing the bridge between IP access and an ATM core. It also provides a means for interoperability between ATM and IP networks. Scalable VPNs (virtual private networks) are enabled and enhanced through the use of MPLS traffic engineering.

MPLS Terminology and Concepts

In MPLS, packets are sent to their destinations by labeling them and forwarding them. Short, fixed-length labels are added to the IP packets when they enter the network. From this point on, instead of using the IP header information, the labels are used to forward the packets to their destinations. To do this, a new protocol is developed for distributing the labels. Extensions to existing protocols are also used to facilitate this.

MPLS and Its Components

Several techniques have been proposed to address the QoS issues in the current Internet, either focusing on the link layer, network layer, or the applications themselves. Most of them helped reserve resources, reduce transit delays, and improve upon bandwidth utilization (e.g., RSVP). MPLS, in trying to alleviate the issue of routing delays, focuses on the switching of traffic flows through the network and thereby improves routing and forwarding of IP packets. MPLS is a means for mapping of IP addresses to fixed-length labels that are used by packet-switching and -forwarding technologies. The labels used are cheap and faster to handle compared with the processes of using the IP packet header fields for packet routing. In addition, MPLS also:

- Provides mechanisms to manage traffic flows of various granularities, such as flows between different applications, machines and hardware, and machines
- Is independent of the layer 2 and layer 3 protocols
- Supports ATM, IP, and Frame Relay
- Works with routing protocols such as RSVP and Open Shortest Path First (OSPF)

Label-Switched Paths

Several label-switching protocols are in existence and they include, for example, ATM, Frame Relay, X.25, and TDM (Time Division Multiplexing). In ATM, the label-switching fabric is the virtual circuit identifier (VCI) and the virtual path identifier (VPI). Frame Relay refers to them as DLCIs (data link connection identifiers) and they travel with the frames. In TDM, they are time slots, an implied time for communication.

MPLS has the responsibility of directing IP packets along a path based on the labels on the packets, and that path is called the label-switched path (LSP). An MPLS label-switched path is akin to the VPI and VCI in an ATM network. Using LSPs provides a clear-cut demarcation between the control of packet transmission (routing) and the act of sending the packets or data to their destination (forwarding), which is undertaken by the MPLS. This function of switching the labeled packets is undertaken by the virtual switched routers (LSRs).

There are two methods for establishing LSPs: the control-driven method and the data-driven method. In the control-driven method, the LSPs are established before forwarding the packets. In the data-driven approach, a detection of a particular type of data flow is used to establish the paths. That is, the condition of the flow forces the decision of path switching. The setup of the path is simplex. That is, for a flow in a particular direction, a distinct LSP is created. If duplex transmission is required, then a different LSP is required for the second direction. An LSP is created by concatenating one or more label-switched hops. This allows a packet to be forwarded from one label-switching router to another label-switching router in the path. When an IP packet enters the ingress router, it examines it and assigns it a label based on its destination. Thus, at the ingress router, the packet is transformed from one that is forwarded by an IP address to one that is forwarded by labels. Therefore, the routers in the core network need not examine the IP address of this packet to know where to send it; rather, they examine its label. This makes their operations faster.

The labels are protocol-specific identifiers and are distributed using the Label Distribution Protocol (LDP) or RSVP. They can also be distributed by combining OSPF and Border Gateway Protocol (BGP) operations. At each router, the IP packet is encapsulated in an MPLS SHIM (header). The SHIM carries the label that is used for forwarding the packet to its destination. By inserting a fixed-length MPLS header at the beginning of the packet, high-speed switching is made

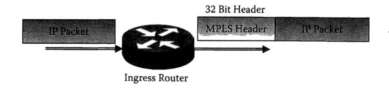

Figure 7.1 MPLS packet encapsulation.

possible. Routing within an MPLS domain is either hop-by-hop or through the use of explicit routing:

- *Hop-by-hop routing:* For this method, each LSR selects the next hop for a given FEC (forward equivalence class). The LSR then uses any available routing protocols, such as OSPF or ATM. This is a distributed control method and is destination-based forwarding with the tree routed at the destination.
- *Explicit routing:* Explicit routing is akin to source routing. The ingress LSR designates the list of nodes through which the LSP (packet) passes. The routing starts at the source, and a path is established first to the destination.

Constraint-based routing (CR) takes into account QoS parameters, such as bandwidth, delay, hop count, and QoS value. Therefore, the LSPs that are created could be CR–LSPs. For such cases, the constraints may be explicit hops or QoS requirements. Using explicit hops forces LSRs to paths that must be taken by packets. QoS requirements also dictate which links, queuing mechanism, or scheduling mechanism should be employed for the FEC. Based on this, it is therefore possible that some packets take longer paths or more costly paths to their destinations. CR, however, increases network utilization but adds complexity to the routing process because paths need to be selected to match the requirements of QoS for the LSP. The IETF through CR–LDP has defined a component to facilitate constraint-based routes.

Label Switching and Label Edge Routers

The routers in an MPLS core network fabric are called label edge routers (LERs) and label-switching routers (LSRs). An LER operates at the edge of the access network and MPLS network. An LSR is a router that supports MPLS-based forwarding and operates in the core of an MPLS network and performs label switching to forward packets toward their destinations using the established paths. This model therefore is like routing only at the edges of the network based on IP and (label) switching in the core network.

LERs can support multiple ports that are connected to networks of different types (for example, ATM, Ethernet, and Frame Relay) to forward the packets from an MPLS network to the dissimilar networks using the established LSPs. An appropriate signaling protocol is used at the ingress router to establish the LSPs.

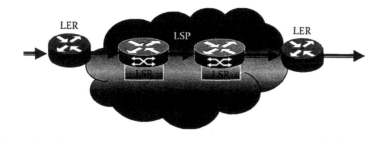

Figure 7.2 MPLS terms.

Figure 7.3 FEC and labels.

Furthermore, the LERs also assign and remove labels as traffic transverses the router (entry and exit).

FEC: Forward Equivalence Class

MPLS groups packets into the so-called forward equivalence class (FEC). An FEC is a description of a group of packets sharing the same transport requirements. Assignment of packets to an FEC is done once at the edge router as they transit to the destination. FECs are assigned based on service requirements and destination address prefix. Packets in the same FEC are given the same consideration or treatment as they are forwarded through to their destination. The FEC is specified as a combination of two elements:

- An IP address prefix of length from 0 to 32 bits
- 2A host address (a 32-bit address)

At each LSR, a table consisting of information on how the packet should be forwarded based on the FEC is constructed. This table is called a label information base (LIB). The entry in the table is a match of FEC-to-label bindings. The FECs help the LSRs specify which packets are mapped to which LSPs. A packet matches an FEC if and only if the IP address prefix of the FEC component matches the packet's destination IP address.

MPLS Packet Header

An MPLS packet header (Figure 7.6) is a 32-bit value used to encapsulate the packet of the underlying delivery protocol and consists of:

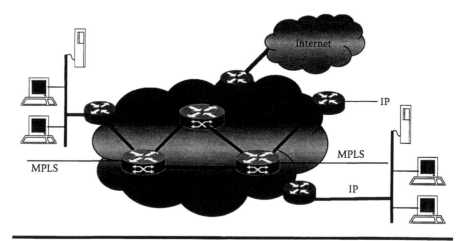

Figure 7.4   An MPLS core network structure.

Figure 7.5   Packets IP1 and IP2 forwarded in the same FEC.

1. The 20-bit label field is used to identify the packet to a particular LSP
2. The experimental bits (EXP), 3 bits, identify the class of service, which is used for priority queuing through the network. At each hop along the path, the EXP value determines which packets receive preferential treatment within the tunnel.
3. The Stacking (S) bit shows that this MPLS packet has more than a label associated with it.
4. TTL (8 bits), or the time-to-live value, gives the number of hops that this MPLS packet is permitted to travel in the network. At each LSR, the TTL is decremented by 1 until the TTL = 0, when the packet is discarded.

MPLS Protocol Stack

MPLS may be considered a layer 2.5 protocol, sitting above the data-link layer and below the network layer (Figure 7.7). As such, it appears as an interworking protocol between the two layers.

The relationship between the MPLS-related protocols and the OSI layers is shown in Figure 7.8.

MPLS is used along with several other protocols. The range of MPLS-related protocols includes:

Figure 7.6   MPLS packet header.

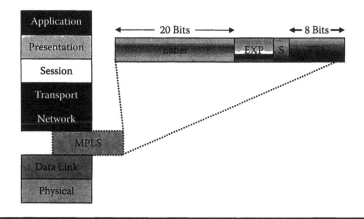

Figure 7.7   MPLS position within the OSI Model.

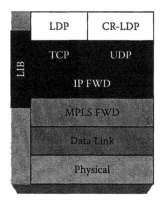

Figure 7.8   MPLS-related protocols.

1. CR-LDP is used for constraint-based LDP and for traffic engineering; resource reservation protocol traffic engineering (RSVP-TE) is another signaling mechanism used for traffic engineering.
2. Internet protocol (IP) FWD is the next hop forwarding based on IP address; longest match forwarding used.
3. LDP, the Label Distribution Protocol.
4. LIB (the label information base), which is a table of labels mapping input port and label to output port and label.
5. MPLS FWD, which is label switching based on MPLS label and LIB lookup.
6. TCP, the Transmission Control Protocol.
7. UDP, the User Datagram Protocol.

Labels and Label Bindings

MPLS is often described as a layer 2.5 protocol as its operation is sandwiched between the data-link layer and the network layer. For operation, it requires a label that identifies the path a packet should traverse. A layer 2 header encapsulates the label along with the packet. Each LSR examines the packet for its label to determine the next hop. On its journey from the ingress LER to the destination through the MPLS core network, the packet is switched just by the label without the IP packet being examined for its address. Each label has local significance and therefore refers to the present hop between the two LSRs. The values of labels are derived from the underlying data-link layers such as Frame Relay (data-link connection identifiers [DLCI]) or ATM's virtual path identifiers (VPIs) and virtual circuit identifiers (VCIs). Labels are bound to an FEC using either some policies or events. Two types of bindings are defined:

- Control-driven binding
- Data-driven binding

The decision on how to assign labels may be a result of the type of forwarding; for example:

- Traffic engineering
- Unicast routing to destination
- Multicast
- Quality-of-service
- Virtual private network (VPNs)

The labels used in LSR for FEC label bindings are of two forms: per platform and per interface. In the per platform case, the values of the labels are unique across the LSR. Because the labels are chosen from a pool, each LSR has distinct and unique labels.

For the per interface case, a range of labels are associated with an interface. Therefore, multiple sets of labels are used for the interfaces and allocated from separate pools. This also means that different interfaces can use the same labels (label reuse).

Creating Labels

Several protocols are used to create labels, including BGP, RSVP, and OSPF.

- Request-based method makes use of RSVP-like requests by processing of the requests control traffic.
- Traffic-based approach depends on the arrival of a packet to trigger the assignment and the distribution of labels.
- Topology-based method processes routing protocols such as BGP and OSPF.

The traffic-based method is data-driven binding and the topology and request-based methods are examples of control-driven label bindings.

Label Distribution Protocol

Apart from the application of existing protocols such as BGP and RSVP, which have been enhanced to provide signaling for label distribution, a dedicated protocol is also mandated for that purpose. The Label Distribution Protocol (LDP) is a specially engineered IETF protocol that provides explicit signaling and management labels. It provides label binding information to LSRs in an MPLS network. The LDP can also be used for explicit routing based on quality of service and class of service. The schemes for distributing protocols are as follows:

1. LDP performs mapping of unicast IP addresses of destinations to labels.
2. RSVP/CR–LDP (constraint-based routing LDP) is used for resource reservation and traffic engineering.
3. Protocol-independent multicast (PIM) is used for mapping multicast states labels.
4. BGP uses external labels (VPN).

The LDP is used to map FECs to labels and later to create LSPs. LDP sessions are established between LDP peers in the MPLS network. Peers need not be adjacent to each other. Four types of messages are exchanged between peers:

1. *Advertisement messages* are used to create, change, and delete FEC label mappings.
2. *Discovery messages* are used to announce and maintain the presence of an LSR in a network.

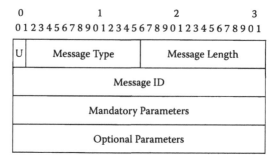

Figure 7.9   LDP message format.

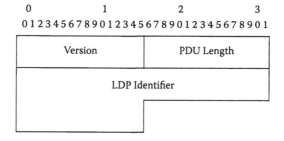

Figure 7.10   LDP PDU.

3. *Session messages* help to establish, maintain, and terminate sessions between LDP peers.
4. *Notification messages* provide signaling error information and advisory information.

LDP Message Formats

The format of the LDP message is shown in Figure 7.9. It consists of the optional parameters, the mandatory parameters, the message ID, message type, length of message, and the U field. The LDP sends messages by using the so-called LDP protocol data units (PDUs). Each LDP PDU consists of a message and a header field. The header is shown in Figure 7.10. The header has the version, length of PDU, and the LDP identifier fields.

Signaling

Signaling is used for requesting labels and for mapping MPLS labels to FECs (see Figure 7.11):

◾ *Label Request:* For label requests, an LSR requests labels from its downstream (toward egress LER) neighbor(s) for binding with a specific FEC. The request

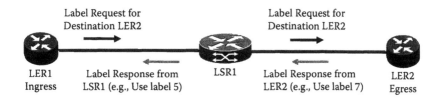

Figure 7.11   Signaling mechanism.

can be propagated downstream in the chain of LSR until it is received by the egress LER.

■ *Label Request Response:* A downstream LSR on receiving the label request sends a label to the upstream LSR requestor using a label mapping framework.

Distribution of Labels

Label distribution ensures that neighboring routers have a common view of the FEC requirements through the label bindings. Because labels are created based on the FECs, then to make label swapping possible, a uniform and clear understanding of which FECs map to which labels must be established between adjacent routers (LDP adjacencies). The communication of label binding information (or the so-called binding of an FEC to a specific label) between LSRs is accomplished by label distribution.

Label distribution can use one of two methods: (1) downstream-on-demand label distribution or (2) the canonical method. The distribution is a relay handover of information on labels in an LDP adjacency. Two LSRs have LDP adjacency if they are neighbors to create a hop between them.

Canonical Label Distribution Method

The canonical method is shown in Figure 7.12. In this method of distribution,

1. LSR1 and LSR2 have an LDP adjacency.
2. LSR2 discovers a "next hop" for a particular FEC.
3. LSR2 generates a label for the FEC and informs LSR1 about the binding.

Figure 7.12   Canonical label distribution.

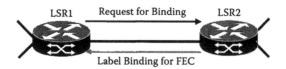

Figure 7.13 Downstream on-demand label distribution.

4. LSR1 inserts the binding in its forwarding tables.
5. If LSR2 is the next hop for the FEC, LSR1 can use that label knowing that LSR2 understands its meaning.

Downstream On-Demand Label Distribution

In this method of label distribution (Figure 7.13):

1. LSR1 recognizes LSR2 as its next-hop for an FEC.
2. A request is made to LSR2 for a binding between the FEC and a label.
3. If LSR2 recognizes the FEC and has a next hop for it, it creates a binding and replies to LSR1.
4. Both LSRs then have a common understanding.

Both of these methods can be supported in the same network. The label bindings shared by the LSRs are of the form **<label, prefix>**.

In Figure 7.14, each LSR downstream advertises label bindings for a particular FEC. In this example, both the routers with prefixes 169.89.10 and 171.69 advertise their bindings using LDP. The advertisement is upstream and the routers upstream

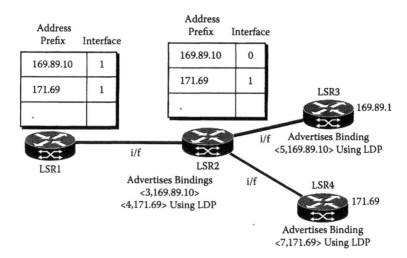

Figure 7.14 Binding advertisement.

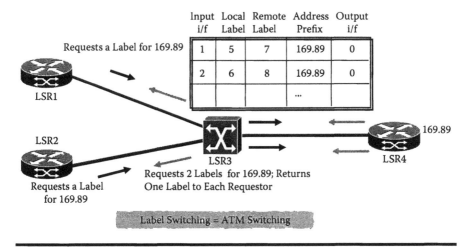

Figure 7.15 Label switching – ATM.

also advertise their FEC binding toward the source. In Figure 7.14, the following label binding advertisements were issued using LDP:

LSR4: advertises the label binding <7, 171.69> to LSR2.

LSR3: advertises the label binding <5, 169.89.10> to LSR2.

LSR2: advertises the label binding pair <3, 169.89.10> and <4, 171.69> to LSR1.
    Each binding in the pair enables LSR2 to use the two interfaces in its downstream for sending packets to LSR3 and LSR4 when needed. If there were more than two LSRs in the downstream of LSR2, there would have been bindings for each of them.

In Figure 7.15, the LSRs upstream have packets with specific FECs to send downstream and each one of them makes a request for label bindings. This request is propagated downstream toward the destinations. The underlying core is ATM based. LSR3 is therefore an ATM switch. LSR4 responds to the request by sending two label bindings with labels 7 and 8 (7 for use for packets from LSR1 and 8 for use for the packets from LSR2).

The ATM switch LSR3 also issues two label bindings, one for each of the requesting LSRs (1 and 2). The labels for the bindings are 5 and 6, respectively. Therefore, the label bindings are <5, 169.89> and <6, 169.89>. Let us trace the path through LSR1, LSR2 to LSR4 and show the states of the routing tables and LIBs (Figure 7.16).

Merging of Labels

Stream merging or aggregation of flows refers to when several streams from different interfaces are destined for the same end-point so that MPLS allows them to

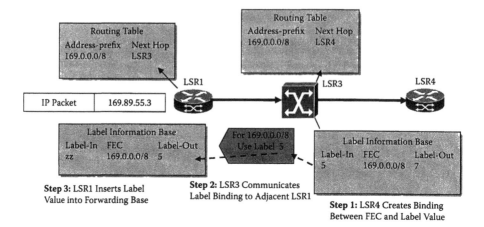

Figure 7.16 LIB and routing table states.

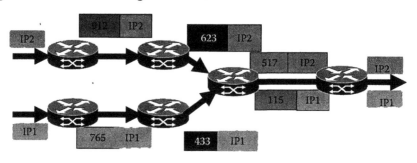

Figure 7.17 Two traffic streams passing through an LSR with different labels.

be merged together at an LSR and be switched using a common label if they are going through the network toward the same final destination. In an ATM core, the merging of multiple streams of traffic can lead to interleaving problems, which arise when multiple streams of traffic are merged. This can be avoided if the LSR uses virtual paths or virtual circuit merging.

In Figure 7.17, two streams are traverse an intermediate LSR and are transported to their destination using different labels (115 and 517). In Figure 7.18, a common label 517 is used to reflect the merging of the two streams.

Operation of MPLS

Because different FECs can be used for different packets and also because of different traffic characteristics and class-of-service requirements, in an MPLS domain, traffic from a source is not necessarily transported through the same path. The process of sending a packet through an MPLS domain involves the following five steps:

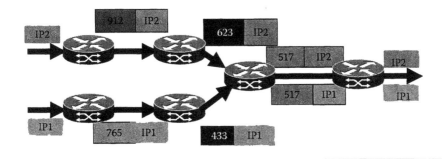

Figure 7.18   Merging of labels.

1. Creating labels and distributing them
2. Creating the FLIB table at each LSR and LER
3. Creating the label-switched path(LSP)
4. Inserting the label and table lookup
5. Packet forwarding

Creating Labels and Distribution

Well before traffic is available to for forwarding, the routers make the decision to bind a label to a specific FEC and build tables for them. The downstream routers also initiate the distribution of labels using LDP, and decisions on label and FEC bindings are taken. Next, traffic characteristics and the capabilities of the MPLS are negotiated using LDP.

Two modes of label control are defined in MPLS and used by LSRs for distribution of labels to neighboring LSRs. They are the independent and ordered modes:

1. *Independent mode:* In this mode, an LSR makes the decision to bind a label to an FEC independently based upon it recognizing a particular FEC and makes the decision to bind a label to an FEC independently to distribute the binding to its peers. Whenever new routes become available to the router, the new FECs are recognized.
2. *Ordered mode:* In the ordered mode, an LSR binds a label to a particular FEC for one of two reasons:
   - If and only if it is the egress router
   - It has received a label binding for the FEC from its next hop LSR. This is the mode recommended for ATM–LSRs.

Creating the Tables

Creating tables happens at each LSR. With the received label bindings, each LSR creates the FLIB. The content of this table shows the mapping between a label and

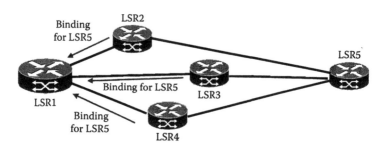

Figure 7.19   Binding for upstream router.

a matching FEC. The tables also show the mapping between the input port of the router and input label, and the output port and output label. This information in the table is refreshed whenever a renegotiation of the label bindings takes place.

Creating the LSP

This step follows the creation of all the LIBs. It is undertaken in the reverse direction, from the egress LER back through the core network to the ingress LER. In Figure 7.19, this is equivalent to starting at LER4 and LSR2, LSR1, and then LER1.

Inserting the Labels and Table Lookup

The ingress router (an LER) (Figure 7.19) uses the LIB table to find the next hop and makes a request for a label for the specific FEC. From then on, the LSRs just use the label to discover the next hop. When the packet reaches the egress LER, the label is removed and the packet is supplied to the destination as an IP packet.

I/f In	Label In	Destination Prefix	I/f Out	Label Out
3	40	128.1	1	50

Label Retention Mechanism

MPLS defines the treatment for label bindings received from LSRs that are not the next hop for a given FEC. Two modes are defined.

An LSR can receive label bindings from multiple LSRs. Some of the received label bindings are not the valid next hop for that FEC. There are two label retention methods: liberal label retention and conservative label retention. Consider the label bindings for LSR5 in Figure 7.19.

Liberal label retention (Figure 7.20): In this mode, the bindings between a label and an FEC received from LSRs that are not the next hop for a given FEC are

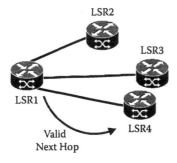

Figure 7.20   Liberal label retention.

retained. This mode allows for quicker adaptation to topology changes and allows for the switching of traffic to other LSPs in case of changes.

- LSR maintains bindings received from LSRs other than the valid next hop.
- If the next hop changes, it may begin using these bindings immediately.
- May allow more rapid adaptation to routing changes.
- Requires an LSR to maintain many more labels.

Conservative label retention: In this mode, the bindings between a label and an FEC received from LSRs that are not the next hop for a given FEC are discarded. This mode requires an LSR to maintain fewer labels. This is the recommended mode for ATM–LSRs.

- LSR only maintains bindings received from valid next hop.
- If the next hop changes, binding must be requested from new next hop.
- Restricts adaptation to changes in routing.
- Fewer labels must be maintained by LSR.

Forwarding of Packets

With reference to Figure 7.21, let us examine the path of a packet as it is transported to its destination from LER1 (the ingress LSR) to LER4 (the egress LSR). Consider a packet being sent through the ingress router LER1 to a destination on LER4.

1. Being the packet in front and sent first, LER1 may not have any labels for this packet. In a typical IP network, it will find the longest address match to discover the next hop. Let LSR1 be the next hop for LER1.
2. LER1 therefore initiates a label request toward LSR1.
3. This request propagates through the network in a manner similar to what happens in RSVP requests.

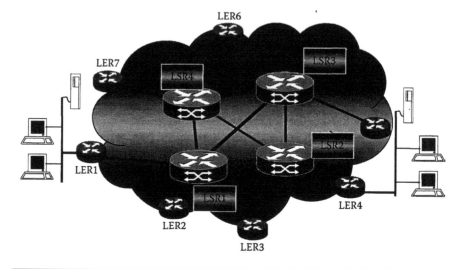

Figure 7.21 Operation of MPLS.

4. The downstream routers send labels to the request to fulfill the request. The green broken path or the LSP is set up using a signaling protocol such as LDP. Occasionally, traffic engineering is required. Then CR–LDP is used to determine the actual path setup. This is to ensure that the requirements for the class-of-service or quality-of-service are met.
5. LER1 inserts the label on the packet and forwards it to LSR1.
6. LSR1 receives the packet, strips it of the label that came with it from LER1, and attaches its own label and then forwards it to LSR2.
7. LSR2 repeats the label stripping and attaches a new label and forwards it to LER4.
8. At LER4, the packet is popped out of the MPLS domain. This means that the label is removed but no new label need be attached because the packet will now be delivered using the IP address of the destination.

The actual path followed by the packet from source to destination is indicated by the broken green lines in Figure 7.21.

Now consider the following example with two input streams entering the MPLS domain at LSR1 of Figure 7.21:

Input Port	Incoming Label	Output Port	Outgoing Label
1	3	3	4
2	8	1	6

1. Assume the packets belong to different FECs (e.g., a file transfer stream and a video content stream).
2. The video content stream requires a quality-of-service parameter.
3. At the ingress LSR, the two packet streams are classified into two separate FECs.
4. The matching label mappings for the streams are 3 and 8, respectively.
5. The input ports at the LSR are 1 and 2.
6. The corresponding output interfaces are 3 and 1, respectively.
7. Label swapping must also be done, and the previous labels must be exchanged for 4 and 6, respectively.
8. The packets are forwarded to LSR2, which also repeats a similar process as in Step 7 before the packets are forwarded to LER4.

Forwarding of Packets Using Labels

The beauty of using labels is that the forwarding of packets is independent of the underlying network and also independent of how labels are assigned. At the ingress, the IP packet is pushed into the network and it attaches a label to the packet. From then on, the following process takes place at each LSR:

1. Extract the label from a packet.
2. Find the LFIB entry with incoming label = label from packet.
3. Replace the label on the packet with outgoing label or labels.
4. Send the packet on through the outgoing interface(s).

Tunneling in MPLS Network Segments

Forwarding of MPLS packets from one network segment to another network segment is a little bit more involved and tricky compared with forwarding packets within the same segment. MPLS can control the whole path spanning the two network segments and forward packets using the so-called tunneling method. Consider for this exercise that two network segments are involved and each network segment consists of two edge routers (LERs) and three label-switching routers (LSRs). To forward packets in this structure, the MPLS defines three label-switching paths. The first path (LSP) spans the core or the LSRs in the first network segment, the second LSP spans the core of the second network segment, and the third LSP is an internetwork segment LSP defined by the edge routers (ingress and egress routers of each) of the two network segments (Figure 7.22). This concept is vital for MPLS VPN architectures.

In Figure 7.22 we have three LSPs defined as:

The LSP2 and LSP3 are internal to the two network segments, and LSP1 works across them. Consider that the BGP is used to create the label-switched paths. For

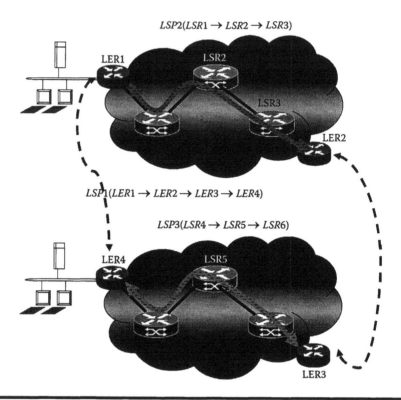

Figure 7.22   Inter-segment packet tunneling.

LSP1, the LER1 is aware that its destination is LER2. LER2 is also aware that its next destination is LER3, and so also LER3 is aware that its next destination is LER4. The LDP is used to receive and store labels from the egress router (LER4) all through to the ingress LER1.

In Figure 7.22 we have two tunnels, the first is between LER1 and LER2 (orange tunnel) and the second is between LER3 and LER4 (green tunnel). The labels used in each tunnel (LSP2 and LSP3) are different from the labels used by the LSP1. To achieve this, label stacking is used for carrying the packets from source to destinations. In label stacking, a packet carries two or more labels at a time—in our case, two labels at a time. The order of labels is:

1. First segment label for LSP1 and LSP2
2. Second segment is label for LSP1 and then LSP3

At the outlet of the first segment, the packet on arriving at LER3 has its LSP2 label removed and is replaced with the LSP3 label and swaps the LSP1 label with the packet within the next hop label. The last edge router LER4

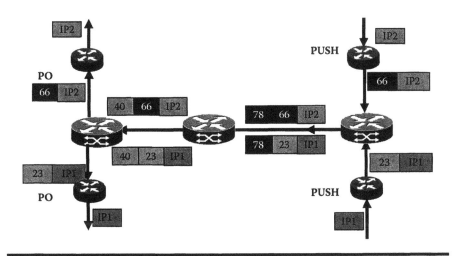

Figure 7.23 Label switching through two segments.

removes both labels before handing the packet over to its destination using the IP address.

Figure 7.23 provides an example of label switching through two segments. The leading segment is swapped at each LSR and the trailing label is retained.

MPLS Core

The MPLS backbone is turning out to be a significant development in WAN services. MPLS enables relatively easy provisioning of virtual private network (VPN) connections among enterprise locations as well as to partner and supplier locations.

Using an MPLS-based WAN service, an enterprise can consolidate Internet access, storage, file sharing, daily point-of-sale transmissions, Web commerce, and other types of data traffic onto a single, enterprisewide private network.

MPLS by itself provides some security but any enterprises considering using an MPLS WAN for services such as VoIP should also insist on a truly private MPLS network that is secured from the public Internet. "Private MPLS networks give service providers and enterprises better capability to manage quality of service and mitigate security risks" [8].

Summary

The need to offer QoS led to protocols such as MPLS. The original reason for introducing MultiProtocol Label Switching (MPLS) was to make routers faster. MPLS enables integration of IP and ATM, so that ATM can be mapped to IP. In conventional IP packet routing, the destination IP address in the packet header is used to forward a packet to the destination. MPLS helps to decouple this IP packet forwarding dependence to the information carried in the IP packet header. MPLS has so far been found to meet the bandwidth management and service requirements for the proposed all-IP backbone next-generation networks. Transforming networks to all-IP means new requirements must be addressed to ensure service quality comparable to circuit-switched networks. MPLS addresses scalability and QoS-based routing as well as service quality measures in IP networks. MPLS improves packet-forwarding performance in the network by simplifying packet forwarding through routers with layer 2 switching fabric. MPLS uses label-switched paths and label-switching routers for forwarding the packets. An MPLS header is a 32-bit value used to encapsulate the packet of the underlying delivery protocol. MPLS is used along with several other protocols, including LDP, RSVP, BGP, and OSPF. The LDP is used to map FECs to labels and later to create LSPs. LDP supports label distribution, which ensures that neighboring routers have a common view of the FEC requirements through the label bindings. The remainder of the chapter demonstrated how MPLS is used.

Practice Set: Review Questions

1. MPLS is a layer 2 protocol.
   a. True
   b. False
2. Labels can be created based on the:
   a. Traffic
   b. Topology
   c. Request
   d. All of the above
   e. None of the above
3. Creating label-switched paths is done only by hop-by-hop routing.
   a. True
   b. False
4. What is label merging?
   a. Merging of traffic destined for different destinations on different interfaces
   b. Merging of traffic for the same destination on different interfaces
   c. Merging of traffic for the same destination on the same interface
   d. None of the above
   e. (a) and (b) above
5. The creation of MPLS tunnels is for controlling the entire path a packet takes to destination:
   a. On the same network segment only and does not require explicitly specifying the intermediate routers

    b. On one or more network segments and does not require explicitly speci-
fying the intermediate routers

    c. Is restricted to the same network segment but needs to explicitly specify
the intermediate routers

    d. Needs to explicitly specify the intermediate routers and can span mul-
tiple network segments

    e. None of the above

6. What is the signaling mechanism in MPLS?

    a. Session Initiation Protocol (SIP)

    b. Discovery and session message

    c. Advertisement and notification message

    d. Label request and label mapping message

    e. (a), (b), and (c) above

7. Because the LDP is used for label distribution, only the downstream routers
initiate the distribution of labels and the bindings of label/FEC.

    a. True

    b. False

8. Only LERs participate in label swapping in MPLS networks.

    a. True

    b. False

9. Only LSRs participate in label swapping in MPLS networks.

    a. True

    b. False

10. Both LSRs and LERs participate in label swapping in MPLS networks.

    a. True

    b. False

11. Which of the following routing protocol can be used for initial route setup?

    a. ATM PNNI

    b. BGP

    c. OSPF

    d. All of the above

    e. None of the above

12. In MPLS, LIBs are maintained by:

    a. LERs

    b. LSRs

    c. Both LERs and LSRs

    d. None of them maintains an LIB

    e. LIBs not required

13. Which of the following is (are) used by the LDP for transmission of control
data between LSRs?

    a. TCP

    b. UDP

    c. SCTP

    d. Both (a) and (b)

    e. None of the above

Exercises

1. What is the role of an ingress router? Given the following figure, draw on it the structure of the packet as it emerges from in ingress router.

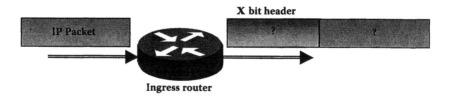

2. Suppose the MPLS network path shown in the following figure has two labels (3 and 7) to use; describe how these labels can be used to service a request for destination LER2. Label the figure correctly to show the label requests and response with the assigned labels.

3. List three methods of routing in an MPLS system. Distinguish them fully with illustrations.
4. Explain the fields of an MPLS packet header.
5. What motivates the decision on how to assign labels in an MPLS system?
6. List and explain the chain of events in an MPLS system operation.

References

[1] <http://www.iec.org/online/tutorials/mpls/>.
[2] Multiprotocol Label Switching Architecture, <http://www.javvin.com/protocol/rfc3031. pdf>.
[3] MPLS Label Stack Encoding, <http://www.javvin.com/protocol/rfc3032.pdf>.
[4] Time To Live (TTL) Processing in Multi-Protocol Label Switching (MPLS) Networks, <http://www.javvin.com/protocol/rfc3443.pdf>.
[5] LDP Specification, <http://www.javvin.com/protocol/rfc3036.pdf>.

[6] Constraint-Based LSP Setup using LDP, <http://www.javvin.com/protocol/rfc3212.pdf>.

[7] RSVP-TE: Extensions to RSVP for LSP Tunnels, <http://www.javvin.com/protocol/rfc3209.pdf>.

[8] T. Kiehn, Select Carefully to Get the Most from Converged Services Networks, Broadwing, <http://www.broadwing.com>.

[9] Applicability Statement for CR-LDP, <http://www.javvin.com/protocol/rfc3213.pdf>.

[10] IETF, <http://www.ietf.org>, RFC 3031 and RFC 3032.

Chapter 8

QoS: Quality-of-Service

The introduction of the Internet as a communication infrastructure has exposed the need for quality-of-service. QoS is much more pressing in real-time traffic (services) such as voice over IP, IPTV, video streaming, and video-conferencing. The term "quality-of-service" is used very loosely in telecommunications to refer to signal qualities that determine the performance of a communication system. Quantities such as signal-to-noise ratio (SNR), bit error rate (BER), and packet (frame or cell) loss ratio are involved. In addition, QoS is some measure of the average transfer rate of information between the sender and receiver. This may include the bandwidth, throughput, information rate, or sustained cell rate. It is also a measure of the maximum transfer rate of information between the sender and receiver. This may be called the peak cell rate, burst rate, or maximum information rate (this parameter only has meaning for a variable bit rate service). Therefore, QoS encompasses many system performance parameters that determine not only the capacity, but also the quality of the signal heard (voice) or seen (video and images).

QoS also includes a measure of the duration of data bursts, or some statistical properties that can approximately describe the probability density function of the transfer rate as a function of time. This can be called the burst size, burst tolerance, burst length, burstiness ratio, etc. This parameter is only valid for variable bit rate services. Furthermore, QoS is also a measure of the average (or perhaps maximum permissible) end-to-end delay. End-to-end (E2E) delay is critical for real-time traffic but may be measured for other traffic in order to characterize performance. Moreover, QoS is also a measure of the variability in the E2E delay (jitter) as a function of time. Again, this is only of significant interest for real-time traffic. For general services, QoS is a measure of the average success that users have in gaining access to desired services (e.g., probability of blocking, probability of buffer overflow, etc.). In a voice over IP system, QoS also includes a measure of average delay

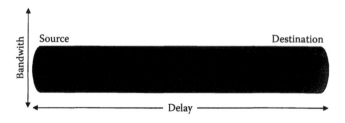

Figure 8.1   QoS characteristics of link medium.

in a service becoming active after a request from a user (post-dial delay, configuration change, etc.). In a nutshell, the definition and variables to include when composing a QoS are varied and depend on the service in hand.

QoS

Many of the parameters cited above are involved in the measurement of the telecommunications system capacity through Shannon's equation.

$$C = B \log_2 \left(1 + SNR\right)$$

In this equation, B is the natural bandwidth of the channel material. The SNR provides not only the noise performance of the system, but also when broken down it includes all the parameters that work together leading to the signal and noise. For example, many phenomena conspire and cause signal (QoS) degradation and they include electronic noise and interference from other communication systems. In general, a packet in transit from source to destination sees a path as having a bandwidth and a delay (Figure 8.1). In Figure 8.1, the delay increases along the horizontal axes and is reflected by the changing color from blue to red. The bandwidth of the "pipe" or channel is along the vertical axis.

Using the pipe analogy, variation in delay is variation in the length of the pipe (jitter). Jitter is more like the length of the pipe dilating (increasing and decreasing), depending on the packet being delivered and the route it has taken to the destination. If there are losses in the channel, this is seen as the pipe leaking, and bandwidth is the width of the pipe.

Bandwidth

Normally, the path between source and destination consists of several links of different capacities, as in Figure 8.2. Several factors contribute to the bandwidth available to the packet in the path. These factors include the forwarding speeds of the

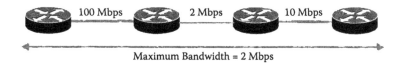

Figure 8.2 Constraint posed by link with lowest bandwidth.

devices in the path, the level of congestion in each hop, and the priority given to the queues.

Along the path, the link with the lowest bandwidth determines the bandwidth of the path. In Figure 8.2, the bandwidth of the path reduces to 2 Mbps despite the fact that the path has two links at 100 Mbps and 10 Mbps.

Throughput

Throughput is the average rate of transfer of an amount of information per unit time. In a nutshell, the proportion of the channel capacity just used for the time instance by an application is the throughput. Usually, it is measured as the average number of IP packets (or bytes) transferred per second. Therefore, the time it takes to carry a certain number of packets from source to destination is important. In slow links, the time required T_s to carry N packets is significantly higher than the time required in fast links (T_f) or $T_s > T_f$. Therefore, throughput is inversely proportional to the time T required to transfer the information. The objective for achieving a high (maximize) throughput is therefore to reduce this time. It can be achieved by deploying faster links, faster servers, and also through mechanisms that reduce the intermediate delays between the source and destination.

The time required to carry the number of packets N consists of various components that add up along the path the packet takes from source to destination. The components are:

- Insertion time (transmission rate)
- Average queuing time
- Processing delay at intermediate nodes
- Buffer size limitations
- Flow control and error control algorithm design

In many situations, it might be necessary to guarantee throughput, and several mechanisms can be used for that purpose. The mechanisms include:

- Using connection-oriented techniques
- Managing connections in real-time (this might mean using connection admission control schemes to ensure that connections that cannot be met are rejected)

- Using congestion control mechanisms to ensure that the impact of congestion in the path between a source and destination is minimized
- Defining different traffic classes and marking packets accordingly
- Servicing traffic according to its needs (rather than first-come-first-served, service queues according to traffic types)
- Using intelligent routing (this can be done at call setup to choose routes that support user requirements)
- Using small packets so that packets do not wait too long in queue to be served (ATM cells are suitable for this)
- Using fixed-length packets or cells and, if possible, using hardware switching to reduce delay

In a nutshell, to guarantee throughput, one needs to also guarantee an E2E delay.

Delay

Guaranteeing E2E delay in a network is very difficult except in single-hop, direct line-of-sight communication. The objective is therefore to minimize all sources of delay. Some of the major delays in a communication network include:

- Propagation delay
- Queuing delay
- Processing delay (switching/routing)
- Physical layer access delay

Microwave signals propagate at the speed of light. The propagation delay between source and destination is proportional to the distance between them. The longer this distance, the larger the delay. There is nothing much anyone can do to minimize the propagation delay. It is more worrisome in trans-international communications and over satellite routes.

Queuing of packets at network switches (queuing delay) before they are served creates significant delay. It is proportional to the service rate of the node and the number of sources of traffic it has to serve. To reduce this delay, service queues *according to need* rather than on a first-in, first-served basis. It is also essential to use small packet sizes and to service the packets according to their priorities so that a low-priority packet being serviced does not hold up higher-priority packets that may have arrived since the low-priority packet was taken up. To reduce queuing delay further, use multiple server processes per switch (node) where necessary.

Switching delay depends to a great extent on the speeds of routers. Using fast processors will reduce this delay. When fixed-sized cells are involved with small fixed header formats, it is useful to implement hardware switching to speed up the transfer of information. Cache routing and optimizing the switch architecture help. Above all, use packet formats that do not need a lot of processing at the switches.

QoS Architectures and Models

Apart from the best-effort model presented by the vanilla Internet, differentiated services and integrated services are two other QoS architectural models. In the best-effort model, all IP packets and traffic are treated equally. No packet is given special attention, and delay/jitter are unpredictable and so is bandwidth. The bandwidth (throughput) obtainable is what is available at the time.

Summary

Claude Shannon's equation is fundamental to most issues in telecommunication as it relates the system capacity to the communication bandwidth and the signal-to-noise ratio. The signal-to-noise ratio also provides a snapshot of what happens in the communication links, including fading, Doppler, inherent system noise, and the effects of the environment. Hence, the equation is fundamental to QoS although most articles on QoS fail to understand this relationship. By reading this chapter and relating the QoS parameters to the equation, deeper insight into telecommunications—and indeed how the system performs under different conditions—will be gained. The chapter clearly explained bandwidth, delay, throughput as opposed to capacity, and quality-of-service architectures.

Practice Set: Review Questions

1. The path between a source S and a destination D consists of four hops of speeds 100, 10, 2, and 30 Mbps. What is the maximum bandwidth of the path?
   a. 100 Mbps
   b. 142 Mbps
   c. 2 Mbps
   d. 30 Mbps
   e. 10 Mbps
2. The quality-of-service metric consists of:
   a. The amount bandwidth a service requires
   b. The amount of capacity available per second
   c. The amount of delay in the network used for the service
   d. The bit error rate of the service
   e. All of the above
3. This delay is deterministic and can be easily computed in a communication system:
   a. Propagation delay
   b. Processing delay

     c. Queuing delay
     d. (a) only
     e. (a), (b), and (c) above
  4. Throughput is:
     a. The average rate of transfer of information
     b. The route through a network
     c. The capacity of the link divided by its width
     d. The average bit error rate
     e. The average packet drop rate

Exercises

1. In a digital system, what can cause an increase in the packet loss or bit error rate?
2. Can the "bandwidth" vary for a circuit-switched system? Why or why not?
3. Can the "bandwidth" vary for a packet-switched system? Why or why not?
4. What could cause a reduction in the available bandwidth for a particular source/destination pair?
5. What causes delay in
     a. A circuit-switched system?
     b. A packet-switched system?
6. Is delay variation applicable to a circuit-switched system? Why or why not?
7. What is jitter?
8. What are the sources of delay variation for a packet network?
9. What could cause a reduction in gaining access to a service to degrade? (Consider this for IP, PSTN, and ATM networks.)
10. What could cause service activation and update delay?
11. What is throughput, and how can we guarantee it?

Chapter 9

SIP: Session Initiation Protocol

This chapter explains SIP in detail. The different networks where SIP can be used are discussed, as are the problems in SIP networks and some of the solutions. A few SIP applications are also discussed.

SIP: A Historical Snapshot

SIP RFC 3261 [1] emerged in the mid-1990s, originally created by Henning Schulzrinne [2] at Columbia University (United States) and his research team. A co-author of the Real-Time Transport Protocol (RTP) [4] for transmitting real-time data via the Internet, Professor Schulzrinne also co-wrote the Real Time Streaming Protocol (RTSP) [6], a proposed standard for controlling streaming audio-visual content over the Web. Hence, Schulzrinne is indeed a pioneer in IP communications.

Schulzrinne's intent was to define a standard for Multi-party Multi-media Session Control (MMUSIC). In 1996, he submitted a draft to the IETF that contained the key elements of SIP. In 1999, Schulzrinne removed extraneous components regarding media content in a new submission [2] and the IETF issued the first SIP specification, designated RFC 2543 [8]. While some vendors expressed concerns that protocols such as H.323 and MGCP could jeopardize their investments in SIP services, the IETF continued its work and issued SIP specification RFC 3261 [1] in 2001.

The advent of RFC 3261 signaled that the fundamentals of SIP were in place. Since then, enhancements to security and authentication, among other areas, have been issued in several additional RFCs. RFC 3262 [9], for example, governs Reliability and Provisional Responses. RFC 3263 [15] establishes rules to locate SIP Proxy Servers, while RFC 3264 [16] provides an offer/answer model and RFC 3265 [17] determines specific event notification.

As early as 2001, vendors began launching SIP-based services. Today, the enthusiasm for the protocol has grown to IP communication devices that support SIP. Organizations such as Sun Microsystems' Java Community Process are defining application programming interfaces (APIs) using the popular Java programming language so developers can build SIP components and applications for service providers and enterprises. Most importantly, increasing numbers of players are entering the SIP marketplace with promising new services, and SIP is on the path to become one of the most significant protocols since HTTP and SMTP.

Overview of SIP

SIP is an application layer signaling and mobility support protocol that can establish, modify, and terminate multimedia sessions such as Internet telephony calls [20]. SIP also is used to "invite" participants to already existing sessions, such as multicast conferences. Although SIP does not carry media itself (that role is reserved for RTP), SIP permits media to be added to (and removed from) an existing session. SIP transparently supports name mapping and redirection services, which support personal mobility [20]—users can maintain a single externally visible identifier regardless of their network location. This advantage means that the SIP user can be anywhere in the world and can be identified by one SIP URI (uniform resource identifier).

SIP supports five facets of establishing and terminating multimedia communications:

1. *User location:* determination of the end system to be used for communication.
2. *User availability (presence):* determination of the willingness of the called party to engage in communications.
3. *User capabilities:* determination of the media and media parameters to be used; this extends to the capabilities of devices and the descriptions of the communication parameters such as data rates and QoS specifications through the Session Description Protocol (SDP).
4. *Session setup:* "ringing," establishment of session parameters at both called and calling party and the reestablishment of sessions when necessary.
5. *Session management:* this includes transfer and termination of sessions, modifying session parameters, and invoking services.

SIP has become a protocol that can be used with other IP protocols to build a complete multimedia architecture. Typically, these architectures support protocols such as the RTP (RFC 1889, [4]) for transporting real-time data and providing QoS feedback, the RTSP (RFC 2326 [6]) for controlling delivery of streaming media, the Media Gateway Control Protocol (MEGACO) (RFC 3015 [7]) for controlling gateways to the PSTN, and the SDP (RFC 2327 [3]) for describing multimedia sessions. Therefore, SIP should be used in conjunction with other protocols to provide complete services to users. However, the basic functionality and operation of SIP does not depend on any of these protocols. SIP only works with them to support real-time delivery of IP services and for interconnectivity of disparate distributed networks so they can support IP communications.

SIP does not provide services. Rather, it provides primitives that can be used to implement different services. For example, SIP can locate a user and deliver an opaque object to that user's current location. If this primitive is used to deliver a session description written in SDP, for instance, the endpoints can agree on the parameters of a session. If the same primitive is used to deliver a photo of the caller as well as the session description, a "caller ID" service can be easily implemented. As this example shows, a single primitive is typically used to provide several different services.

SIP does not offer conference control services such as floor control or voting, and does not prescribe how a conference should be managed. This function is reserved for protocols such as RTSP and RSVP. SIP can be used to initiate a session that uses some other conference control protocol. Because SIP messages and the sessions they establish can pass through entirely different networks, SIP cannot—and does not—provide any kind of network resource reservation capabilities. This function is also reserved for the RSVP.

The nature of the services provided makes security particularly important. To that end, SIP has been enhanced to provide a suite of security services that includes denial-of-service prevention, authentication (both user to user and proxy to user), integrity protection, and encryption and privacy services. These security services are discussed in Chapter 10. SIP works with both IPv4 and IPv6 and in the future will be more wholly integrated to working with IPv6 when it becomes dominant.

SIP Protocol Stack

SIP is structured as a four-layer protocol, as shown in Figure 9.1; this means that its behavior is decoupled in terms of a set of fairly independent processing stages with only a loose coupling between each stage. The protocol behavior is described in layers for the purpose of presentation, allowing the description of functions common across elements in a single section. It does not dictate an implementation in any way. When we present a protocol and describe it as containing layers of operation, we mean that it is compliant with the set of rules defined by that layer.

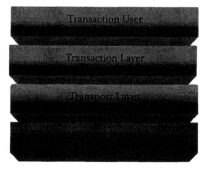

Figure 9.1   SIP stack.

Not every element specified by the protocol contains every layer. Furthermore, the elements specified by SIP are logical elements, not physical ones. A physical realization can choose to act as different logical elements, perhaps even on a transaction-by-transaction basis.

The lowest layer of SIP is its syntax and encoding (Figure 9.1). Its encoding is specified using an augmented Backus-Naur Form grammar.

The second layer in Figure 9.1 is the transport layer. It defines how a client sends requests and receives responses, and how a server receives requests and sends responses over the network. All SIP elements contain a transport layer.

The third layer shown in Figure 9.1 is the transaction layer. Transactions are a fundamental component of SIP. A transaction is a request sent by a client (using the transport layer) to a server transaction, along with all responses to that request sent from the server transaction back to the client. The transaction layer handles application layer re-transmissions, matching of responses to requests, and application layer timeouts. Any task that a user agent client (UAC) accomplishes takes place using a series of transactions. User agents (UAs) contain a transaction layer, as do stateful proxies. Stateful proxies are proxies that maintain state information [22]. Stateless proxies do not contain a transaction layer. Stateless proxies are proxies that do not keep any state information [22]. The transaction layer has a client component (referred to as a client transaction) and a server component (referred to as a server transaction), each of which is represented by a finite state machine that is constructed to process a particular request.

The layer above the transaction layer, the fourth layer in Figure 9.1, is called the transaction user (TU). Each of the SIP entities, except the stateless proxy, is a TU. When a TU wishes to send a request, it creates a client transaction instance and passes the request along with the destination IP address, port, and transport to which to send the request. A TU that creates a client transaction can also cancel it. When a client cancels a transaction, it requests that the server stop further processing, revert to the state that existed before the transaction was initiated, and generate a specific error response to that transaction. This is done with a cancel request, which constitutes its own transaction but references the transaction as cancelled.

The SIP elements—that is, user-agent-clients and servers, stateless and stateful proxies, and registrars—contain a core that distinguishes them from each other. Cores, except for the stateless proxy, are transaction users. While the behavior of the UAC and user agent server (UAS) cores depend on the method, there are some common rules for all methods. For a UAC, these rules govern the construction of a request; for a UAS, they govern the processing of a request and generation of a response. Because registrations play an important role in SIP, a UAS that handles a register is given the special name "registrar."

Certain other requests are sent within a dialogue. A dialogue is a peer-to-peer SIP relationship between two user agents that persists for some time. The dialogue facilitates sequencing of messages and proper routing of requests between the user agents. The "invite" method is the only way defined in this specification to establish a dialogue. When a UAC sends a request that is within the context of a dialogue, it follows the common UAC rules and also the rules for mid-dialogue requests.

The most important method in SIP is the "invite" method, which is used to establish a session between participants. A session is a collection of participants, and streams of media between them, for the purposes of communication.

Characteristics

SIP is described as a control protocol for creating, modifying, and terminating sessions with one or more participants. These sessions include Internet multimedia conferences, Internet (or any IP network) telephone calls, and multimedia distribution (Figure 9.2). Members in a session can communicate via multicast or via a mesh of unicast relations, or via a combination of the two. SIP supports session descriptions that allow participants to agree on a set of compatible media types. It also supports user mobility by proxying and redirecting requests to the user's current location. SIP is not tied to any particular conference control protocol.

SIP-Enabled Functions

Name translation and user location ensure that the call reaches the called party wherever the called party is located and carries out any mapping of descriptive location information. SIP ensures that the details of nature of the call (session) are supported.

Feature negotiation allows the group involved in a call to agree on the features supported—recognizing that not all the parties can support the same level of features.

During a call, a participant can bring other users onto the call or cancel connections to other users. This is called "call participant management." In addition, users could be transferred or placed on hold.

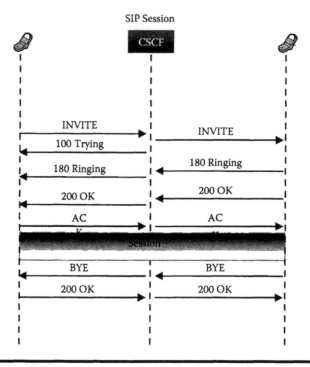

Figure 9.2 Typical SIP session.

A user should be able to change the call characteristics during the course of the call. This feature is called "call feature changes."

SIP Networks and Services

SIP for Wireless Networks

SIP is the key to high-value wireless services [21]. The emerging high-value, high-margin wireless services are about peer-to-peer communications. SIP was designed to support these types of services. SIP can run on any wireless network.

SIP is used in packet-based IP communications networks, so in the mobile field it will come into play with the launch of new high-speed 3G systems and through existing technologies such as wireless LAN and GPRS.

SIP for Fixed-Line Networks

The majority of SIP deployments to date have been in fixed-line IP networks. Gradually, SIP is becoming the preferred protocol over legacy protocols such as H.323. This is due to SIP's ease of application development, openness, lightweight design, and total integration with Web technologies [21]. VoIP technologists agree that all significant VoIP network deployments going forward will be based on SIP.

SIP Mobility

SIP supports personal mobility; that is, a user can be found independent of location and network device (PC, laptop, IP phone, etc.). The step from personal mobility to IP mobility support is basically the roaming frequency and the ability of a user to change location (IP address) during a traffic flow [20]. Therefore, to support IP mobility, we need to add the ability to move while a session is active. It is assumed that the mobile host belongs to a home network, on which there is a SIP server that receives registrations from the mobile host each time it changes location. This is similar to home agent registration in Mobile IP. The host does not need to have a statically allocated IP address on the home network.

When the correspondent host sends an "invite" to the mobile host, the redirect server has current information about the mobile host's location and redirects the invite there (see Figure 9.3).

If the mobile host moves during a session, it must send a new "invite" to the original call setup, as shown in Figure 9.4. It should put the new IP address in the Contact field of the SIP message, which tells the correspondent host where it wants to receive future SIP messages.

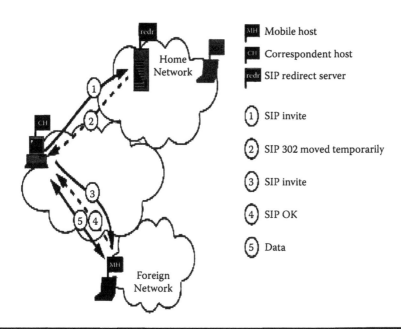

Figure 9.3　SIP mobility: setting up a call.

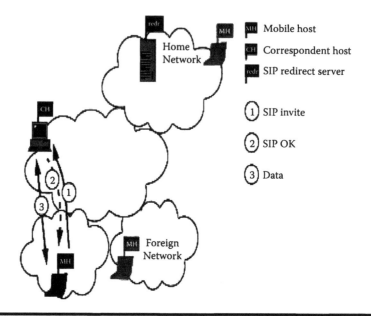

Figure 9.4 A new "invite" to original call setup due to mobility of mobile host during session.

Application Layer Mobility Using SIP

More needs to be done to support mobile Internet multimedia applications beyond the ability to maintain connectivity across subnet changes [18]. Schulzrinne and Wedlund [18] describe how SIP can help provide terminal, personal, session, and service mobility to applications ranging from Internet telephony to presence and instant messaging. A lot of effort has been made over the years to allow computers and communication devices to continue communication, even when mobile.

System design should make it easier for devices to move between different wireless networks, depending on population density, speed of movement, and propagation characteristics. Schulzrinne and Wedlund [18] also describe an architecture that allows the support of a full range of mobility options, independent of the underlying technology. The focus was primarily on the provision of multimedia services but alluded briefly to "data" services. This was taken further to include how application layer support is necessary to offer more than just handoff between base stations and subnets, as well as how it can, under certain circumstances, compensate for the lack of deployment of Mobile IP. SIP was used at the heart of the effort as the protocol.

Application layer mobility can either partially replace or complement network layer mobility [18]. For interactive sessions, SIP-based mobility can be used to provide all common forms of mobility, including terminal, personal, and service mobility. For terminal mobility, an IPv6-based solution is preferable, as it applies to all IP-based applications, rather than just Internet telephony and conferencing. When the home agents are absent, SIP-based mobility can provide mobility services to the most important current application telephony systems.

Error Recovery

If the correspondent host for some reason has an outdated address of the mobile host, it must have a fallback mechanism to break the error situation. To avoid situations like this, a host can send retransmissions of information invitations also to the SIP server on the mobile host's home network. Because the SIP server has a fixed address, the mobile host can always send registrations to it. In this fashion, the correspondent host can relocate a mobile host that has been lost.

Problems with SIP in a Network

SIP can be used in any network. Some of these SIP networks are discussed in this chapter. Because SIP is receiving much attention and seems to be the most promising candidate as a signaling protocol for the current and future IP telephony services, it is worth knowing and even solving the problems that still persist in SIP networks. Two of the most severe problems found in SIP services are delay and security. This section provides a discussion of the problems within SIP networks.

The common problems of SIP in different networks include security, delay, packet loss, and jitter. Undesirable delay causes packet loss, which is detrimental to applications such as VoIP or streaming video with stringent QoS requirements [13].

Security

In the SIP specifications, support for both authentication and encryption of SIP messages were described, using either challenge-response or private/public key cryptography.

SIP provides an elegant application layer mobility support that solves the problems associated with lower layer mobility protocols in next-generation heterogeneous wireless access networks [19]. Because sessions can be between domains, securing the process will involve inter-realm authentication and authorization, which gives rise to most issues such as user privacy and authorization granularity.

A solution by Sicker et al. [19] for the security problem is based on three modular functions:

1. *Resource registration:* Allows a user to register within local domain.
2. *Resource discovery:* Allows a user to locate other users from within the same domain as well as other domains.
3. *Call initiation:* Allows a user to set up a session with another user.

Each of these modular functions is independent of each other. They are protected separately. The goal is to make use of practical security functions while providing a robust level of privacy to the end user.

Rao and Li [14] solved the security and privacy issues by securing the ubiquitous computing (UC) system and, second, by securing the user:

1. *Securing the UC system:* Protects the UC system from unauthorized access and ensures the possibility of tracking down "who does the evil" [14].
2. *Securing the user:* UC systems based on similar designs will not be accepted for wide deployment, as almost every user action might be captured by the system.

Quality-of-Service and SIP

A proposed SIP-based architecture supports soft handoff for IP-centric wireless networks, thus alleviating the problem of packet loss [13]. The architecture ensures that there is no packet loss and the E2E delay jitter is kept under control, thus maintaining two important parameters dictating the QoS for streaming multimedia applications.

Delay, Jitter, and Packet Loss

Although SIP-based mobility management solves the problem posed by Mobile IP route optimization, in some cases it introduces unacceptable handoff delays for multimedia applications with stringent QoS requirements. Moreover, SIP entails application layer processing of the messages, which may introduce additional delay [13]. To alleviate this problem, a SIP-based soft handoff mobility architecture for next-generation wireless networks is given in [13]. The architecture ensures zero packet loss and controlled delay and jitter.

Although security, delay, jitter, and packet loss problems exist in SIP networks, SIP provides an elegant application layer mobility support that solves the problems associated with lower layer mobility protocols in next-generation heterogeneous wireless access networks—and other networks for that matter.

SIP-Based Applications and Services

SIP supports name mapping and redirection services, which support personal mobility. Because SIP can be used with other IETF protocols, a complete multimedia architecture can be built out of SIP. SIP itself does not provide services but does provide primitives that can be used to implement different services.

SIP can create, modify, and terminate sessions. This function is part of SIP's definition. The sessions include Internet multimedia conferences, Internet or any IP network telephone calls, and multimedia distribution.

Mobile Agent Communication Using SIP

Reliable communication protocols can integrate SIP as a signaling protocol—for example, a Synchronous Invite Model (SIM) and Asynchronous Subscribe & Notify Model (ASNM) to establish a framework for inter-agent communication [12]. Both of them can solve "communication failure and message chasing" problems [12]. Two communication problems are inherent in the mobile agent communication:

1. *Communication failure:* Defined as a message delivered to the host that the receiver agent stays; receiver agent has migrated to the next host [12]. Thus, the receiver agent receives unsuccessful messages. It is not guaranteed that the receiver host can get all messages sent to it because messages cannot be delivered instantly, even in an error-free network.
2. *Message chasing:* Defined as when a receiver host migrates frequently and each time a message reaches the host, the receiver host has moved and advanced to the next host, resulting in a race condition. This may corrupt the collaboration of mobile agent systems.

A solution to this problem integrates SIP to location agents and delivers messages in order to improve communication fault rate and thus enhanced system reliability. SIP specifies the architecture of user agents and servers to support agent tracking, call routing, and call redirection. SIM and ASNM were used to solve communication failure and message chasing problems. SIP handles normal inter-agent communication [12]. ASNM is invoked as agents are moving and messages are forwarded indirectly and distributive.

The communication cost of SIM is [12]:

$$\text{Cost(SIM)} = T_{inv} + T_{msg}$$

And the communication cost of ASNM is given by:

$$\text{Cost(ASNM)} = T_{inv} + T_{sub} + T_{mig} + T_{mig}(R) + T_{not} + T_{msg}$$

where T_{inv} is the duration from the moment an agent sends an "invite" to the moment it receives either moved temporarily or temporarily unavailable, T_{msg} is the duration from the moment an agent sends a message to the receiver to the moment it gets an OK response, T_{not} is the time needed to send "notify" to an agent from the home server, T_{sub} is the duration from the moment an agent sends "subscribe" to home server to the moment it gets an OK, and T_{mig} is the moment an agent leaves a node to the moment it reaches another node.

The two models integrate SIP to exploit its signaling feature.

An Advanced Architecture for Push Services

Push technology allows for new, appealing services that foresee the distribution of information useful to mobile users, such as weather forecasts, traffic news, sport news, or simply advertising. A reference architecture to support advanced Push services, based on presence, location, instant messaging (IM), and build-over SIP, is given in [10].

IM is one of the possible outstanding applications for mobile data services, while Presence and Push are considered very interesting enabling services. The architecture uses the Presence concept and IM solution, which is based on SIP and is integrated with the Wireless Application Protocol (WAP). The architecture has been designed with the IP Multimedia Subsystem (IMS) architecture in view.

Context-Aware Communication Application

Pervasive computing applications are usually engineered to provide unprecedented levels of flexibility to reconfigure and adapt in response to changes in computing resources and user requirements. By using a disciplined, model-based approach to engineer a context-aware, SIP-based communication application, a rich set of contextual information can be provided in order to self-configure and adapt, including the location of people and devices, activity types, device network status, and power. In addition, system policies and user preference information help evaluate the suitability of available communication channels for the current context of use. Policies and preferences can be modified on-the-fly to support highly flexible behavior and accommodate an evolving set of system resources and user requirements [5].

Generally, other SIP services include advanced telephony services and ubiquitous computing. Ubiquitous computing makes computational resources or communication more widely available.

Service Deployment and Development in H.323 versus in SIP

Comparisons at the system level between SIP and H.323 show that H.323 provides better functionality, interoperability, and internetworking with respect to supplementary services. SIP has advantages with respect to the design of low-cost terminals.

Summary

This chapter provided a historical and technical snapshot of the Session Initiation Protocol. The discussions extended to the SIP stack, its functions, and mobility support. The SIP supports personal mobility. For interactive sessions, SIP-based mobility can be used to provide all common forms of mobility, including terminal, personal, and service mobility. For terminal mobility, an IPv6-based solution is preferable, as it applies to all IP-based applications, rather than just Internet telephony and conferencing. How security features have been added to SIP was discussed, in addition to quality-of-service. SIP services were explained, including push and context-aware services.

Practice Set: Review Questions

1. SIP is not a routing protocol.
   a. True
   b. False
2. SIP has no real value to an IP network.
   a. True
   b. False
3. SIP was designed with the following security features:
   a. AAA
   b. AES cryptographic features
   c. DES cryptographic features
   d. All of the above
   e. None of the above
4. SIP security is an afterthought.
   a. True
   b. False
5. SIP supports mobility through:
   a. Establishing mobile sessions
   b. Modifying mobile sessions
   c. Terminating mobile sessions
   d. All of the above
   e. None of the above
6. SIP supports several facets for establishing and terminating multimedia sessions:
   a. Seven ways of doing it
   b. Five ways of doing it
   c. Nine ways of doing it
   d. None
   e. Three ways of doing it
7. Transactions are fundamental elements of SIP.
   a. True
   b. False
8. The transaction layer in SIP is:
   a. Layer 3
   b. Layer 4
   c. Layer 5

     d. Layer 2

     e. Layer 1

9. What is a transaction user?

     a. All stateless proxies in SIP

     b. All SIP entities

     c. Both (a) and (b) are true

     d. None of the above

     e. Only human users of SIP

10. SIP elements include:

     a. User-agent clients and servers

     b. Stateless and stateful proxies

     c. SIP registrars

     d. All of the above

     e. None of the above

Exercises

1. List the methods that the SIP uses for establishing and terminating multimedia sessions. Discuss them fully.
2. Transactions are fundamental elements of the SIP. What are SIP transactions, and which layer of the SIP protocol stack handles this function? Discuss.
3. How does a SIP transaction user send a request?
4. How does SIP support application layer mobility and QoS? Discuss these within the context of a multimedia user.

References

[1] J. Rosenberg, H. Schulzrinne, G. Camarillo, A. Johnston, J. Peterson, R. Sparks, M. Handley, and E. Schooler, SIP: Session Initiation Protocol, RFC 3261, June 2002.

[2] SIP Overview, Available: <http://www.sipcenter.com/sip.nsf/html/WEBB5YNVK8/$FILE/Ubiquity_SIP_Overview.pdf>.

[3] M. Handley and V. Jacobson, SDP: Session Description Protocol, RFC 2327, April 1998.

[4] H. Schulzrinne, S. Casner, R. Frederick, and V. Jaconson, RTP: A Transport Protocol for Real-Time Applications, RFC 1889, January 1996.

[5] T. McFadden, K. Hendricksen, J. Indulska, and P. Mascaro, Applying a Disciplined Approach to the Development of a Context-Aware Communication Application, *Proceedings of the 3rd IEEE International Conference on Pervasive Computing and Communications (PerCom 2005)*, March 2005, pp. 300–306.

[6] H. Schulzrinne, R. Rao, and R. Lanphier, Real Time Streaming Protocol (RTSP), RFC 2326, April 1998.

[7] F. Cuervo, N. Greene, A. Rayhan, C. Huitema, B. Rosen, and J. Segers, Megaco Protocol Version 1.0, RFC 3015, November 2000.

[8] M. Handleym H. Schulzrinne, E. Schooler, and J. Rosenberg, SIP: Session Initiation Protocol, RFC 2543, March 1999.

[9] J. Rosenberg and H. Schulzrinne, Reliability of Provisional Responses in the Session Initiation Protocol (SIP), RFC 3262, June 2002.

[10] D. Tosi, An Advanced Architecture for Push Services, Proceedings of the Fourth International Conference on Web Information Systems Engineering Workshop (WISEW'03), December 2003, pp. 193–200.

[11] J.Y. Zhou, Z.Y. Jia, and D.X. Chen, Designing Reliable Communication Protocols for Mobile Agents, *Proceedings of the 23rd International Conference on Distributed Computing Systems Workshops (ICDCSW'03)*, May 2003, p. 484.

[12] H. Tsai and F. Leu, Mobile Agent Communication Using SIP, *Proceedings of the 2005 Symposium on Application and the Internet (SAINT'05)*, January 2005, pp. 274–279.

[13] N. Banerjee, S.K. Das, and A. Acharya, SIP-based Mobility Architecture for Next Generation Wireless Networks, *Proceedings of 3rd IEEE International Conference on Pervasive Computing and Communications (PerCom 2005)*, March 2005, pp. 181–190.

[14] W. Rao and W. Li, Design of an Open and Secure Ubiquitous Computing System, *Proceedings of the IEEE/WIC/ACM International Conference on Web Intelligence (WI'04)*, September 2004, pp. 656–659.

[15] J. Rosenberg and H Schulzrinne, SIP: Locating SIP Servers, RFC 3263, June 2002.

[16] J. Rosenberg and H. Schulzrinne, An Offer/Answer Model with SDP, RFC 3264, June 2002.

[17] A.B. Roach, Session Initiation Protocol (SIP) — Specific Event Notification, RFC 3265, June 2002.

[18] H. Schulzrinne and E. Wedlund, Application-Layer Mobility Using SIP, *ACM SIGMOBILE Mobile Computing and Communications*, 4(3), July 2000, 47–57.

[19] D.C. Sicker, A. Kulkarni, A. Chavali, and M. Fajandar, A Federated Model for Securing Web-Based Videoconferencing, *Proceedings of the International Conference on Information Technology: Computers and Communications (ITCC'03)*, April 2003, p. 396.

[20] R. Padya, Emerging Mobile and Personal Communication Systems, *IEEE Communications Magazine*, 33, June 1995, 44–52.

[21] SIP for Wired Network, Available: <http://www.dynamicsoft.com/innovation/sip-4wire.php>.

[22] S. Berger, H. Schulzrinne, S. Sidiroglou, and X. Wu, Ubiquitous Computing Using SIP, *IEEE Communications Magazine*, 41(11), November 2003, 128–135.

[23] J. Glasmann, W. Kellerer, and H. Müller, Service Development and Deployment in H.323 and SIP, *IEEE Communications Surveys & Tutorials*, 5(2), Fourth Quarter 2003, 32–47.

Chapter 10

SIP Security

SIP security threats fall into one of four different categories. The first set of threats results from the vulnerabilities of SIP. These threats take advantage of the flaws in the definition of SIP in terms of its characteristics, functionality, and description. The second set of threats results from its interoperation with other protocols. This relates to SIP interfacing with other protocols. The third group is threats that exploit the operational details of SIP, including mandatory, nonmandatory, and/or optional requirements of SIP. The fourth type of threat results from the inherent security risks associated with the system under which SIP is used. These four categories of threats are well discussed in [1] and are therefore summarized only.

Threats from Vulnerability of SIP

SIP is not an easy protocol to secure. Its use of intermediaries, its multifaceted trust relationships, its expected usage between elements with no trust at all, and its user-to-user operation make security far from trivial. Security solutions are needed that are deployable today, without extensive coordination, in a wide variety of environments and usages. To meet these diverse needs, several mechanisms applicable to different aspects and usages of SIP are required.

Security Loopholes in SIP

Although security and privacy should be mandatory for IP telephony architecture, most of the attention during the initial design of the IETF IP telephony architecture and its signaling protocol (SIP) has focused on the possibility of providing new dynamic and powerful services, and simplicity. Less attention has been paid to security features.

189

IP networks using SIP can replace PSTN provided the same basic telephony service with comparable levels of QoS and network security are part and parcel of the IP network. The following security characteristics should therefore be guaranteed:

1. High service availability
2. Stable and error-free operation
3. Protection of the user-to-network and user-to-user network traffic

SIP security should cover the overall communication chain. This includes the source, destination, links, signaling, packet headers, and messages. In addition, protocols that interface with SIP should also be secured.

Unlike traditional mobile phone text messages, which are protected by the network and therefore are forced to remain unmodified from sender to destination, SIP text messages can be modified in transit because no security is used.

SIP messages may contain information a user or server wishes to keep private. The headers can reveal information about the communication patterns and individual content information, or other confidential information. The SIP message body may also contain user information (media type, codec, address and ports, etc.) that should not be revealed.

Messages Not Encrypted

Traditionally, SIP messages are not encrypted and are therefore open to interception in the links or at any point in the network. The "To" and "via" fields of the headers are visible so that SIP messages can be routed normally. This makes it possible for an attacker to send spoofed INITIATE messages with deceitful IP addresses. Furthermore, it is possible to use a spoofed BYE message to end SIP sessions prematurely. While this might require capturing SIP messages and holding them for use, it is not a difficult task.

Several approaches are used to protect SIP messages. They include using security mechanisms such as IPSec, TLS/SSL (Transport Layer Security/Secure Socket layer), and S/MIME (Secure/Multipurpose Internet mail Extensions). Another approach is to deploy firewalls at the perimeter of the network to protect and provide application layer security as well as authentication. This also protects against transport layer and protocol attacks.

Securing SIP header and overall information is implemented for the following reasons:

- Maintaining private user and network information to guarantee some form of user privacy
- Avoiding SIP sessions being established or charged by someone faking the identity of another person

Threats in the SIP Communication Chain

Threats exist at all levels of SIP operations: at the servers, proxies, SIP addresses, messages and requests, and message headers:

- Sending ambiguous requests to proxies
- Exploitation of forking proxies
- Listening to a multicast address
- Population of "active" addresses
- Sending messages to multicast addresses
- Contacting a redirect server with ambiguous requests
- Throwaway SIP accounts
- Misuse of stateless servers
- Anonymous SIP servers and back-to-back UAs
- Exploitation of messages and header fields structure

Threats from SIP's Optional Recommendations

This threat exploits the registrar servers. A registrar server is responsible for receiving REGISTER requests and updating location information into the location service for the domain it controls. Actually, it is the front end to the location service for a domain. As such, it could query the location service to gather information for specific registrations. The registrar server could be fooled. A spitter could issue a dictionary attack, or submit a query with special characters to a registrar and get a list of URIs. The authentication of a user to any proxy server, including the registrar, is not defined as a mandatory requirement. Therefore, this form of attack can be mounted. More specifically, the SIP specification does not dictate proxy servers to authenticate each user. Therefore, without any authentication between a UA and a proxy server, several vulnerabilities might occur, including impersonation of a legitimate user.

Attacks and Threat Models

SIP security mechanisms must be defined and combined properly to obtain a trusted network scenario. It is anticipated that SIP will frequently be used on the public Internet. Attackers on the network may be able to modify packets (perhaps at any compromised intermediary). Attackers may wish to steal services, eavesdrop on communications, or disrupt sessions.

SIP communications are susceptible to several types of attacks:

- The simplest attack in SIP is snooping, which permits an attacker to gain information on users' identities, services, media, and network topology. This information can be used to perform other types of attacks.

- Modification attacks occur when an attacker intercepts the signaling path and tries to modify SIP messages in an effort to change some service characteristics. This kind of attack depends on the kind of security used.
- Snoofing is used to impersonate the identity of a server or user to gain some information provided directly or indirectly by the attacked entity.
- Finally, SIP is especially prone to denial-of-service attacks that can be performed in several ways, and can damage both servers and UAs. This attack technique may cause memory exhaustion and processor overload.

Although the security mechanisms provided with SIP can reduce the risks of attacks, there are limitations in the scope of the mechanisms that must be considered. These limitations are discussed later in the chapter.

The following security protection requirements should be considered a priority. They are to first ensure the protection of the registration process by ensuring confidentiality and authentication. Providing confidentiality is a harder requirement than authentication. End-to-end authentication ensures that man-in-the middle security breaches are reduced. This can also limit denial-of-service attacks. This should be done for both random access and for repeat services. In addition to end-to-end authentication, end-to-end message confidentiality will create the required confidence in subscribers for SIP-based services. One open source of problems is the possibility of numerous INVITE messages that could be unsolicited. To limit this, SIP security should ensure that the INVITE messages are actually from valid IP addresses.

Security Vulnerabilities and Mechanisms

The security mechanisms in SIP can be classified as end-to-end or hop-by-hop protection. The first provides long-term and long-reach security protection. The second provides intermediate- and short-distance protection.

End-to-End Security Mechanisms

The end-to-end mechanism involves the caller and/or callee SIP user UAs and is realized by features of SIP specifically designed for this purpose (e.g., SIP authentication and SIP message body encryption). Hop-by-hop mechanisms secure the communication between two successive SIP entities in the path of signaling messages. SIP does not provide specific features for hop-by-hop protection and relies on network-level or transport-level security. Hop-by-hop mechanisms are needed because intermediate elements may play an active role in SIP processing by reading and/or writing some parts of the SIP messages. End-to-end security cannot apply to those parts of messages that are read and written by intermediate SIP entities.

Security Mechanisms

Two main security mechanisms are used with SIP: authentication and data encryption. Data authentication is used to authenticate the sender of the message and to ensure that critical message information is not modified in transit. This is to prevent an attacker from modifying and/or replaying SIP requests and responses. Data encryption is used to ensure the confidentiality of SIP communications, so that only the intended recipient decrypts and reads the data. This is usually done using encryption algorithms such as Data Encryption Standard (DES) and Advanced Encryption Standard (AES).

Four fundamental security services are required for SIP: (1) preserving the confidentiality and integrity of messaging, (2) preventing replay attacks or message spoofing, (3) providing for the authentication and privacy of the participants in a session, and (4) preventing denial-of-service attacks. Entities within SIP messages separately require the security services of confidentiality, integrity, and authentication. Rather than defining new security mechanisms specific to SIP, SIP reuses, wherever possible, existing security models derived from the HTTP (Hyper Text Transfer Protocol) and SMTP (Simple Mail Transfer Protocol) space.

Full encryption of messages provides the best means to preserve the confidentiality of signaling—it can also guarantee that messages are not modified by any malicious intermediaries. However, SIP requests and responses cannot be naively encrypted end-to-end in their entirety because message fields such as Request-URI, Route, and VIA must be visible to proxies in most network architectures so that SIP requests are routed correctly. Note that proxy servers also need to modify some features of messages (such as adding VIA header field values) for SIP to function. Therefore, the trust levels on proxy servers must be high. For this purpose, low-layer security mechanisms for SIP are recommended. Low-layer security mechanisms encrypt the entire SIP requests or responses on the wire on a hop-by-hop basis. They also allow endpoints to verify the identity of proxy servers to whom they send requests.

SIP components also need to identify one another in a secure manner. When a SIP endpoint provides the identity of its user to a peer UA or to a proxy server, that identity should be verifiable. A cryptographic authentication mechanism is used in SIP to address this requirement. An independent security mechanism for the body of an IP message provides an alternative means of end-to-end mutual authentication. It also provides a limit on the degree to which UAs must trust intermediaries.

Network Layer Security Mechanisms

In transport or network layer security, the signaling traffic is encrypted, guaranteeing message confidentiality and integrity. Often, certificates are used for establishing lower layer security. The certificates may also be used to provide a means of authentication.

Two popular alternatives for providing security at the transport and network layer are, respectively, TLS [4] and IPSec [5]. IPSec is a set of network layer protocol tools that collectively can be used as a secure replacement for traditional IP. IPSec is most commonly used in architectures in which a set of hosts or administrative domains have an existing trust relationship with one another. IPSec is usually implemented at the operating system level in a host, or on a security gateway that provides confidentiality and integrity for all traffic it receives from a particular interface (as in a VPN architecture). IPSec can also be used on a hop-by-hop basis. In many architectures, IPSec does not require integration with SIP applications; IPSec is perhaps best suited to deployments in which adding security directly to SIP hosts would be arduous. UAs that have a preshared keying relationship with their first-hop proxy server are also good candidates for IPSec use. Any deployment of IPSec for SIP would require an IPSec profile describing the protocol tools that would be required to secure SIP.

Transport Layer Security Mechanisms

TLS (Transport Layer Security) provides transport layer security over connection-oriented protocols (for the purposes of this document, TCP); "tls" (signifying TLS over TCP) can be specified as the desired transport protocol within a "Via" header field value or a SIP URI. TLS is most suited to architectures in which hop-by-hop security is required between hosts with no preexisting trust association. For example, Alison trusts her local proxy server, which after a certificate exchange decides to trust the local proxy server, which Bob trusts; hence, Bob and Alison can communicate securely. TLS must be tightly coupled with a SIP application. Note that transport mechanisms are specified on a hop-by-hop basis in SIP; thus, a UA that sends requests over TLS to a proxy server has no assurance that TLS will be used end-to-end. The TLS_RSA_WITH_AES_128_CBC_SHA cipher suite [4] must be supported at a minimum by implementers when TLS is used in a SIP application. For purposes of backward compatibility, proxy servers, redirect servers, and registrars should support TLS_RSA_WITH_3DES_EDE_CBC_SHA [4]. Implementers may also support any other cipher suite.

The most commonly voiced concern about TLS is that it cannot be compliant with UDP; TLS requires a connection-oriented underlying transport protocol, which for the purposes of this document means TCP.

It may also be necessary for a local outbound proxy server or registrar to maintain many simultaneous long-lived TLS connections with numerous UAs. This introduces scalability concerns, especially for intensive cipher combinations. Maintaining redundancy of long-lived TLS connections, especially when a UA is solely responsible for their establishment, is cumbersome.

TLS only allows SIP entities to authenticate servers to which they are adjacent. TLS offers strictly hop-by-hop security. Neither TLS nor any other mechanism allows clients to authenticate proxy servers to which they cannot form a direct TCP connection.

SIP URI Scheme

The SIP URI scheme adheres to the syntax of the SIP URI, although the scheme string is "sips" instead of "sip." However, the semantics of SIP are very different from the SIP URI. SIP allows resources to specify that they should be reached securely.

A SIP URI can be used as an address-of-record for a particular user—the URI by which the user is canonically known (on their business cards, in the "From" header field of their requests and in the "To" header field of register requests). When used as the Request-URI of a request, the SIP scheme signifies that each hop over which the request is forwarded, until the request reaches the SIP entity responsible for the domain portion of the Request-URI, must be secured with TLS. Once it reaches the domain in question, it is handled in accordance with local security and routing policies. When used by the originator of a request (as in the case when a SIP URI is employed as the address-of-record of the target), SIP dictates that the entire request path to the target domain be secured.

The SIP scheme is applicable to many of the other ways in which SIP URIs are used in SIP today in addition to the Request-URI, that is, inclusion in addresses-of-record, contact addresses (the contents of Contact headers, including those of register methods), and Route headers. In each instance, the SIP URI scheme allows these existing fields to designate secure resources. The manner in which a SIP URI is dereferenced in any of these contexts has its own security properties, which are detailed in [6].

The use of SIP in particular entails that mutual TLS authentication should be employed, as well as the cipher suite TLS_RSA_WITH_AES_128_CBC_SHA. Certificates received in the authentication process should be validated with root certificates held by the client. Failure to validate a certificate should result in the failure of the request.

Note that in the SIP URI scheme, transport is independent of TLS, and thus "sip:nkosi@tut.com;transport=tcp" and "sip:nkosi@tut.com;transport=sctp" are both valid (however, note that UDP is not a valid transport for SIP). The use of "transport=tls" has consequently been deprecated, partly because it was specific to a single hop of the request. This is a change since RFC 2543. Users who distribute a SIP URI as an address-of-record may elect to operate devices that refuse requests over insecure transports.

HTTP Authentication

SIP provides a challenge capability, based on HTTP authentication, that relies on the 401 and 407 response codes as well as header fields for carrying challenges and credentials. Without significant modification, the reuse of the HTTP Digest authentication scheme in SIP allows for replay protection and one-way authentication.

HTTP Digest

One of the primary limitations in using HTTP Digest in SIP is that the integrity mechanisms in Digest do not work very well for SIP. Specifically, they offer protection of the Request-URI and the method of a message, but not for any of the header fields that UAs would most likely wish to secure [8].

The existing replay protection mechanisms described in RFC 2617 also have some limitations for SIP. The next-nonce mechanism, for example, does not support pipelined requests. The nonce-count mechanism should be used for replay protection.

Another limitation of HTTP Digest is the scope of realms. Digest is valuable when users want to authenticate themselves to a resource with which they have a preexisting association, such as a service provider of which the user is a customer (which is a common scenario and thus Digest provides an extremely useful function). "By way of contrast, the scope of TLS is inter-domain or multi-realm, since certificates are often globally verifiable," so that the UA can authenticate the server with no preexisting association [8].

Application Layer Security Mechanisms

Encrypting entire SIP messages end-to-end for the purpose of confidentiality is not appropriate because network intermediaries (like proxy servers) need to view certain header fields in order to route messages correctly; and if these intermediaries are excluded from security associations, then SIP messages will essentially be nonroutable. However, "S/MIME allows SIP UAs to encrypt MIME bodies within SIP, securing these bodies end-to-end without affecting message headers" [9]. S/MIME can be used to provide end-to-end confidentiality and integrity for the bodies of messages, as well as mutual authentication. Furthermore, S/MIME can be used to provide some integrity and confidentiality for SIP header fields with SIP message tunneling.

A significant defect of the S/MIME mechanism is the absence of a prevalent public key infrastructure for end users. When self-signed certificates (or certificates that cannot be verified by a participant in a dialogue) are employed, the SIP-based key exchange mechanism is open to a man-in-the-middle attack. Therefore, an attacker can potentially inspect and modify S/MIME bodies [9]. The approach is for the attacker to intercept the first exchange of keys between the two communicating parties, extract the existing CMS (Cryptographic Message Syntax)-detached signatures from the request and response, and put a different CMS-detached signature containing the attacker's certificate (which appears to be a certificate for the correct address-of-record). Both parties will think they have exchanged keys with each other, when in fact they have the public key of the attacker.

SSH (Secure Shell) is open to similar man-in-the-middle attacks on the first exchange of keys; however, it is widely accepted that although SSH is not perfect, it improves the security of connections. The use of key fingerprints can provide some help in SIP applications, just as for SSH. For example, if two endpoints use SIP to establish a voice communications session, each party reads off the fingerprint of the key they received from the other party, which may be compared against an original. It is, however, more difficult for the man-in-the-middle to emulate the participants' voices than their signaling (this practice is used with the Clipper chip-based secure telephone).

The S/MIME mechanism allows UAs to send encrypted requests without preamble "if they possess a certificate for the destination address-of-record on their key ring" [8]. However, it is possible that any particular device registered for an address-of-record will not hold the certificate that was previously employed by the device's current user, and that it will therefore be unable to process an encrypted request properly, which could lead to some avoidable error signaling. This is especially likely when an encrypted request is forked.

The keys associated with S/MIME are most useful when associated with a particular user (an address-of-record) rather than a device (a UA). When users move between devices, it may be difficult to transport private keys securely between UAs.

Another difficulty with the S/MIME mechanism is that it can result in very large messages, especially when SIP tunneling is used. For that reason, it is recommended that TCP should be used as a transport protocol when S/MIME tunneling is employed [8].

SIP URIs

Using TLS on all segments of a request path requires that the terminating UAs must be reachable over TLS (perhaps registering with a SIP URI as a contact address) [5]. This is the preferred use of SIP. Much valid architecture, however, uses TLS to secure part of the request path but relies on a different mechanism for the final hop to a UA. Thus, SIP cannot guarantee that using TLS will be truly end-to-end. In essence, because many UAs will not accept incoming TLS connections, even those UAs that do support TLS may be required to maintain persistent TLS connections.

Ensuring that TLS is used for all the request segments up to the target domain is complex. There is the possibility that noncompliant, cryptographically authenticated proxy servers along the route that are compromised may choose to disregard the forwarding rules associated with SIP. Hence, malicious intermediaries could, for example, retarget a request from a SIP URI to a different SIP URI in an attempt to downgrade security.

Alternatively, an intermediary might legitimately retarget a request from a SIP to a SIP URI. Recipients of a request whose Request-URI uses the SIP URI scheme thus cannot assume on the basis of the Request-URI alone that SIP was used for the entire request path (from the client onward) [2].

To address these concerns, it is recommended that recipients of a request whose Request-URI contains a SIP or SIP URI inspect the "To" header field value to see if it contains a SIP URI (however, note that it does not constitute a breach of security if this URI has the same scheme but is not equivalent to the URI in the "To" header field). Although clients may choose to populate the Request-URI and "To" header field of a request differently, when SIP is used, this disparity could be interpreted as a possible security violation, and the request could consequently be rejected by its recipient. Recipients may also inspect the "Via" header chain in order to double-check whether or not TLS was used for the entire request path until the local administrative domain was reached. S/MIME may also be used by the originating UAC to help ensure that the original form of the "To" header field is carried end-to-end.

If the UAS has reason to believe that the scheme of the Request-URI has been improperly modified in transit, the UA should notify its user of a potential security breach.

As a further measure to prevent downgrade attacks, entities that accept only SIP requests may also refuse connections on insecure ports. End users will undoubtedly discern the difference between SIP and SIP URIs, and they may manually edit them in response to stimuli. This can either benefit or degrade security. For example, if an attacker corrupts a DNS cache, inserting a fake record set that effectively removes all SIP records for a proxy server, then any SIP requests that traverse this proxy server may fail. However, when a user sees that repeated calls to a SIP AOR (address-of-record) are failing, they could, on some devices, manually convert the scheme from SIP to SIP and retry. Of course, there are some safeguards against this (if the destination UA is truly paranoid, it could refuse all non-SIP requests)—but it is a limitation worth noting. On the bright side, users might also divine that "SIP would be valid even when they are presented only with a SIP URI" [2].

SIP Processing Cost

Starting from the SIP-based telephony service scenario, eight procedures/scenarios have been identified to compare their processing costs [7]. Tables 10.1 and 10.2 provide the basic procedure/scenarios, which are:

1. SIP call setup with no authentication, where the proxy server is stateless and always uses UDP communication
2. Authentication and corresponds exactly to the call flow
3. Has been considered in order to see the difference between the use of UDP or TCP between the proxy and tester UAs. The motivation for considering TCP-based SIP communication is to obtain an incremental analysis toward a TLS-based SIP communication setup. Scenarios (5 to 8) replicate (1 to 4) as far as authentication and UDP/TCP/TLS are employed, by considering a call stateful proxy server.

Table 10.1 reports the call throughput for the procedure/scenario (1) for different numbers of active threads. Starting from two threads in parallel, the capacity of the server is saturated: each of the N threads takes 1/N of the server capacity, so its throughput is 1/N of the maximum throughput.

Table 10.2 gives the results of the evaluation by Salsano [7]. The third column reports the theoretical maximum throughput in terms of calls per second that the

Table 10.1 Experimental Results (Threads and Throughput)

Active Threads	1	2	3	4	5	6
Single thread throughput (s⁻¹)	27.8	15.9	10.6	7.9	6.4	5.3
Total throughput (s⁻¹)	27.8	31.7	31.9	31.7	31.8	31.7

Source: From S. Salsano, L. Veltri, and D. Papalilo, *Network IEEE*, November/December 2002, 16, 6, 38-44. With permission.

Table 10.2 Experimental Results

	Procedure/Scenario	Total Throughput (s^{-1})	Relative Processing Cost	
1	No authentication, stateless server, UDP	34.8	100	
2	Authentication, stateless server, UDP	19.6	177	
3	Authentication, stateless server, TCP	19.4	180	
4	Authentication, stateless server, TLS	19.2	182	
5	No authentication, call stateful server, UDP	21.0	166	100
6	Authentication, call stateful server, UDP	14.6	239	144
7	Authentication, call stateful server, TCP	13.8	253	152
8	Authentication, call stateful server, TLS	13.6	256	154

Source: From S. Salsano, L. Veltri, and D. Papalilo, *Network IEEE*, November/December 2002, 16, 6, 38–45. With permission.

proxy server can accommodate. This includes all the processing that the proxy performs for a call, from its setup to its termination or teardown. The two rightmost columns are the most important ones; they report the throughput values converted into relative processing cost. In the first row, a reference value of 100 has been assigned to procedure/scenario (5). The results show that the introduction of SIP security accounts for nearly 80 percent of the processing cost of a stateless server and 45 percent of a stateful server. This increase can be explained with the increase in the number of exchanged SIP messages and with the actual processing cost of security. Salsano [7] estimated that 70 percent of the additional cost identified was for message processing and 30 percent for actual security mechanisms.

SPIT

SPam over Internet Telephony (SPIT) is emerging as a major security problem brought to the fore because of the reduced cost of VoIP services compared with traditional voice calls over PSTN. The deterrent that was posed by the high cost of voice calls over PSTN has been removed and in doing so a new problem has emerged (SPIT), or has it? From experience, there is no significant spam through short messaging service (SMS), which suggests that SPIT may not be as major problem as anticipated. SPIT consists of two components: signaling and media. Three different forms of SPIT can be identified:

1. Voice or call SPIT: Defined as "a bulk unsolicited set of session initiation attempts in order to establish a multimedia session" [1].
2. Instant Message SPIT: Defined as an unsolicited set of bulk instant messages.
3. Presence SPIT: Defined as an unsolicited set of bulk presence requests made with the objective of the initiator to gain membership of the "address book of a user or potentially of multiples users" [1].

From these definitions, SPIT threats are more potent that spam threats because three sources of SPIT attack can be mounted, compared with just a single direct spam attack through bulk e-mails.

Summary

SIP security problems were discussed in detail in this chapter. SIP is not an easy protocol to secure. Its use of intermediaries, its multifaceted trust relationships, its expected usage between elements with no trust at all, and its user-to-user operation make security far from trivial. There are security loopholes in SIP, and it is essential to guarantee high service availability, stable and error-free operation, and protection of the user-to-network and user-to-user network traffic. SIP as defined originally did not provide security for the communication chain. SIP security should cover the overall communication chain. This includes the source, destination, links, signaling, packet headers, and messages. In addition, protocols that interface with SIP should also be secure. SIP text messages can be modified in transit because no security is used. Traditionally, SIP messages are not encrypted and therefore are open to interception in the links or at any point in the network. Threats exist at all levels of SIP operations—at the servers, proxies, the SIP addresses, messages and requests, and the message headers. SIP communications are susceptible to several types of attacks. The simplest attack in SIP is snooping, which permits an attacker to gain information on users' identities, services, media, and network topology. This information can be used to perform other types of attacks. Modification attacks occur when an attacker intercepts the signaling path and tries to modify SIP messages in an effort to change some service characteristics. This kind of attack depends on the kind of security used. Snoofing is used to impersonate the identity of a server or user to gain some information provided directly or indirectly by the attacked entity. SIP is especially prone to denial-of-service attacks that can be performed in several ways, and can damage both the servers and user agents. Security mechanisms that can be used in the network, transport, and application layers were explained in detail. The chapter concluded with SIP processing overhead, which adds to the traditional system processing overhead, and discussed SIP spam (SPIT).

Practice Set: Review Questions

1. SIP:
   a. Is a routing protocol.
   b. Is a resource reservation protocol.
   c. Is for initiating the point at which a media stream starts.
   d. Is a signaling protocol.
   e. Is required to provide quality-of-service.
2. SIP is only suitable for:
   a. ATM networks.
   b. Supporting session mobility.
   c. Establishing and tearing down sessions.
   d. None of the above.
   e. All of the above.
3. Traditionally, are all SIP messages encrypted?
   a. Only the "To" field is not encrypted.
   b. All the fields are not encrypted.
   c. All SIP messages are encrypted.
   d. Only the "From" field is not encrypted.
   e. The "Via" field is not encrypted.
4. Do threats exist in SIP proxies?
   a. Sending ambiguous requests to proxies.
   b. Exploitation of forking proxies.
   c. No threats exist in the proxies.
   d. Only (a).
   e. Both (a) and (b).
5. Do threats exist in SIP addresses?
   a. Population of "active" addresses.
   b. Listening to a multicast address.
   c. Sending messages to multicast addresses.
   d. All of the above.
   e. No threats exist in SIP addresses.

Exercises

1. List the methods that the SIP uses for establishing and terminating multimedia sessions. Discuss them fully.

2. List and discuss all the possible threats that exist in the SIP communication chain.
3. What is a dictionary attack? Discuss dictionary attacks with respect to the SIP servers.
4. Why is it possible to attack the signaling path in SIP? Do you think snooping attacks are possible in SIP?
5. What are some of the security mechanisms that can be used in the SIP communication chain?
6. Enumerate and discuss the security mechanisms in the network and transport layers.

References

[1] S. Dritsas, J. Mallios, M. Theoharidou, G.F. Marias, and D. Gritzalis, Threat Analysis of the Session Initiation Protocol Regarding Spam, in *Proc. of the 3rd IEEE International Workshop on Information Assurance (in conjunction with the 26th IEEE International Performance Computing and Communications Conference (IPCCC-2007)*, pp. 426–433.
[2] E.T. Lakay, SIP-Based Content Development for Wireless Mobile Devices with Delay Constraints, M.Sc. thesis, Department of Computer Science, University of the Western Cape, Bellville, Cape Town, South Africa, 2006.
[3] T. Dierks and C. Allen, The TLS Protocol Version 1.0, RFC 2246, January 1999.
[4] P. Chown, Advanced Encryption Standard (AES) Cipher Suites for Transport Layer Security (TLS), RFC 3268, June 2002.
[5] S. Kent and R Atkinson, Security Architecture for the Internet Protocol, RFC 2401, November 1998.
[6] J. Rosenberg and H Schulzrinne, SIP: Locating SIP Servers, RFC 3263, June 2002.
[7] S. Salsano, L. Veltri, and D. Papalilo, SIP Security Issues: The SIP Authentication Procedure and Its Processing Load, *Network IEEE*, 16(6), November/December 2002, 38–44.
[8] J. Rosenberg, H. Schulzrinne, G. Camarillo, A. Johnston, J. Peterson, R. Sparks, M. Handley, and E. Schooler, RFC 3261 — SIP: Session Initiation Protocol, <http://www.faqs.org/rfcs/rfc3261.html>, 2002.
[9] S. Duanfeng, L. Qin, H. Xinhui, and Z. Wei, Security Mechanisms for SIP-Based Multimedia Communication Infrastructure, in *Proceedings of the International Conference on Communications, Circuits and Systems, ICCCAS 2004*, 1, June 27–29, 2004, 575–578.

Chapter 11

SIP Quality-of-Service Performance

The Session Initiation Protocol (SIP) is a new application level protocol that was developed to support the exchange of information that is required for the establishment of media sessions in IP networks. It supports mobility at the application layer through its capability to establish and tear down sessions. Its main function is to create, establish, and tear down media sessions between one or more SIP end-points. These end-points have valid SIP addresses and form part of the information that is exchanged during SIP signaling. The end-points may reside in either the IPv4 or IPv6 domain, and application level gateways (ALGs) have been developed for the translation of the addresses inside the SIP messages.

SIP targets the enabling of emerging IP services, including voice over IP (VoIP), Internet telephony, Presence, IPTV, and instant messaging (IM). Although SIP does not carry the media itself, it works closely with other protocols such as Real-time Transport Protocol (RTP), which is used to transport the media; RTP Control Protocol (RTCP), which is used for the management and control of the data stream; Session Description Protocol (SDP), which is used for specifying the parameters of the media sessions (equivalent to the H.245 Call Control in H.323 Protocol); TCP; and UDP.

Securing SIP by itself is not easy because each session that is established must be secured. In a highly mobile network environment when either the terminal or the network is mobile, the problem is compounded. In essence, securing SIP will mean securing the dynamic links, signaling, and data involved.

The traditional PSTN is relatively secure and provides a level of QoS to which subscribers are accustomed. Hence, when SIP is used in IP networks, subscribers will demand as a minimum the same levels of security and QoS. However, SIP is not mature enough to support that form of security and QoS. No real similar solutions exist yet for providing the security mechanisms that subscribers have become accustomed to over the PSTN and Internet. There are, however, several proposals that have been published; they are aimed at addressing the identified security concerns in SIP. This chapter therefore provides a summary of the security issues in SIP and the proposed solutions. It also includes the QoS performance of SIP.

QoS Performance of SIP

SIP is never used in isolation. It must work in consonance with other QoS-determining protocols such as DiffServ and real-time protocols such as RTP, RTCP, RTSP, and RSVP. The IETF has provided several QoS solutions in IntServ, RSVP, and DiffServ. The RSVP is an end-to-end QoS solution but it is relatively hard to design due to its high cost. RSVP cannot provide guaranteed QoS in a wireless environment because of the dynamic links and significant bit error rates (BERs) that result. Therefore, the QoS performance of SIP is intricately woven into the QoS performances of these protocols. The QoS performance of SIP alone can be measured by how fast it establishes and maintains sessions through its signaling mechanisms under TCP (UDP) and either IPv4 or IPv6. This section summarizes these two approaches to the QoS performance of SIP. SIP was not developed originally with QoS considerations in mind. Therefore, to handle QoS, extension to SIP proxy server is normally undertaken [1].

Measuring Session-Based QoS Performance

The key QoS performance measures for SIP sessions include communication time, handoff delay when mobile nodes are involved in and between neighboring subnets, data volume exchanged during the SIP session time, jitter, and BERs. For real-time traffic, these parameters refer to end-to-end performance. In non-real-time traffic, guaranteed bit rates are used. A SIP session normally consists of several ON and OFF periods. Three data types are considered: audio, data, and video. For real-time traffic such as VoIP, the ON time refers to talk periods and the OFF time refers to silence periods. In the case of data, the ON time refers to the download time and the OFF time refers to data reading time. When video is considered, only the ON time is necessary. To provide a means for comparing different implementations and SIP-based platforms, this section provides a set of parameters and expressions. Based on these definitions, in one flow of a multimedia TCP session, the following parameters apply in defining SIP QoS performance following the example in [1]. Consider that arrival time for the traffic can be modeled as a Poisson process with the following parameters [1]:

D: Session rate per second (session/s)
N: Session load (number of simultaneous sessions)
T: Average mean duration of one session (s)
ρ: Average rate of one session (kbps)
L: Average rate of one flow (kbps)
N_p: Average number of ON-OFF patterns in one session
T_{on}: Average duration of ON period (s)
T_{off}: Average duration of OFF period (s)
R_{on}: Average rate in ON period (kbps)
Q: Average file size in ON period (kb)

The ON period for data transfer is a function of the data rate and the average file size used during the ON period. This is given by the expression:

$$T_{on} = \frac{Q}{R_{on}}$$

The ON and OFF durations for Web and FTP sessions are known to have Pareto distributions.

Normally, the download rate is a function of the quality of the network or the level of activity in the network. At some point, the download speed will be very high and at other times could grind to a very low value when the network is busy. Hence, it is necessary to use only the average download rate. The average download rate can be calculated based on a statistical average of per-minute rates accumulated over a length of time. Consider that a session consists of several ON and OFF patterns. That being the case, the average data transferred during a session is given by the product of the data transfer rate and the number of ON and OFF patterns:

$$\alpha = Q * N_p$$

Let the number of active sessions be given by:

$$N = D * T$$

and the duration T of one session is:

$$T = N_p \left(T_{on} + T_{off} \right)$$

The flow rate has an average value L given by:

$$L = N * d$$

The session data transfer rate can therefore be estimated using the expression:

$$\rho = R_{on} * \frac{T_{on}}{T_{on} + T_{off}}$$

Apart from the level of activity in a network that determines the return trip time (RTT), the rate at which packet acknowledgment is performed impacts the average rate of a TCP connection. The approach used in [1] is adopted below. It considers that acknowledgment is provided for every two packets received. This leads to the expression:

$$T(Q) = RTT * \log_{1.57} Q$$

When long TCP connections are involved, this expression should be substituted with the maximum of the TCP rate permitted by the source.

In the next sections we discuss the QoS performance of SIP from various delay perspectives.

Quality-of-Service and SIP

"Quality-of-service" is a term that has been loosely used in telecommunications to refer to network parameters that impact the performance of the system. In telecommunications, this term refers primarily to network performance variables such as delay, BER, packet loss rate, interference, bandwidth, system capacity, and throughput. In a dynamic network environment, the definition extends to the velocity of the nodes, the length of the links between nodes, and the characteristics of the links. In a multimedia situation, it extends to the quality of the data, such as the quality of the audio content; the mean opinion scores in VoIP systems; and for IPTV where video is involved, to the quality of the video (images), compression quality, and the frame rate. Thus, the definition of QoS includes multiple system parameters and depends on the application in question. In this chapter, a subset of these system parameters that typically affects SIP performance is considered. This includes delay in its various forms in SIP networks, including setup delay, signaling delay, call setup delay, queuing delay, message transfer delay, session setup delay, handoff delay, disruption time, and throughput. The SIP-T signaling mechanism is used to facilitate the interconnection of PSTN with carrier-class VoIP networks. Its major attributes are the promises for scalability, flexibility, and interoperability with PSTN. It also provides the call control function of a media gateway controller required for setting it up, tearing it down, and managing VoIP calls in carrier-class VoIP networks. The performance of the SIP-T signaling system has a direct impact on optimizing network QoS. The QoS performance attributes include the queue size, the mean of queuing delay, and the variance of queuing delay. To represent the SIP-T signaling system, Wu [2] has assumed an M/G/1 queue with non-preemptive priority assignment. Formulae for queuing size, queuing delay, and delay variation for the non-preemptive priority queues from queuing theory, respectively, were presented and readers are urged to check this work. The number of SIP-T signaling messages in the queue is called the queuing size, and the queuing delay is the average delay of messages in the queue. Kueh [3] also evaluated the performance of SIP-based session setup over s-UMTS (satellite-Universal Mobile Telecommunications System), taking into account the larger propagation delay over a satellite link as well as the contribution of the UMTS radio interface. In the next paragraphs we summarize other similar works in more detail.

Queuing Delay Performance

Queuing delay is a result of queuing size and is one of several sources of delay in SIP operations. This was analyzed in [2] using imbedded Markov chain and semi-Markov processing, while the queuing delay and delay variation were analyzed using the Laplace-Stieltjes Transform (LST) [4]. Queuing size is defined as the number of SIP-T messages in the system. Tables 11.1 and 11.2 are the comparative

Table 11.1   Mathematical Results

SIP-T Message Queue Size (messages/s)	Mean (ms)	Std. Dev. (ms)	Buffer Size
50	0.49	0.49	0.12
100	0.62	0.68	0.23
150	0.78	0.89	0.37
200	1.00	1.14	0.52
250	1.29	1.46	0.72
300	1.71	1.90	0.97
350	2.37	2.57	1.33
400	3.57	3.75	1.92
450	6.39	6.41	3.17
500	25.00	18.57	8.20

Source: From J.S. Wu and P.Y. Wang, 17th International Conference on Advanced Information Networking and Applications (AINA'03), *IEEE* March 2003, 39. With permission.

Table 11.2   Simulation Results

SIP-T Message Queue Size (messages/s)	Mean and 95th Percent (ms)	Std. Dev. (ms)	Buffer Size	Sample Size (SIP-T message)
59.2	0.49 ± 0.01	0.47	0.12	4925
100.7	0.61 ± 0.01	0.65	0.25	10074
150.4	0.79 ± 0.01	0.88	0.39	15044
201.8	0.98 ± 0.02	1.10	0.54	20178
251.4	1.27 ± 0.02	1.40	0.76	25138
301.1	1.70 ± 0.02	1.91	1.03	30115
348.9	2.36 ± 0.03	2.57	1.40	34891
399.5	3.62 ± 0.04	3.98	1.97	39955
450.7	6.49 ± 0.06	6.67	3.49	45067
502.5	21.08 ± 0.18	20.12	8.85	50251

Source: From J.S. Wu and P.Y. Wang, 17th International Conference on Advanced Information Networking and Applications (AINA'03), *IEEE* March 2003, 40. With permission.

views between the mathematical and simulation results, including buffer size, mean queuing delay, and standard deviation of queuing delay for error probability of a SIP-T message $p_e = 0.004$. The mean queuing delay and standard deviation of queuing delay vary slowly as a function of the queue size at SIP-T messages arrival rates of less than 450. However, these values increase dramatically when the arrival rate of SIP-T messages exceeds 450. It is reasonable to suggest that this phenomenon is due to the SIP-T message arrival rate approaching the processing capability of the system for heavy traffic intensity [2]. It can be observed that the theoretical estimates are in excellent agreement with simulation results. Therefore, we can determine the cost to performance and the planning and design compromises needed to meet the requirements of a carrier-class VoIP network.

Call Setup Delay

Call setup delay consists of post-dialing delay, answer-signal delay, and call-release delay. The post-dialing delay occurs when signaling takes place through the network to the receiver. A measure of these delays was undertaken by Cursio and Landan [5] using a 3G network emulator. These delays happen during the lifetime of a SIP call. The results account for local, national, international, and overseas intranet LAN calls.

- *Post-dialing delay (PDD)* is also called post-selection delay or dial-to-ring delay. This is the time elapsed between when the caller clicks the button of the terminal to call another caller and the time the caller hears the callee's terminal ringing.
- *Answer-signal delay (ASD)* is the time elapsed between when the callee picks up the phone and the time the caller receives indication of this. This delay is obviously a function of the depth and extent of the network infrastructure between the caller and callee. In a local call scenario, this time is relatively short. For international and long-distance calls, it can be significant.
- *Call-release delay (CRD)* is the time elapsed between when the releasing party (the caller) hangs up the phone and the time the same party can initiate/receive a new call.

Table 11.3 provides a summary of these delays. The values in Table 11.3 are relative to the point of call and the destination, and therefore are subject to wide variation. For example, a SIP call between Europe and Australia will indicate significantly larger PDD, ASD, and CRD. Tests in the case of bandwidth limitations were also performed in [5]. The call success rate was always 100 percent; thus, bandwidth limitation does not prevent a successful SIP call. The PDD for 2 kbps bandwidth was less than 1 s. This is significantly large compared with the PDD for 5 kbps bandwidth, which was approximately 420 ms. The values of PDD decrease with increasing bandwidth. At a bandwidth of 254 kbps, the PDD was around 50 ms. In realistic 3G network configurations, the bearer allocated for signaling is a few kbps [5], requiring very small time. ASD values recorded in [5] were constantly 45 ms for channels of at least 5 kbps, but increased for very narrow channels (2 kbps) to 166 ms. This could be attributed to message queuing and congestion at lower channel bandwidths and the possibility of retransmissions [5]. The CRD could not be measured because the media packets queued up in the simulator and blocked the channel for a long time [5].

Table 11.3 3G SIP Calls versus Intranet Calls Delay Results

	Local SIP Call (ms)	National SIP Call (ms)	International SIP Call (ms)	Overseas SIP Calls (ms)	3G SIP Calls (ms)
PDD	24	38	153	24	62
ASD	23	31	147	237	45
CRD	11	30	138	230	50

Source: From I.D.D. Curcio and M. Lundan, Seventh International Symposium on Computers and Communications (ISCC'02), *IEEE* July 2002, 835. With permission.

Note: PDD = post-dialing delay; ASD = answer-signal delay; and CRD = call release delay.

Globally, these SIP signaling delays are well within the grade of service (GoS) bounds proposed by the ETSI TIPHON QoS classes [6].

Losses in the air interface and narrow channels have a significant impact on the overall SIP call setup time. They result in large call setup times. However, it is expected that UDP/IP header compression and SIP message compression algorithms will reduce significantly the SIP call setup delay over 3GPP networks.

Message Transfer Delay

Because the INVITE method is considered the most important method in SIP, as it is used to establish a session between participants and normally contains the description of the session to be set up, it is interesting to determine its performance in general. This was done by Kueh [3] for message transfer delay at block error rates (BLERs) of 0 percent, 10 percent, and 20 percent. It was found that the message transfer delay increases as the message size (and hence the session description) increases. This is expected. It was found that the transfer delay of the "invite" request is substantially reduced compared to when no link layer retransmission is employed; whereby the delay reduction increases as the message size and the BLERs increase. This is because without retransmission at the link layer, a segment that is lost means that the whole message cannot be recovered at the receiver side and thus the whole message must be retransmitted at the session layer, according to the SIP reliability mechanism.

The tests also show that the delay is lower with the unsolicited and solicited status report option set, compared to only having the solicited feedback. This is because, by incorporating unsolicited feedback on top of solicited feedback, the missing protocol data units (PDUs) can be recovered faster because retransmissions of missing PDUs can be performed before polling; also, the reduction in delay is greater at a higher BLER.

Session Setup Delay

Handley and Jacobson [1] have shown that session setup delay and call blocking probability for a simple call setup sequence consist of sending of "invite" request and the 180 ringing response. Two status report trigger settings present a bound on the delay. Tgood status report trigger settings range between 0.5 s and 10 s, and Tbad settings are equal to 0.5 s, 2 s, and 4s. Comparing both schemes, it can be seen that when the channel is good (i.e., for a low Tbad value or a high Tgood value), there is hardly any difference in performance; but as the channel gets worse, combining both unsolicited and solicited feedback options gives a lower delay and blocking probability.

SIP signaling is transaction based and generous in size. Therefore, merely transporting packets over the radio interface is not sufficient. Consideration must be given to the delay values and the errors in the process. When passing over the error-prone wireless link plus a larger satellite propagation delay, the session establishment delay can be rather large. At the moment, SIP over Satellite is not advisable. From a study [3], it was shown that with the presence of radio link control-acknowledgment mode

(RLCAM), the session setup performance can be substantially improved. Also, it was shown that the combination of unsolicited and solicited status report triggers gives better performance than just solicited alone in terms of delay and blocking probability in a more hostile environment.

Mobility and Handoff Delay

The mobility of a user can be described by three distinct terms: roaming, micro-mobility, and macro-mobility. Roaming is the mobility of a user in the absence of or independent of Internet access. Roaming is then initiated or triggered when the mobile user gains Internet access. Semantically, this is no different from roaming in cellular networks. Micro- and macro-mobility are the changes in Internet access within a domain (intra-domain) and between domains (inter-domain), respectively. If ad-hoc networking exists within the domain, micro-mobility results in hand-off between nodes in the same domain or subnet. Otherwise, in inter-domain or macro-mobility, handoff is required for movement of nodes between two neighboring domains or subnets.

Two approaches are used for handling mobility when IP services are involved. Mobility is handled at the network layer using Mobile IP, or in the application using SIP. The most prominent problem with mobility management at the network layer with Mobile IP is the so-called triangular routing that is inherent in the protocol. Triangular routing leads to more delay in the routing process. Second, because Mobile IP depends on tunneling to forward packets to the mobile node's foreign network, it increases overhead specifically in terms of the encapsulation using the new care-of-address. This increases the IP header. This is not suitable for narrowband wireless links because it consumes the limited capacity. Although the SIP-based approach using UDP offers several advantages over Mobile IP by removing the triangular routing problem and the overhead associated with tunneling, it suffers from several drawbacks, of which the most prominent are the handoff delay and handoff disruption time. Disruption can occur if the new SIP session is not completely created within the overlapping coverage regions of the two nodes involved in handover. The time required to acquire DHCP (Dynamic Host Configuration Protocol) address renewal while in the overlap region can be significant and can cause disruption in calls (VoIP) and silence periods during handoff. To understand the mechanisms that lead to the delays and the performance of mobility management at the network and the application layers, this section summarizes the two mobility management schemes and compares them.

Mobility Management Using Mobile IP

The well-known Mobile IP triangular routing scheme is shown in Figure 11.1. Two agents (routers) are involved. The home agent (HA) is at the home network (subnet

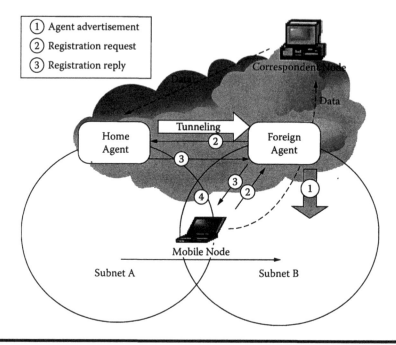

Figure 11.1 Pre-call mobility in SIP. (From J.W Jung, R. Mudumbai, D. Montgomery, and H.K. Kahng, Global Telecommunication Conference 2003 (GLOBECOM'03), *IEEE* 2003, 3, 1190-1194. With permission.)

A) and the foreign agent (FA) is at the foreign network (subnet B). The correspondent node (CN) seeks to establish communication with a mobile terminal originally attached to the HA. In this mobility management scheme, while the mobile node is within the intersecting region of the two neighboring subnets, it must acquire a new address using DHCP from the FA. The FA advertises its availability using an agent advertisement (1) to all the nodes within its coverage region. The mobile node, upon hearing this advertisement, sends a request for a registration message (2) to the FA. This message contains, in part, the identities of the mobile node and of its HA. If a DHCP address is available, it is issued to the mobile node and it is registered at subnet B. The FA also informs the HA of the new registration (2). Data sent by the CN for the mobile node originally through the HA can therefore now be sent to the mobile node at the foreign network (subnet B). To do this, tunneling is used. From this point on, the mobile node can start to communicate directly with the CN through the FA (Figure 11.1).

When mobile nodes move, they make, break, and reestablish connections with their CNs. This mobility affects the performance of SIP. Using IPv6, SIP can support terminal mobility without increasing the SIP payload but incurs delay penalties resulting from the desire to hand off. In terminal mobility, the terminal changes access between two nodes to maintain communication. This change could

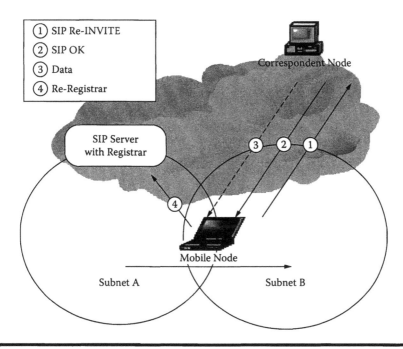

Figure 11.2 Mid-call mobility in SIP. (From J.W Jung, R. Mudumbai, D. Montgomery, and H.K. Kahng, Global Telecommunication Conference 2003 (GLOBECOM'03), *IEEE* 2003, 3, 1190-1194. With permission.)

be a result of better received signals from the new node or better channel conditions to the new node. Whichever one of these is involved, there are delays to consider during handoff.

The initial design of SIP did not consider mobility management of the end nodes an issue and hence mobility management using SIP is an afterthought. However, to support real-time applications, it is essential to address the mobility of nodes to ensure that there is no disruption of communication due to lost attachment and poor link quality. In the mobility management scheme described in Figure 11.2, during an active communication session, the mobile node obtains a new address through a DHCP server and sends a new session invitation (SIP re-INVITE) to the CN (1). This invitation message contains details of the identity of the mobile node, its new IP address, and the new subnet to which it is attached. Usually this is done by updating the session description. The CN therefore knows where to send data to the mobile node first by issuing a SIP OK message (2) and sending the data (3). A re-registrar message is sent to the SIP server to complete the signaling.

There are two crucial problems with the SIP mobility management scheme. First, SIP does not offer a solution for the mobility management scheme for long-term TCP connections. Second, disruption can occur during handoff when the mobile node is within the overlap region if the acquisition of the new SIP session is

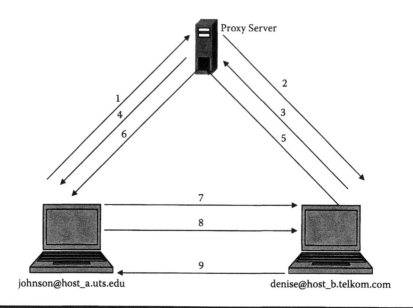

Figure 11.3 Experimental test bed for handoff measurement in IPv6 mobile networks. (From N. Nakajima, A. Dutta, S. Das, and H. Schulzinne, *International Conference on Communications (ICC'03)*, 26, May 2003, 1085–1089. With permission.)

not completed before the node moves out of the overlap region. Because the mobile node must acquire in this case a new address through DHCP, there are delays associated with completing the acquisition of the new IP address. This handoff delay can be significant and is a function of the implementation of SIP and the mobility scheme. Some versions of DHCP can cause a delay of about 2 s during address renewal. This delay is reduced to about 0.1 s if the duplicate address detection scheme is removed from the DHCP. Delays in SIP mobility-based communication can further be distinguished and analyzed.

Figure 11.3 shows that in trying to establish sessions, SIP operations lead to delays associated with the signaling. In the scenario described in Figure 11.3, the SIP proxy is assumed to have Denise's location address (host_b.telkom.com) and Johnson therefore invites Denise through that address. The sequence of messages sent is as follows. Johnson sends an INVITE message through the proxy server (1), and the proxy server relays the INVITE message to Denise at host_b.telkom.com (2). While Denise is being alerted, a 180 Ringing message is sent back to Johnson via the proxy server (3, 4). After Denise has answered the call, a 200 OK response is sent back to Johnson to notify him that Denise accepts the call (5, 6). Then Johnson sends an ACK (7) message to Denise to confirm session establishment. Because Johnson knows the contact address of Denise, denise@host_b.company.com, via 200 OK, Johnson can send an ACK to Denise directly without going through the

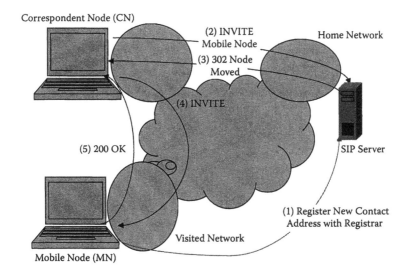

Figure 11.4   Handoff process in SIP in IPv6 environment.

proxy server. When Johnson wants to terminate the call, he sends a BYE request to Denise (8), which is confirmed by another 200 OK (9) issued by Denise.

Apart from establishing sessions, which causes delays, two types of terminal mobility can be identified: pre-call terminal mobility and mid-call terminal mobility. The pre-call terminal mobility ensures that a CN can reach a mobile node (MN). Figure 11.4 illustrates this [7]. The pre-call terminal mobility scenario is akin to Mobile IP. When a SIP-enabled MN moves to a foreign network (Visited Network), it registers and has a new contact address with the SIP proxy server (1). Therefore, an INVITE request from a CN through the SIP server (2) is used by the server to notify the new contact address of the MN to the CN. This provides the means for the CN to send an INVITE request to the MN directly (4) with an OK message sent back by the MN (5).

The second type of mobility is mid-call mobility. It allows a node to maintain an ongoing session with its peer during handoff, as in Figure 11.5 [7]. As shown, an MN sends a re-INVITE request with a new IP address to CN (1), and the CN sends packets to the MN directly at the new point of attachment to the network (2).

Handoff Delay

Handoff duration is composed of two broad delays: link layer establishment and signaling delays. Link layer establishment delay is normally negligible compared with signaling delay. Handoff delay as a result of mobility was measured in [7] using a small network testbed of three mobile nodes, one correspondent node, a home network, and a visited network, each consisting of a router within an IPv6 domain. The network used in [7] is shown in Figure 11.6.

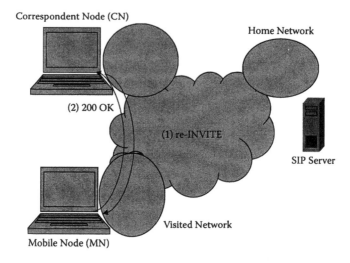

Figure 11.5   Handoff delay and packet loss instances.

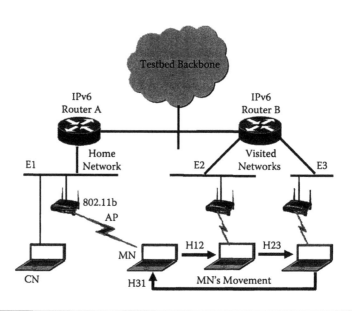

Figure 11.6   Handoff delay measurement architecture.

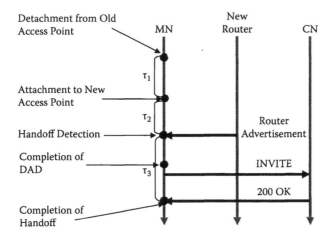

Figure 11.7 Disruption time as a function of wireless link delay for MIP and SIP.

Total handoff delay (τ) is defined [18] as the time between detachment from the old access point and establishment of communication with a new CN. It consists of three components:

1. The time for switching a lower-layer medium to access network (τ_1)
2. The time for detecting a new router and a new link (τ_2)
3. The time for recovery of communication with a CN after detecting a new link (τ_3)

Therefore, $\tau = \tau_1 + \tau_2 + \tau_3$. Nakajima et al. [18] concentrated on τ_3. There are two main factors that contribute to delay τ_3: duplicate address detection (DAD) and router selection. The purpose of DAD is to confirm the uniqueness of the IPv6 address on the link. DAD imposes a delay between receiving a router advertisement (RA) and sending a packet out of the interface with an auto-configured IPv6 address.

Nakajima et al. [18] also measured the handoff delay of SIP terminal mobility in the IPv6 testbed (Figure 11.7). Two different scenarios were considered:

1. SIP mobility without kernel modification
2. SIP mobility with kernel modification

Measurements were performed for the two scenarios above. Table 11.4 shows the handoff delay τ_3, which is related to signaling: H12 is the handoff time between mobile nodes 1 and 2, H23 is the handoff time between mobile nodes 2 and 3, and H13 is the handoff delay between mobile nodes 1 and 3, as shown in Figure 11.7.

Table 11.5 shows the results of another handoff delay related to voice communication using UDP.

Table 11.4   Handoff Delay of Signaling

Handoff Case	(a) (ms)	(b) (ms)
H12	38290.0	171.4
H23	3932.2	161.6
H31	1934.7	161.1

Source: From N. Nakajima, A. Dutta, S. Das and H. Schulzinne, *International Conference on Communications (ICC'03)*, 26, May 2003, 1085–1089. With permission.

Table 11.5   Handoff Delay of Media UDP Packet

Handoff Case	(a) (ms)	(b) (ms)
H12	38546.3	420.8
H23	4187.7	418.6
H31	1949.4	408.4

Source: From N. Nakajima, A. Dutta, S. Das and H. Schulzinne, *International Conference on Communications (ICC'03)*, 26, May 2003, 1085–1089. With permission.

From Tables 11.4 and Table 11.5, the modified kernel has reduced handoff delay although the delay figures are unacceptable for real-time multimedia communications. For example, a delay of 420 ms is likely to result in echo if other delays in the network add to it. This could lead to packets being dropped or played back with significant delays, and hence render the playback unsuitable for listening. Nakajima et al. [7] showed that by integrating MIPL MIPv6 in the network for SIP mobility, they observed that the handoff delay for signaling is about 2 ms and the handoff delay for media UDP packet is less than 31 ms. This is acceptable in most voice communications. The results also showed that MIPL MIPv6 with modified kernel outperforms SIP mobility with modified kernel. Hence, application layer mobility based on SIP mobility has potential for real-time applications provided MIPL and MIPv6 and other similar speed-up processes are integrated.

Disruption Time

The amount of disruption time for processing handoff and interconnections is a concern in VoIP services. Handoff between terminals or nodes could disrupt speech

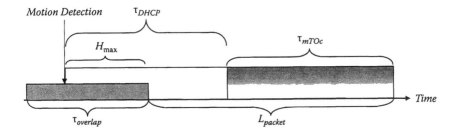

Figure 11.8   SIP proxies routing SIP and fax packets. (From T.T. Kwon, M. Gerla, S.K. Das, and S. Das, *IEEE Wireless Communications,* IEEE, 9, October 2002, 66–75. With permission.)

communications if speech packets are lost or delayed too much. Lost packets lead to lost words and sentences, and hence make the communication annoying and uncomfortable. Delayed packets also lead to long delays between sections of words or between words, and hence lead to the same uncomfortable speech communication. Therefore, handoff has potential to lead to an annoying listening experience in VoIP and IPTV services. Roaming users are interested in staying connected with the network while moving from one network to another network with multiple network interfaces. Hence, an efficient mobility management scheme is necessary for handling micro-mobility, macro-mobility, and roaming. The objective of the scheme should be to reduce handoff delay and disruption effects. Micro-mobility (intra-domain mobility) protocols aim to handle local movement (e.g., within a domain) of mobile hosts without interaction with the Mobile IP enabled Internet [8]. A macro-mobility (or inter-domain mobility) protocol based on Mobile IP manages mobility between domains [8]. In [4] it was found that the disruption for handoff of the Mobile IP approach is smaller than that of the SIP approach in most situations; however, SIP shows shorter disruption when the MN and CN are close because the need for handoff is reduced in such situations. While the SIP-based approach offers several advantages over a corresponding MIP-based solution for typical UDP-based VoIP streams, it continues to suffer from several drawbacks. These include the absence of mobility management for long-term TCP connection. Second, it can cause call disruption if the new SIP session is not created completely while the mobile terminal is in the overlapped area. As opposed to an MN using Mobile IP, an MN using SIP-mobility always needs to acquire an IP address via DHCP, which can be a major part of the overall handoff delay.

Several delays or disruption factors [9] affect the handoff delay and packet loss (Figure 11.8) over wireless networks, including:

1. τ_{mTOc}: The one-way delay from the mobile node to correspondent node.
2. τ_{hTOn}: The delay in sending a message between the new foreign agent and the home agent.

3. τ_{DHCP}: IP address renewal time through DHCP.
4. H_{max}: The maximum time for seamless handoff. It is the time difference between when motion is detected and when the time the node escapes from the old cell.
5. $\tau_{overlap}$: The time over which the mobile node is within the overlap area of two adjacent cells.
6. L_{packet}: Packet loss time during handoff process. This is the length of time over which the node will lose packets when a handoff process is ongoing.
7. $\tau_{overtime} = (H_{max} - \tau_{DHCP})$: The time difference between H_{max} and T_{DHCP}. H_{max} is the maximum time for seamless handoff time difference between move detection and the escaping time from old cell. T_{DHCP} is DHCP IP address renewal time.

Tests were performed by Jung et al. [9] to assess some of these disruption times, and the results are reflected in Tables 11.6, 11.7, and 11.8. Three factors are considered in this section.

Table 11.6 Handoff Disruption Time as Function of Delay τ_{hTOn}

	Disruption Time (ms)		
τ_{hTOn}	MIP	SIP	MIP-SIP
10	42	100	41
20	70	99	71
30	90	107	95
40	110	95	95
50	131	100	98
60	151	100	100
70	170	92	91
80	191	93	92
90	210	93	92
100	230	94	94

Source: From J.W Jung, R. Mudumbai, D. Montgomery, and H.K. Kahng, Global Telecommunication Conference 2003 (GLOBECOM'03), *IEEE* 2003, 3, 1190–1194. With permission.

Table 11.7   Disruption Time versus τ_{mTOc}

	Disruption Time (ms)		
τ_{mTOc}	MIP	SIP	MIP-SIP
10	90	49	48
20	90	70	59
30	90	90	91
40	90	120	91
50	90	125	91
60	90	143	91
70	90	180	91
80	90	200	90
90	90	205	90
100	90	230	90

Source: From J.W Jung, R. Mudumbai, D. Montgomery, and H.K. Kahng, Global Telecommunication Conference 2003 (GLOBECOM'03), *IEEE* 2003, 3, 1190–1194. With permission.

1. Table 11.6 illustrates the handoff disruption time as the delay τ_{hTOn} increases. Here, the τ_{mTOc} are set at 30 ms.
2. Table 11.7 shows the disruption time as the delay τ_{mTOc} increases. Here, the τ_{hTOn} is assumed to be 30 ms. As SIP mobility only depends on the distance between the MN and CN, the disruption time of SIP-mobility increases according to the delay τ_{mTOc}.
3. In Table 11.8, τ_{mTOc} and τ_{hTOn}, respectively, are set at 30 ms and 100 ms. Also, H_{max} is fixed at 500 ms. Because the handoff in SIP over Mobile IP does not need IP renewal time, its disruption time is the same as that in disruption time versus τ_{hTOn}. However, as the SIP-based approach always needs a new IP address, its disruption time increases in proportion to the rate of increase for $\tau_{overtime}$. Simulation results show that the proposed approach outperforms the existing approach in most cases. This shows that VoIP can be supported over wireless Internet.

Handoff Delay Disruption of SIP in IP Services

The performance of SIP in IP services can better be appreciated if we adopt similar metrics and measures for comparing them. One such approach uses the so-called

Table 11.8   Disruption Time versus $\tau_{overtime}$

	Disruption Time (ms)		
$\tau_{overtime}$	MIP	SIP	MIP-SIP
0	0.1	0.1	0.21
0.1	0.2	0.2	0.21
0.2	0.21	0.3	0.21
0.3	0.21	0.4	0.21
0.4	0.21	0.5	0.21
0.5	0.21	0.6	0.21
0.6	0.21	0.7	0.21
0.7	0.21	0.8	0.21
0.8	0.21	0.9	0.21
0.9	0.21	1	0.21
1	0.21	1.1	0.21

shadow registration proposed in [4]. In shadow registration, the MN establishes security association with the neighboring AAA server *a priori* before handoff is initiated. This can be done in a distributed manner with many known AAA servers or selectively with only the neighboring server. (This section does not discuss how this is done.)

Handoff Delay Disruption of SIP in VoIP Services

Handoff delay for SIP-based VoIP services was measured in [4] for both Mobile IP and SIP using the setup in Figure 11.9. Following [4], the variables are defined as:

1. t_s: The time for message transfer in the wireless link between the mobile node and the foreign agent router (RFA) or DHCP server.
2. t_f: The delay between the mobile node and the foreign AAA server (AAAF). This is the time required to send a message in the foreign network.
3. t_h: The delay between the MN and the communicating entities in its home network, which are its home registrar (HR), AAAH, or home agent (HA). It is therefore the time for the MN to send messages to its home network. Generally, $t_s < t_f < t_h$.
4. t_{mc}: The delay between the MN and the CN.
5. t_{hc}: The delay between the home network and the CN. These delays are shown in Figure 11.9.

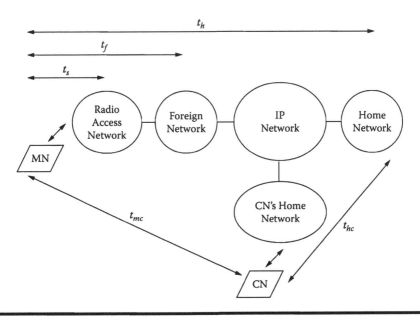

Figure 11.9 Separate paths for SIP messages and fax images. (From T.T. Kwon, M. Gerla, S.K. Das, and S. Das, *IEEE Wireless Communications,* IEEE, 9, October 2002, 66–75. With permission.)

6. t_{arp}: The time required for the ARP (Address Resolution Protocol).
7. t_{no}: The time to send a message between the new foreign agent (NFA) and the old foreign agent (OFA).

Figure 11.9 assumes that a simple VoIP application unaware of mobility is involved in the QoS parameter measurements and the MN caches the address of the callee's CN. Furthermore, it assumes that there was no *a priori* Internet connection. Hence, the mobile node must initiate the connection when it requires a VoIP application. Therefore the initial delay is the sum of the time required for Internet connection and VoIP signaling. These times are different for the MIP and SIP situations.

Initial Registration and Session Setup

The time required by Mobile IP for initial registration and session setup is given by the sum of the delay for the Router Solicitation message and Router Advertisement messages $(2t_s)$ as a round-trip delay, plus the round-trip registration message with the home agent $(2t_h)$ and the time for the simple VoIP application (SVA) to initiate a VoIP service request with the CN. This time consists of the request and the OK response from the CN $(2t_{mc})$. For the Mobile IP case, the total time is:

$$t_{MIP\_init} = 2t_s + 2t_h + 2t_{mc}$$

When SIP is used instead of MIP, the time required to acquire a new address through the DHCP ($4t_s$) and for the ARP (t_{arp}) must be added to the time for the MIP case. The total time is:

$$t_{SIP\_init} = 4t_s + t_{arp} + 2t_h + 2t_{mc}$$

When the same networks and nodes are used, the time required to initiate the IP service using SIP is distinctly higher than for MIP.

Intra-Domain Handoff Delay

In the case MIP, the MN detects a new access point and then the new IP subnet. It subsequently sends the Router Solicitation message to the RFA and receives a response in terms of the Router Advertisement message. This round-trip time is ($2t_s$). Because intra-domain does not involve AAA, the registration time is also ($2t_f$), and the total intra-domain handoff is:

$$t_{MIP\_init} = 2t_s + 2t_f$$

If SIP is used instead of MIP, the time to acquire a new IP address using a DHCP ($4t_s$) and the ARP time (t_{arp}) are needed. The remaining time is for the MN to register for the SVA with the foreign registrar ($2t_f$). The total time for intra-domain handoff in this case is:

$$t_{SIP\_init} = 4t_s + t_{arp} + 2t_f$$

Generally in the example given in [4], t_s is about 10 ms and t_f is about (t_s + 2) ms.

Inter-Domain Handoff Delay

In inter-domain, two subnetworks are involved. Therefore, the delays involve communication and signaling between the elements of the two subnetworks and the MN. Using MIP, for handoff to take place, the MN must first detect the new IP subnetwork, select it, and initiate handoff. The delays involved include the time for the MN to initial Router Solicitation and Router Advertisement messages ($2t_s$). The next delays are the time for the MN to register with the NFA ($2t_h$), and cache the message from the HA ($2t_h - t_s$). During the message caching process, two signaling flows take place: route optimization and smooth handoff. As the NFA caches the registration message, it sends a Binding Update message to the OFA. The OFA, upon receiving the Binding Update message (t_{no}), starts sending packets destined to the MN to the NFA ($t_s + t_{no}$), thus starting the inter-domain handoff. This results in the total time for starting the inter-domain handoff or smooth handoff:

$$h_{MIP\_smb} = \left(2t_s + \left(2t_b - t_s\right) + t_{no} + \left(t_s + t_{no}\right)\right) = 2t_s + 2t_b + 2t_{no}$$

During route optimization, other delays are involved in the handoff process, including the time when the OFA updates its binding cache with the care-of-address (CoA) of the MN ($t_b - t_s$) and sends a Binding Warning message to the HA of the MN. The HA then sends the Binding Update message to the CN (t_{hc}). Finally, the CN sends the packets for the MN to the NFA, which are then subsequently delivered to the MN with time (t_{mc}). The total time for route optimization is:

$$h_{MIP\_ropt} = \left(2t_s + \left(2t_b - t_s\right) + t_{no} + \left(t_b - \dot{t}_s\right) + t_{hc} + t_{mc}\right) = 3t_b + t_{no} + t_{hc} + t_{mc}$$

After route optimization, handoff is triggered. It is essential to understand the two signaling flows involved in route optimization in terms of the time packets from the CN reach the MN and the time the packets forwarded by the OFA reach the NM. There is a blackout period until packets from the OFA arrive at the MN. This blackout period is about $2t_s + 2t_b + 2t_{no}$, after which the VoIP communication resumes without a disruption until packets from the CN reach the MN. The MIP inter-domain handoff time becomes:

$$t_{MIP\_inter} = 3t_b + t_{no} + t_{hc} + t_{mc}$$

Note that when the MN is located in its home network, $t_{mc} = t_b + t_{hc}$.

With SIP, the inter-domain handoff time occurs after the times for acquiring the new IP address through DHCP and an ARP ($4t_s + t_{arp}$) and the time for the MN to send the SIP REGISTER message to the home registrar (HR), which is ($2t_b$). At this point, Internet connectivity is enabled. Then, the MN re-invites the CN by sending to it an INVITE message. This takes time ($2t_{mc}$). Hence, the inter-domain handoff time in this case is:

$$t_{SIP\_inter} = 4t_s + t_{arp} + 2t_b + 2t_{mc}$$

In general in the above equation, the choice of t_{arp} is between 1 and 3 ms, and t_{no} is about 5 ms. The disruption time increases as the delay distance between the MN and CN increases. In the case of MIP, the MN the disruption time is $2t_s + 2t_b + 2t_{no}$ and t_f is about ($t_s + 2$) ms; therefore, the disruption time is about 54

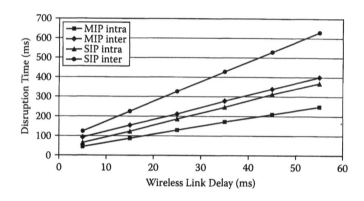

Figure 11.10 Protocol stack for SIP security. (From T.T. Kwon, M. Gerla, S.K. Das, and S. Das, *IEEE Wireless Communications,* IEEE, 9, October 2002, 66–75. With permission.)

ms, which is large enough (for speech sampled at 8 kHz) to take out more than 432 contiguous speech samples—a significant portion of the speech communication— and make it annoying to listen to. However, MIP outperforms SIP because the mobile terminal starts to receive speech after only about $2t_s + 2t_b + 2t_{no}$ ms, and the terminal can start to play back some of the speech until the overall handoff is completed. In both inter-domain and intra-domain handoffs, MIP performance is superior. The intra-domain disruption times for both MIP and SIP are relatively flat with increasing delay distance between the MN and home network, and also always smaller than the inter-domain disruption times. The inter-domain disruption times increase linearly for both MIP and SIP. At smaller delay separations, the MIP outperforms the SIP performance but after some time, SIP is advantageous as its disruption time becomes smaller than that for MIP over larger delays.

For both MIP and SIP, the inter-domain handoff disruption time also increases with the delay separation between the CN and MN, and their intra-domain handoff disruption times remain flat, with MIP always outperforming SIP.

Over low-bandwidth wireless links, the disruption times resulting from inter- and intra-domain handoffs increase linearly as the wireless link delay increases, and the intra-domain disruptions are always the smaller (i.e., MIP smaller than SIP) (Figure 11.10).

Disruption Time with Shadow Registration

When shadow registration is included in the operations of MIP and SIP, the disruption times change. Because AAA resolution for the MN can be performed within the foreign network (AAAF), the disruption time for the MIP case is a function of the Router Solicitation and Advertisement message exchange time ($2 t_s$), the registration message time from the MN to the NFA and the reply time to the MN $(2t_f - t_s)$, and the route optimization signaling flow time $(t_b - t_s) + t_{no} + t_{bc} + t_{mc}$. Therefore, the total disruption time is given by the sum:

$$t_{MIP\_int\,er\_sdw} = 2t_s + \left(2t_f - t_s\right) + \left(t_b - t_s\right) + t_{no} + t_{bc} + t_{mc}$$

$$= 2t_f + t_b + t_{no} + t_{bc} + t_{mc}$$

In case of SIP, the REGISTER message is also handled locally in the foreign network DHCP acquisition and the ARP consumes the time $(4t_s + t_{arp})$. Furthermore, the AAA resolution and REGISTER are performed in the foreign network with time $(2t_f)$. This is followed by the MN re-inviting the CN by sending the re-INVITE message with time $(2t_{mc})$. The total disruption time is therefore given by:

$$t_{SIP\_int\,er\_sdw} = 4t_s + t_{arp} + 2t_f + 2t_{mc}$$

Note that the disruption time now has t_f replacing t_b. Hence, shadow registration has decreased the disruption time in both cases of MIP and SIP. Therefore, shadow registration is useful when the MN is far from the home network.

Performance of SIP in Fax over IP

The performance of fax over IP (FoIP) needs to rival its performance in legacy fixed line at user premises. Fax over the PSTN is very robust and reliable. However, the reach and popularity of the Internet, combined with the fact that its use is almost free, is a major driver for VoIP networks and for similar data applications that exist on the PSTN. The major application is fax transmission.

Fax transmission has special requirements. First, while the loss of a packet during a human conversation is not likely to significantly affect a voice call, it can easily affect a fax call. This is because fax transmission requires far more signaling and handshaking than a regular telephone call. This includes negotiating details such as speed, paper size, and delivery confirmation. Apart from the signaling in a fax call, the sending and receiving of fax documents are mostly done and interpreted by automated fax machines. Therefore, errors in either signaling or the actual transmission are likely to result in lengthy recovery times.

A study was performed in [10] on the performance of fax over an IP network. The performance parameters for fax over an IP network are significantly different compared with VoIP. In fax transmissions, no roaming or handoffs are required. However, end-to-end delays, link utilization, and the performance of the wireless links are essential.

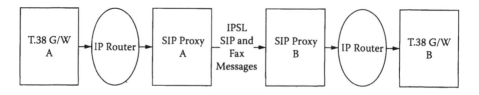

Figure 11.11 Mobile IP mobility management scheme. (From U. Choudhary, E. Perl, and D. Sidhu, 26th Annual IEEE Conference on Local Computer Networks (LCN'01), *IEEE*, November 2001, 74. With permission.)

Experimental Network Models

With three network models, Choudhary et al. [10] measured the link utilization of the inter-proxy-server link (IPSL), the link utilization of the auxiliary link (AUXL), the average end-to-end delay of the SIP signaling packets, the average end-to-end delay of fax data packets, the average SIP call setup time, and the average fax call setup time.

1. In Network Model 1, calls are generated from the T.38 gateway, and all messages are sent to the SIP proxy server on its network. The path between the T.38 gateway and the SIP proxy server contains an IP router. Although this is not necessary, it is however done to maintain compatibility with the next two network models where IP routers play an important role. The SIP proxy server of the sender's network communicates with that of the recipient's network and initially sends SIP messages to set up a call. Once that is done, the T.38 gateway starts fax transmission. The originating T.38 gateway starts sending fax packets to the SIP proxy server. That is, the SIP proxy server routes and interprets SIP messages. It also routes Internet Fax Protocol (IFP) packets without actually interpreting them. This scenario is likely if the SIP proxy server is implemented on a router within the network. In such a case, it interprets SIP messages, possibly translating them and maintaining the state for them. All other messages are not interpreted by it, but merely routed. This model is shown in Figure 11.11 [10].

2. Network Model 2 is the same network as Model 1 but a different setup. It not only includes the components and links of the first model, but also has a direct link between the IP routers (the so-called proxy-bypass approach). That is, all the SIP signaling is carried out on the path that traverses the two SIP proxy servers. However, once the call is set up, all fax data packet transmission is done through IP routers and proxy bypass links only. This mimics a network where signaling travels on separate links, and data is sent across another set of links. This is possibly because data does not need to go through SIP proxy servers or other network entities. This frees up resources at such entities and segregates signaling from data transmission, much like PSTN

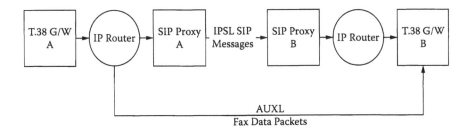

Figure 11.12 Mobility management scheme using SIP. (From U. Choudhary, E. Perl, and D. Sidhu, 26th Annual IEEE Conference on Local Computer Networks (LCN'01), *IEEE,* November 2001, 74. With permission.)

networks where the Signaling System Number 7 (SS7) links are distinct from the trunks. Figure 11.12 gives a view of this model.

3. Network Model 3 uses the same network as Network Model 2. However, not all SIP messages travel between the SIP proxy servers in this case. In general, all SIP terminals are configured to know their network's SIP proxy server. Hence, to set up a SIP call, they contact their SIP proxy server first, not having to know the entire route to a receiver themselves. They only need to know how to route SIP calls they initiate to their designated SIP proxy server. The SIP proxy server then handles all the signaling for the remainder of the call. The AUXL is used for fax data packets (as in Network Model 2).

Table 11.9 shows the performance measures recorded by Choudhary et al. [10]. In terms of link utilization values on IPSL, Network Model 3 is the most suitable. In terms of just link utilization values on AUXL, Network Model 2 is only

Table 11.9 Experimental Results

	Model 1	*Model 2*	*Model 3*
Average link utilization of IPSL (Mbps)	77.50	9.223	0.9546
Average link utilization of AUXL (Mbps)	–	69.03	75.96
Average end-to-end fax data packets delay (s)	0.2023	0.2017	0.2017
Average end-to-end signaling delay (s)	0.2014	0.2015	0.2009
Average SIP call setup times (s)	3.4023	3.4042	3.4020
Average fax call setup time (s)	4.2086	4.2106	4.2060

Source: From U. Choudhary, E. Perl and D. Sidhu, 26th Annual IEEE Conference on Local Computer Networks (LCN'01), *IEEE* November 2001, 74. With permission.

marginally better. In terms of end-to-end fax data packet delay, Network Models 2 and 3 are, on average, equally good. When just packet end-to-end delay is considered, Network Model 3 is the best. The average SIP call setup time and average fax call setup for Network Model 3 are the best.

SIP appears to be a powerful and useful signaling protocol supporting mobility for wireless IP networks but it has inherent weaknesses and dangers. These weaknesses and dangers are discussed next.

Effects of Security Protocols on Performance of SIP

SIP has been in use for setting up secure calls in VoIP. For this, SIP can employ several security protocols, such as TLS, DTLS, or IPSec, combined with TCP, UDP, or SCTP as a security protocol in VoIP (Figure 11.13). These security mechanisms introduce additional overheads into SIP performance. From a privacy point of view, VoIP delivered over an IP network is vulnerable to message spoofing and eavesdropping. Impersonation of a legal user could also result in false accounting and avoidance of payment. Hence, many vendors resort to encrypting VoIP messages. SIP therefore assumes an additional role of authenticating users who send and receive VoIP messages with the potential to negatively affect the performance of SIP. An obvious effect of a security protocol on the performance of SIP is the call setup delay. Traditionally, SIP was not developed with its own security features but the standard recommends that existing security mechanisms could be used instead.

Two factors are responsible for the increased call setup delay: authentication of messages and security handshake. A security channel is established during the handshake. During this process, a key is generated and messages are exchanged. This results in additional setup delay. Also, during authentication, the SIP messages are encrypted, resulting in increased packet size or payload, thereby causing additional delay.

There are two types of handshakes: full handshake and abbreviated handshake. During full handshake for normal session initiation, the user proxy or servers

Figure 11.13   Basic SIP procedure. (1) INVITE sip:denise@telkom.com; (2) INVITE sip:denise@hos_b.telkom.com; (3) 180 Ringing; (4) 180 Ringing; (5) 200 OK; (6) 200 OK; (7) ACK sip:denise@host_b.telkom.com; (8) BYE sip:denise@host_b.telkom.com; (9) 200 OK.

authenticate themselves and also create an encryption session key. During abbreviated handshake (Phase 2 re-key), a previous security channel could be reused for session resumption. This is called abbreviated security handshake because the information of the previous security channel is still available. Fewer messages are exchanged during the abbreviated handshake because the authentication process may be skipped and consequently the delay involved is also smaller.

Furthermore, additional setup delays are introduced by the processes required for establishing the confidentiality and integrity of messages. These processes include encryption and adding message authentication code (MAC) to the message. For example, the SHA-1 MAC process results in 25 bytes of extra data on top of the original message. In addition, every encryption step introduces its own processing delay, which worsens if multiple levels of encryption and decryption are involved.

Different security mechanisms have different effects on SIP, particularly in terms of delays and overheads. This section summarizes the performance of SIP when combined with IPSec, TLS, and DTLS (Datagram Transport Layer Security) within the domains of transport protocols such as TCP, UDP, and SCTP. Noticeable performance degradation was observed during SIP call setup [4]. A relatively larger call setup delay is observed with the combination of TLS and SCTP than in other combinations (e.g., TLS/TCP or TLS/UDP). The best performance is achieved with the combination of UDP and other security mechanisms (IPSec/UDP and DTLS/UDP). However, there is no congestion control involved with SIP over IPSec/UDP or DTLS/UDP. A high level of congestion increases the call setup delay with TLS/TCP. At low network congestion, there is no significant difference between the call setup times for the combinations of IPSec/TCP and TLS/TCP. In general, the overhead associated with message authentication is very small when TLS and IPSec are involved.

Summary

SIP was not designed initially with security as a major objective. Security features are, however, used to enhance its performance and the features have benefits and disadvantages. They provide authentication while increasing delays in the services. The QoS performance of SIP was presented in this chapter, setting up the foundation for a better understanding of what to expect in SIP applications and services. The key QoS performance measures for SIP sessions include communication time, handoff delay when mobile nodes are involved in and between neighboring subnets, data volume exchanged during the SIP session time, jitter, bit error rates, handoff delay, registration, and setup time. This chapter not only explained these in detail, but also presented how to measure and model them. Delays involved in SIP telephony are quantified through experimental data and can also be estimated mathematically, as was clearly shown in the chapter.

Practice Set: Review Questions

1. If SIP is a signaling protocol, why are we worried about QoS?
   a. It takes time to set up an SIP session.
   b. SIP communication takes some time.
   c. There could be jitter associated with SIP packets.
   d. Handoff delays exist.
   e. All of the above.

2. A SIP session involves a file of size 50 kbps. If the average rate ON period for the session is 12.5 kb, what is the average duration of the ON period?
   a. 5 milliseconds
   b. 4 milliseconds
   c. 4 seconds
   d. 2 seconds
   e. 1 second

3. Using the answer in Question 11.2 above, if the average duration of OFF period is 250 milliseconds and the average number of ON-OFF patterns in one session is 20, what is the total duration of one session?
   a. 85 seconds
   b. 42.5 seconds
   c. 250 milliseconds
   d. 5 seconds
   e. 2.5 seconds

4. QoS is a loose term in telecommunication that is used to assess the performance of networks. Which one of the following is used to assess QoS?
   a. Range of the access point
   b. Throughput
   c. Length of the wireless link
   d. Line of sight
   e. Height of the antenna

5. QoS is a loose term in telecommunication that is used to assess the performance of networks. Which of the following are not used directly to assess QoS?
   a. Length of a communication link
   b. Throughput
   c. Bit error rate
   d. Network delay
   e. Height of the antenna

6. QoS is a loose term in telecommunication that is used to assess the performance of networks. Which ones of the following delays are not used directly to assess QoS?
   a. Network delay
   b. Message transfer delay

c. SIP signaling delay
d. Call setup delay
e. All of the above
7. Several disruption delays affect SIP handoff. What are they?
   a. One-way delay from the mobile node to the correspondent node
   b. Delay in sending a message between the New Foreign Agent and the Home Agent
   c. Packet loss time during the handoff process
   d. All of the above
   e. Only (a) and (b)

Exercises

1. List all the possible delays in SIP communication session setup.
2. List all the possible delays in the SIP communication chain.
3. List and discuss the associated delays in SIP call setup. Which one is likely to be more costly in terms of offending SIP network users? Why?
4. SIP supports application layer mobility. Discuss the delays associated with handoffs in SIP communication.
5. Discuss the terminal mobility scheme using SIP.
6. A mobile node using SIP-mobility always needs to acquire an IP address via DHCP, which can be a major part of the overall handoff delay. Discuss all the disruption factors that can affect SIP handoff and packet losses.
7. An MIP setup was observed to incur a total delay of 500 milliseconds. If the router solicitation and advertisement cost 100 milliseconds of time, and the time to contact the CN and receive a response is 150 milliseconds, what is the total round-trip delay for the registration message? If this is a wireless link, what is the physical distance between the devices on registration? (Assume that the speed of light is 3×10^8 meters per second.)
8. Given the values in Question 11.14 and your answers, if the address resolution time t_{arp} is 3 milliseconds, what is the inter-domain handoff delay time?
9. List the most prominent sources of delay in a SIP voice call.
10. What is post-dialing delay?
11. What is answer-signal delay?
12. What is call-release delay?
13. Why is the INVITE method considered the most important in SIP?
14. Distinguish between macro mobility, micro mobility, and roaming.
15. How is mobility handled in the:
    a. Network layer?
    b. Application layer?
    c. Transport layer?
16. What is terminal mobility? Is it a problem or an asset in IP networks?

17. What is handoff delay? Mention its constituent delay categories.
18. What is likely to happen if the handoff disruption time is very long in voice communications? How would it affect mobile IPTV services?
19. What happens during a SIP security handshake? Distinguish between full handshake and abbreviated handshake.

References

[1] H. Hassan, J.M. Garcia, and O. Brun, Session Based Quality of Service in SIP Networks, *Information and Communication Technologies, ICTTA'06,* 2006, pp. 3251–3256.

[2] J.S. Wu and P.Y. Wang, The Performance Analysis of SIP-T Signalling System in Carrier Class VoIP Network, *17th International Conference on Advanced Information Networking and Applications (AINA'03),* IEEE, March 2003, pp. 39.

[3] V.Y.H Kueh, R. Tafazolli, and B. Evans, Performance Evaluation of SIP Based Session Establishment over Satellite-UMTS, *Vehicular Technology Conference 2003 (VTC 2003-Spring). The 57th IEEE Semi-annual,* IEEE, April 22–25, 2003, Vol. 2, pp. 1381–1385.

[4] T.T. Kwon, M. Gerla, S.K. Das, and S. Das, Mobility Management for VoIP Services: Mobile IP vs SIP, *IEEE Wireless Communications,* Vol. 9, October 2002, pp. .66–75

[5] I.D.D. Curcio and M. Lundan, SIP Call Setup Delay in 3G Networks, *Seventh International Symposium on Computers and Communications (ISCC'02),* IEEE, July 2002, pp. 835.

[6] ETSI Telecommunications and Intranet Protocol Harmonization Over Networks (TIPHON), End to End Quality of Service in TIPHON System. Part 2: Definition of Quality of Service (QoS) Classes, TS101 329-2 v.1.1.1, July 2000.

[7] N. Nakajima, A. Dutta, S. Das, and H. Schulzinne, Handoff Delay Analysis and Measurement for SIP Based Mobility in IPv6, *International Conference on Communications (ICC'03),* 26, May 2003, 1085–1089.

[8] A.T. Campbell, J. Gomez, S. Kim, A.G. Valko, and C. Wan, Design, Implementation, and Evaluation of Cellular IP, *IEEE Personal Communications,* 7(4), August 2000, 42–49.

[9] J.W Jung, R. Mudumbai, D. Montgomery, and H.K. Kahng, Performance Evaluation of Two Layered Mobility Management Using Mobile IP and Session Initiation Protocol, *Global Telecommunication Conference 2003 (GLOBECOM'03),* IEEE, 3, 2003, 1190–1194.

[10] U. Choudhary, E. Perl, and D. Sidhu, Using T.38 and SIP for Real-Time Fax Transmission over IP Networks, *26th Annual IEEE Conference on Local Computer Networks (LCN'01),* IEEE, November 2001, pp. 74.

Chapter 12

Softswitch

Switching has evolved from circuit switching (CS), to packet switching (PS), Internet Protocol (IP) switching, and multiprotocol label switching (MPLS).

The traditional public switched telephone network (PSTN) and the Internet (or other Internet protocol-based networks) require specific paths for the traditional or VoIP call to take from the source to the destination. Duplex channels are required for transmit and receive operations. The traditional fixed-network Internet requires three basic functions: routing, transmission, and billing. These three requirements must be translated for IP communications.

For the PSTN, most of the intelligence required to accomplish those functions resides in the central office (CO) switches. These are placed at strategic optimal locations in suburbs and around the country at their best proximity points to the populations being served. The CO switches themselves contain the switching building block and the switching logic. The switching fabric creates the physical connections between the endpoints, and the switching logic provides the call routing functions, calling features, and interfaces to other parts of the telephone system, including billing and OSS (operations support systems). From the inception of analog telephony, it was normal to have dedicated human operators who performed the interconnectivity for calls based on requests from subscribers and availability of trunk lines. Other functions such as billing were also done manually. These early switching fabrics have evolved, first as relay-based mechanical switches well known for their problems and huge manpower that is often required in times of maintenance and system upgrading. It was not unusual in those days for upgrading several thousands of lines to take months.

Hard and Soft Switching

Advances in computer software development and fast electronic switches and computers have, however, provided a means for replacing mechanical switching and routing fabric with software, and for replacing the mechanical PBX with a pure software PBX. These new software telephony components make it easy to add intelligence to the basic telephony infrastructure and also for operations and maintenance, including upgrading. Hence, softswitches have emerged to replace the old-generation hardware telephone switches. Therefore, instead of large, dedicated hardware for switching, it is normal to have a computer-based softswitch replacing the hardware. "Short for software switch, a softswitch is an API that is used to bridge a traditional PSTN and VoIP by linking PSTN to IP networks and managing traffic that contains a mixture of voice, fax, data, and video. Softswitches are able to process the signaling for all types of packet protocols. Softswitch is a software-based switching platform, which is opposed to traditional hardware-based switching center technology" [5].

Softswitch is also known by many other names, including proxy gatekeeper, call server, call agent, media gateway controller, and switch controller. It is software used to bridge a PSTN and IP services. Softswitch performs call control functions such as protocol conversion, and authorization, accounting, and administration operations.

The softswitch performs all the functions that used to be handled by hardware switching, and much more. With this approach, it is therefore possible to easily decouple the control logic function as software from the switching logic, which is also software. This means that these software components in the softswitch can be located in different locations and be distributed. Furthermore, new functionalities such as interconnectivity between existing circuit-switched (PSTN) and packet-switched (VoIP) services can be provided and enabled with easier evolution and migration paths.

The softswitch, also popularly referred to as a media gateway controller, consists of switching software and control software. The switching architecture decomposes into two major parts: a call agent and a media gateway. The call agent contains most of the network intelligence and is often referred to as the media gateway controller or the call controller. Its main functions include network signaling, call routing, and billing, to name a few. The call agent may control one or more media gateways and may be dispersed geographically, depending on the locations of the networks that it helps to control and the format of transmission. The media gateway functions to provide the end-to-end path or the physical connectivity. Therefore, it incorporates a mixed format of transmissions of IP, ATM, LAN, WLAN, and T1-T3.

The softswitch deals mainly with the integration of VoIP systems with legacy circuit-switched networks (such as private branch exchange (PBX) switches that are used in medium- to large-sized enterprises, or the CO switches) that are deployed throughout the PSTN. Integration of the VoIP network with a cellular network uses the IP Multimedia Subsystem (IMS), and many softswitch vendors are beginning to include the IMS within their own product architectures. The IMS is discussed in Chapter 13.

Architecture of the Softswitch

The softswitch decouples the major functions of a legacy central office switch into logical software functional components and allows the physical switching functions to reside in one device and the logic functions in another device [1]. By decoupling the functional components, they can be distributed into several devices and also into disperse geographical locations. This approach is clearly different from the old legacy central office switching system in which the functions are resident in one large infrastructure and located in one place. In practical terms, because there is no real reason for centralizing them, the operator is given the freedom of deciding where best they fit to optimize network performance. In addition, two new network elements were introduced: the call agent, which provides the call routing, signaling, and billing, and the media gateway, which provides the physical interconnectivity to the wide area network (WAN) or LAN.

The fundamental architecture of a softswitch is designed to promote greater interoperability between service providers and vendor products. The description here is in terms of functional planes from the top to the bottom. Four major function groups or functional planes are defined: transport, call control and signaling, service, and application and management. These planes also define the major functions of a VoIP network and its softswitches. The four planes can be further subdivided into other domains with the following functions:

1. *Transport plane:* This plane consists of the transport media, call setup, and call signaling messages across the VoIP network. The underlying physical transport technology may vary, based upon the requirements for the media being carried. "Three domains are defined:
   - The *IP Transport Domain* transports packets across the VoIP network, and includes devices such as routers and switches.
   - The *Interworking Domain,* which interacts with networks that are external to the VoIP network, includes signaling gateways, media gateways, and interworking gateways.
   - The *Non-IP Access Domain* allows non-IP terminals, such as ISDN telephones or mobile telephones, to connect to the VoIP network through access gateways or residential gateways" [2].
2. *Call control and signaling plane:* SIP signaling is the norm. This plane controls the major devices in the VoIP network, and also handles the establishment and teardown of media connections. The call agent (also known as the media gateway controller) operates in this plane.
3. *Service and application plane:* This plane provides the control and logic functions for services and applications that are available on the VoIP network, and includes devices such as application server and feature servers.
4. *Management plane:* This plane provides the services required for subscriber and service provisioning, and support and billing functions. This plane may

interact with the other three planes using the Simple Network Management Protocol (SNMP) or proprietary protocols [2].

Media Gateway Controller

Although the media gateway operates within the transport plane, the media gateway controller (MGC) operates within the call control and signaling plane. Whether through a VoIP network or a PSTN, telephone calls require reliability and the ability to establish and disconnect end-to-end. The MGC provides these functions and therefore requires significant network intelligence to achieve that. The intelligence is required to undertake the functions of call routing, control of the connections, and control of the network resources. Figure 12.1 shows that the MGC and the media gateways sit at the edge of the network. Hence, as opposed to legacy networks where the intelligence resides inside the network, IP networks locate the intelligence hub at the edges of the network. This affords the network elements to proactively manage and condition the calls at one point in the network before they reach the overall network.

The specific operations of the MGC include [3]:

■ Originating and terminating signaling messages to and from end-user stations, other MGCs, or networks that are external to the IP network.
■ Maintaining call state information for every call through the media gateway.
■ Serving as a conduit for the negotiation of media parameters. This includes, for example, the selection of an audio or video codec that could impact the bandwidth consumption for that particular call, and on a larger scale, the bandwidth consumption of the network as a whole. "These parameters are typically

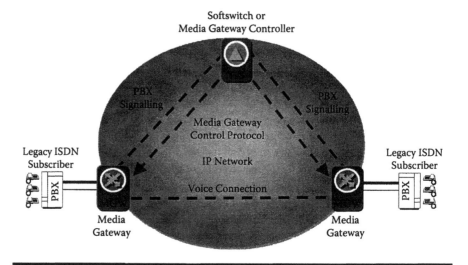

Figure 12.1   Locations of MGC and media gateway in IP networks.

negotiated between the end stations when the call is established, which may require the involvement of the MGC" [3]. Where necessary, format transcoding such as voice transcoding and image/video transcoding are undertaken as well so as to match the capabilities of the end-user stations. To be effective and to maintain QoS, such transcoding processes must be lossless.

■ Management of network resources, such as ports on the media gateway, the available network bandwidth, etc.

■ Providing policy functions, such as access capabilities and permissions for endpoints.

■ Interfacing with other services and/or servers, such as the application, accounting, or routing functions that are required in support of the call [3].

■ Participation in network management tasks.

In addition, the architecture defines two other functions that are subsets of the MGC. The first is the *call agent function,* which provides the call control and call state maintenance operations. This function may involve a protocol such as SIP or H.323. The second is the *interworking function.* This is required when a connection is needed to/from a network that deploys another type of networking protocol. Examples of interworking functions include H.323, SIP, or IP/ATM network connections. While SIP and H.323 interwork in terms of signaling for legacy networks and IP networks, IP and ATM interworking is required when there is the need to transform IP packets into ATM cells for the connected ATM network to function properly with an underlying IP network.

Architecture of Media Gateway [4]

The call agent handles the vast majority of logical functions in an IP network but the media gateway is responsible for providing the physical connection to the WAN or LAN. The media gateway operates within the transport plane. It is therefore responsible for the physical transport of the IP service or VoIP messages and the required access to external networks and associated terminals.

According to the Open Systems Interconnection (OSI) networking reference model, a *gateway* is defined by the International Organization for Standardization (ISO) as a device that converts protocols for one system to the protocols for another system, and operates at all seven layers of the OSI model. The two networks that sit on the two sides of the gateway are most likely based on different technologies (e.g., PSTN/UMTS), such as circuit switching on one side and packet switching on the other [4], or a cellular telephone network on one side and a packet-switching network on the other (e.g., GSM/GPRS or all-IP network). With this type of definition, a key attribute of the gateway will be as an interface to allow seamless communication with each network. In practical terms, a voice caller does not want to know when her call transverses from the cellular network she is using to the fixed network number she is calling or where the callee resides. For example, the gateway may have been used to connect a PSTN with an IP-based network, possibly with a T-1 interface, operating at 1.544 Mbps on the PSTN side, and an Ethernet interface operating at 10 or 100 Mbps on the IP network side. Many other combinations are possible, including cellular wireless, high-speed optical, cable television, ATM, and others [4]. Creating the physical connection and the necessary interface and rate conversions is a fraction of the responsibilities of the gateway. Depending on how the gateway is used and the networks involved, it may also perform the following functions:

- Converting ATM cells into packets used with IP networks
- Performing media processing functions, such as canceling line echoes, managing the jitter, and compensating for packet loss, to thus ensure the overall QoS of the connection
- Transcoding of data from one standard to the other, such as converting voice from ADPCM to LD-CELP and vice versa, and video from proprietary standards to MPEG-4
- Inserting appropriate signals into the media, such as dialing and call progress tones and adding comfort (background) noise
- The ability to detect specific call events, such as on/off hook states and voice activity

Thus, the media gateway consists of three key elements:

1. A physical interface to the first network, and another physical interface to the second network.
2. Signal processing functions that are required to make all the protocol conversions between these two networks.
3. Depending on the network type, media gateways are frequently located as edge devices close to the community of end users. Intelligence at the edge of the networks further reduces connection costs, and provides even more efficiencies as it handles all rate and format conversions before the data enters the destination network.

Media Gateway Control Protocol [6]

For the media gateway and MGC to fulfill their responsibilities to the networks they interface to, communication between them is required. This allows them to interconnect to the physical and logical sides of the softswitch architecture. This communication is in the form of a master/slave relationship with the MGC as master and the media gateway as the slave. The MGC sends commands to the media gateway using two protocols:

1. Media Gateway Control Protocol (MGCP)
2. MEGACO/H.248 standard (also called the Media Gateway Control)

Two major differences can be specified for the two protocols. They differ in the commands used for communication with each other and also in the transport mechanisms for the two protocols. MGCP is defined for UDP/IP transport. MEGACO is independent of the underlying transport, supporting UDP/IP, TCP/IP, or ATM

MGCP was developed by the Internet Engineering Task Force (IETF) as RFC 2705 and updated its information in RFC 3435. RFC 3435 describes a broad spectrum of gateway types that could use such a protocol. The following is a short listing of such gateways:

- Access gateways, which provide a traditional analog (RJ11) or digital PBX interface to a VoIP network
- Residential gateways, which provide a traditional analog (RJ11) interface to a VoIP network, such as cable modems/cable set-top boxes, xDSL modems, and broadband wireless devices
- Trunking gateways, which interface between the telephone network and a VoIP network; they typically manage a large number of digital circuits
- Voice over ATM gateways, which interface to an ATM network
- Network access servers, which can attach a modem to a telephone circuit and provide data access to the Internet
- Circuit switches, or packet switches, which offer control interface to an external call control element

Two forms of information are communicated between the MGC and the media gateway: signaling and events. An example of an event is a telephone going off-hook, while an example of a signal may be the application of dial tone to an endpoint. "These events and signals are grouped into packages, which are then supported by a particular type of endpoint such as a telephone or a video system" [6].

The communication between MGCs (call agents) and media gateways uses MGCP commands. There are nine defined MGCP commands. The commands are constructed using a command verb followed by a set of parameter lines.

Seven of the nine commands are sent by the MGC and the remaining two by the media gateway. The commands have the ability to clearly name the notified entity and are [8]:

1. EndpointConfiguration (EPCF) is sent by the call agent to a media gateway and is used to instruct the gateway about the coding characteristics expected by the line-side of the endpoint.
2. NotificationRequest (RQNT) is sent by the call agent to a media gateway and is used to instruct the gateway to watch for specific events such as hook actions or DTMF tones on a specified endpoint.
3. CreateConnection (CRCX) is sent by the call agent to a media gateway and creates a connection that terminates in an endpoint inside the gateway.

4. ModifyConnection (MDCX) is sent by the call agent to a media gateway and changes the parameters associated with a previously established connection.
5. DeleteConnection (DLCX) is sent by the call agent to a media gateway and deletes an existing connection, or indicates that a connection can no longer be sustained.
6. AuditEndpoint (AUEP) is sent by the call agent to a media gateway and queries the status of a particular endpoint (sent from call agent to the gateway).
7. AuditConnection (AUCX) is sent by the call agent to a media gateway and queries the status of a particular connection.
8. Notify (NTFY) is sent by the media gateway to the call agent and the gateway informs the call agent when the requested events occur.
9. RestartInProgress (RSIP) is sent by the media gateway to the call agent and the gateway notifies the call agent that the gateway is being taken out of service, or is being placed back in service.

MEGACO/H.248 Protocol [7]

The second protocol developed for communicating between the call agent and the media gateway is called the MEGACO/H.248 standard. This standard was jointly developed by the IETF and the International Telecommunications Union—Telecommunications Standards Sector (ITU-T) and therefore has two names: MEGACO being the IETF designation as explained in RFC 3525 and Recommendation H.248.1 designated by the ITU-T.

The text for these two documents is virtually identical. (As an aside, MEGACO was the designation given by the IETF Working Group that developed the protocol, and that name has therefore been associated with the protocol since its early development. The new title of RFC 3525 reflects a more generic protocol name: Gateway Control Protocol. Many industry documents still refer to the protocol as MEGACO. MEGACO and MGCP were developed based on the notion of distributed gateway architecture by assuming that the intelligence to process calls and have access to the media stream and therefore should reside at the edge of the network and hence at the media gateway controller.

There are, however, key differences between the protocols. First, they differ in the abstractions used in the connection model. The commands in MGCP apply to the connections. The commands in MEGACO apply to "Terminations" that are related to a "Context."

■ A Termination is a source and/or sink of one or more streams of information. For example, in multimedia conferences, the Termination could be multimedia and the sources and/or sinks could be multiple media streams.
■ A Context is an association between a collection of Terminations. A Context describes the topology (who hears/sees whom) and the media mixing and/ or switching parameters for the cases where more than two Terminations

are involved with this association. Contexts are modifiable using the Add, Subtract, and Modify commands. These are described subsequently. "A Connection is created when two or more Terminations are placed in a common Context" [7].

Packages are also defined as in MGCP to specify the characteristics of the Termination. Descriptors called *properties* and *statistics* are added to the events and signals as are found in MGCP. Furthermore, Annex E of RFC 3525 defines different packages from those found with MGCP. They are Tone Generator, Tone Detection, Basic DMTF Generator, DTMF Detection, Generic, Base Root, Call Progress Tones Generator, Call Progress Tones Detection, Analogue Line Supervision, Network, RTP, and TDM Circuit and Basic Continuity.

There are eight MEGACO commands [7]. Some of them are used to modify a Context and others are used by the media gateway to communicate with the MGC. The commands are:

1. Add: Adds a Termination to a Context. The Add command on the first Termination in a Context is used to create a Context.
2. Modify: Modifies the properties, events, and signals of a Termination.
3. Subtract: Disconnects a Termination from its Context, and returns statistics on the Termination's participation in the Context.
4. Move: Atomically moves a Termination to another Context.
5. AuditValue: Returns the current state of properties, events, signals, and statistics of Terminations.
6. AuditCapabilities: Returns all possible values for Termination properties, events, and signals allowed by the media gateway.
7. Notify: Allows the media gateway to inform the MGC of the occurrence of events in the media gateway.
8. ServiceChange: Allows the Media Gateway to notify the MGC that a Termination or group of Terminations is about to be taken out of service, or has just been returned to service. A number of ServiceChangeReasons have been defined that provide further details [7].

Summary

Recent developments in telephone switching and software engineering have led to replacing the mechanical switching and routing fabrics in PSTN with software and replacing the mechanical private branch exchange (PBX) with a pure software PBX. Software telephony components make it easy to add intelligence to the basic telephony infrastructure and also for operations and maintenance (including upgrading). Hence, softswitches have emerged as replacements for the old-generation hardware telephone switches. The softswitch performs all the functions that

used to be handled by hardware switching and much more. The switching architecture decomposes into two major parts: a call agent and a media gateway. The
softswitch architecture is segmented into planes. The *media gateway* (MGW) operates within the transport plane, and the *media gateway controller* (MGC) operates
within the call control and signaling plane.

Practice Set: Review Questions

1. Which of the following are not methods of switching in telecommunication?
   a. Circuit switching
   b. ATM switching
   c. Packet switching
   d. Cell switching
   e. Bit switching
2. The architecture of softswitches has:
   a. Five planes
   b. Four planes
   c. Three planes
   d. Two planes
   e. One plane
3. In the call control and signaling plane, the main signaling method is:
   a. SS7
   b. SS7 and SIP
   c. mSCTP
   d. SIP
   e. RSVP
4. The softswitch operates in:
   a. The transport plane
   b. The call control and signaling plane
   c. The service and application plane
   d. None of the above
   e. All of the above
5. The key element(s) of the media gateway is (are):
   a. Physical interface
   b. Signal processing
   c. Edge device
   d. All of the above
   e. None of the above
6. How does the MGC communicate with the media gateway?
   a. Using the Media Gateway Control Protocol (MGCP).
   b. Using the MEGACO/H.248 standard.
   c. Both (a) and (b) are applicable.

7. Which of the following is not another name for softswitch?
   a. Proxy gatekeeper
   b. Proxy server
   c. Call server
   d. Call agent
   e. Media gateway controller
8. The media gateway controller sends commands to the media gateway using the following protocols:
   a. MGCP
   b. SIP
   c. MEGACO
   d. MPLS
   e. (a) and (c)
9. Which of these is not a type of gateway?
   a. Access gateway
   b. Trunking gateway
   c. Residential gateway
   d. Highway gateway
   e. Voice over ATM gateway

Exercises

1. Distinguish between hard and soft switching.
2. What is the role of the transport plane?
3. Discuss the main function of the MGC.
4. Discuss the roles of MEGACO and MGCP. List and explain the MGCP commands.
5. List and discuss the MEGACO and MGCP commands.
6. What are the major functions of the softswitch?
7. Discuss the functions of a media gateway.
8. The media gateway consists of three elements. What are they and their functions?
9. The architecture of a softswitch is separated into planes. What are the names of these planes, and what does each plane do?

References

[1] M.A. Miller, Softswitches—Part I: Getting There from Here, Oct. 11, 2005, <http://www.voipplanet.com/backgrounders/article.php/355296>.
[2] M.A. Miller, Softswitches—Part II: Functional Planes of the Softswitch Architecture, Oct. 19, 2005, <http://www.voipplanet.com/backgrounders/article.php/3557421>.

[3] M.A. Miller, Softswitches—Part III: The Media Gateway Controller, <http://www.voipplanet.com/backgrounders/article.php/>.

[4] M.A. Miller, Softswitches—Part IV: Media Gateway Architecture, <http://www.voipplanet.com/backgrounders/article.php/>.

[5] <http://webopedia.com/TERM/s/softswitch.html>.

[6] M.A. Miller, Softswitches—Part V: Media Gateway Control Protocol, <http://www.voipplanet.com/backgrounders/article.php/>.

[7] M.A. Miller, Softswitches—Part VI: MEGACO/H.248 Protocol, <http://www.voipplanet.com/backgrounders/article.php/>.

Chapter 13

IMS: IP Multimedia Subsystem

The Internet Protocol Multimedia Subsystem (IMS) was defined by the Third Generation Partnership Program (3GPP/3GPP2) as a generic standard architecture for offering IP multimedia services and also voice over IP. IMS was created initially as a standard for wireless networks. However, the wireline community, in the search for a unifying standard, has since realized the potential of IMS for fixed communication as well. Hence, it is being adopted for that purpose. As an international standard, it provides the framework and the support for fixed-to-mobile convergence [1] and hence the migration of all traffic from a circuit-switched domain to IP packet-switched domain in an all-IP network. The IMS inherently has the glue for working with the SIP for signaling and setting up IP sessions. As a 3GPP and subsequently international standard, it supports multiple access types, including GSM/GPRS, UMTS, CDMA2000, and HSDPA, broadband wireline access, and WLAN.

Intelligent Multimedia Subsystem

The IMS aims to put an end to fractured technologies by standardizing the network interfaces. It also provides open network interfaces and open platforms. The IMS provides operators and users with specific advantages. For operators, it provides the basis for a horizontal architecture, thereby taking the concept of layered architecture a step forward. This enables reuse of service enablers and common functions for offering multiple applications. It offers a common means for interoperability and roaming, and better control of bearer, billing, and security. Furthermore, it is integrated with existing data and voice networks and adopts significant benefits of the IT domain. Therefore, it is a major enabler of fixed-to-mobile convergence in a seamless manner. Thus, new services can be offered across the board in a standardized and well-structured manner. In doing so, new service creation is facilitated and enhanced, and it enables legacy interworking across multiple disparate access types. Because it offers a horizontal architecture, overlapping of functionalities in different access types is avoided, thereby eliminating costly complex layered network architectures. Specifically, it eliminates overlapping charging, presence, group and list management, routing, and provisioning functionalities. Hence, a secure migration path is provided toward an all-IP architecture.

For users, the IMS provides and enables person-to-content and person-to-person communications for voice, text, image, and video content, and their hybrids. With subscribers using so many different types of terminals and with so many vendors, they are not interested in the dividing borderlines between networks and different access techniques. Subscribers really demand a seamless offering and access to multimedia content irrespective of the type of terminal they own and use or the hosting vendor. IMS (Figure 13.1) provides the basis for this rich communication experience and freedom of access. By eliminating interoperability problems, the user is more prone to venture into buying new terminals and subscribing to new and different multimedia content types irrespective of where they originate. This makes reaching each person easier and less prone to disruptions.

As an advantage, the IMS provides the enabler for field force automation in which on-the-road and remote office workers are able to work from anywhere in the world, at home, on the road, and in transit, and yet are able to access their office services and data without encumbrance by the underlying technology. Hence, a

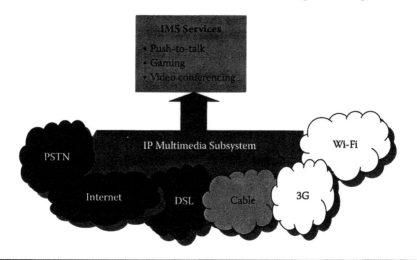

Figure 13.1   IMS Vision. (From H. Mowafy and Z. Zeb, Telcordia, Presentation, October 4, 2006. With permission.)

single user can be reached with a single name or address irrespective of different terminal type, access types, or location. The IMS will therefore provide the basis for richer social networking communities and offer the levels and quality of experience commensurate with or close to face-to-face interactions in meetings, conferences, and presence.

With a standardized platform and common functionality, operators can respond quickly to developing, implementing, and offering new services that are transparently the same to all other operators. This also permits enterprises to interoperate more easily based on new technologies and hence reduce the costs of ownership and subscriptions. Furthermore, an IMS provides a savvy operator the basis for creating new services and/or integrating third-party services into its service and product offerings, which enables its revenue base to grow quickly. Therefore, the IMS offers operators the basis for differentiation and service interoperability.

IMS Architecture

A three-layered architecture is used to implement an IMS: the connectivity layer, the control layer, and the services layer (Figure 13.2).

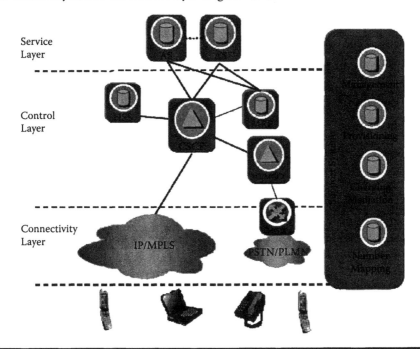

Figure 13.2   Overview of IMS architecture. AS: application server, HSS: home subscriber server, MGCF: media gateway control function, MGW: media gateway, MPLS: multiprotocol label switching, MRF: media resource function, PLMN: public land mobile network, SG: signaling gateway.

The first layer is the *connectivity layer,* which consists mainly of routers and switches that compose the access network and backbone infrastructure. The IP network may also use MPLS and permit interoperation of different access types such as PSTN and other network types.

The second layer is the *control layer.* The control layer consists mainly of control platforms for managing session or call setup, including modification, reconnections, and release. This layer therefore contains the call session control function (CSCF) as its core element. The CSCF is also known as the SIP server. In a nutshell, the IMS is a SIP platform! As an infrastructure of SIP proxies, it is responsible for providing the SIP functionalities and establishing of sessions, negotiation of session descriptions, and teardown of sessions. This layer also contains components that enable provisioning, charging, and operations and management (O&M). Interoperation with other networks and access types is managed in this layer by border gateways. It also handles user sign-on and authentication.

There are three types of CSCF (or SIP proxies) inside the IMS: the P-CSCF, S-CSCF, and I-CSCF [3].

1. The (Proxy) P-CSCF is the IMS contact point for SIP signaling and usually there will be several of them in a domain and also located in the visited domain. Its functions are to secure SIP messages, compress and decompress SIP messages, and ensure the correctness of SIP messages. SIP terminals must know the identity and location of this proxy, and this is achieved through the use of DHCP.
2. The (Serving) S-CSCF controls the user's SIP sessions. There will be at least one or more of these in a domain. It is located in the home domain and serves as the SIP registrar and also as proxy.
3. The (Interrogating) I-CSCF proxy is the domain's contact point for inter-domain SIP signaling. There will be at least one or more of them in a domain. When there is more than one S-CSCF in a domain, the I-CSCF locates which S-CSCF is serving a user.

Figure 13.3 shows how they are located and used for a simple case of one domain without roaming in a 3G setting.

The third layer is the *service* or *application layer.* This layer consists mainly of service platforms or servers to offer and execute value-added services to end users. This layer provides services such as presence, voice over IP, group list management that as SIP application servers. Understand that this concept is uniquely different from the existing legacy service offering concept in which service platforms and services are deployed as islands of services with interactions only made possible through protocols. Under this concept it is not uncommon for each service to be created from scratch and operate independently of others and therefore do not interoperate. What the IMS has offered is the opportunity to reuse some of the

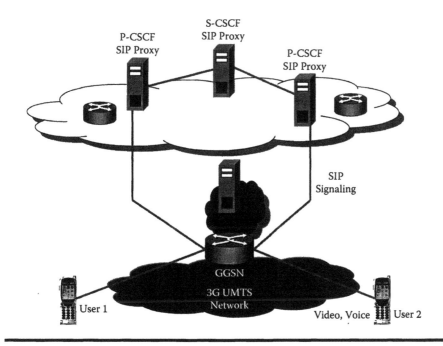

Figure 13.3   SIP signaling path in IMS.

functionalities across different service types, thereby enabling fast service creation. Therefore, the IMS defines through the application layer how the services are composed, requested, routed, the protocols supported, and billing. With this type of definition, a single application server may host multiple services (e.g., VoIP, instant messaging, and presence). By collocating many services, software components can be shared and reused by the services, and the locating and loading of the service nodes can be optimized. This helps reduce the workload of the CSCF of the control layer.

IMS services are primarily intended to address mass markets incorporating telecom-grade quality-of-service. To cater to a large number of users, the IMS also needs to provide support for a complex service mix or different service bundles and offerings that meet specific customer needs. These service bundles most likely have different numbers of users and different user behaviors. This will affect the network dimensioning to provide capacity and functionality. Therefore, the IMS network architecture offers a significant advantage because it is designed to scale independently of the traffic mix. Hence, CSCF capacity can scale in proportion to the number of subscribers, and the number of application servers can also scale in proportion to how the different services are used. "In addition, the amount of inter-domain (for example, VoIP to PSTN) interworking capacity can grow as services that utilize these capabilities are introduced" [1].

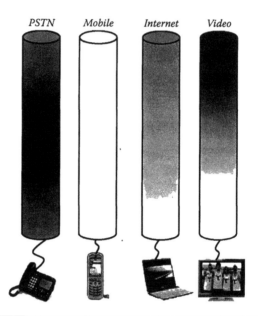

Figure 13.4 Vertical integration service offering. (From H. Mowafy and Z. Zeb, Telcordia, Presentation, October 4, 2006. With permission.)

Vertical versus Horizontal Integration Services

There are two distinct paradigms for provisioning telecommunication services in industry: vertical and horizontal integrated services. These two concepts are related. Traditionally, almost anyone who has used fixed telephony and the Internet has had to use services through vertically integrated service provisioning.

Vertically Integrated Services

Legacy service offerings in traditional networks integrate and offer services in a vertical manner (Figure 13.4). In this structure, each service has its unique functionality for charging, presence, group and list management, routing, and provisioning. There is no reuse of service components and this is very costly, complex, and difficult to build and maintain. In essence, for each service, each layer must be built, and this structure is repeated across the network, the core network, and the user terminals. Thus, it is not uncommon to find that each new service requires new user terminal and new network and core network specifications and definitions.

Horizontal Integration in IMS

The IMS framework replaces this expensive exercise with a horizontal specification that is shared across multiple services. As an extension to the concept of layered

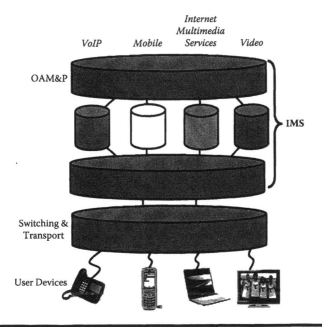

Figure 13.5 Horizontal integration service provisioning with IMS. (From H. Mowafy and Z. Zeb, Telcordia, Presentation, October 4, 2006. With permission.)

architecture, the IMS defines a horizontal architecture that enables service enablers and common functions to be reused across multiple applications (Figure 13.5). The IMS also defines roaming and interoperability including bearer control, charging, and security. The following common functions are defined by the horizontal IMS architecture: provisioning, presence, charging, operations and management, directory, and deployment. These functions are common and shared by all the services.

It is obvious that the IMS simplifies the retention and deployment of expertise and makes it generic rather than specific and unique to services. Hence, it saves or minimizes both CAPEX (capital expenditures) and OPEX (operational expenditures) in areas such as O&M, customer care, provisioning, and billing. Training and retention of skilled staff are facilitated as it focuses on well-known common technology.

Pictorially, the overall paradigm change in service offering is captured in Figure 13.6, which shows a migration from vertical integrated services without component reuse to horizontal integration of services with component reuse.

In the vertical integration, the application logic, common functions, routing, and service discovery are replicated but not reusable, while in the horizontal integration, the common functions, routing, and discovery process are reused, thereby saving costs and facilitating the speed to service creation and ease of maintenance (Figure 13.7).

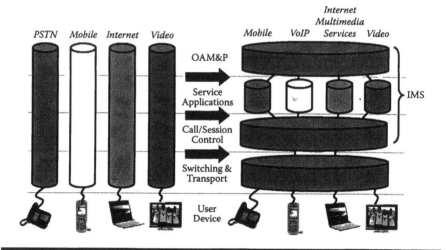

Figure 13.6   PSTN to IMS migration. (From H. Mowafy and Z. Zeb, Telcordia, Presentation, October 4, 2006. With permission.)

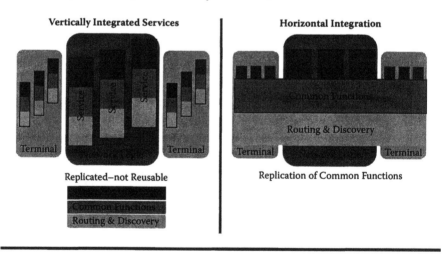

Figure 13.7   Vertically and horizontally integrated services in legacy and IMS services.

Service Enablers in IMS

Let us discuss horizontal integration in IMS further. The IMS makes use of key elements or multiple service enablers. Service enablers are generic and reusable building components for service creation. They are designed to be built once and reused many times across services. Hence, they can be included in new services and applications without modifications. Two of the most important service enablers in the IMS are presence and group list management.

Presence

A presence service enabler allows users to be aware of the availability and means of communication with the other group members. Therefore, users can see when a person is present or connected (active) or has disconnected (inactive). They can also receive alerts when other group members are online or are available. This service enabler is prevalent in instant messaging and social networking systems. "In IMS, presence is sensitive to different media types, users (requestors), and user preferences. IMS presence function is also aware of what terminals the user can be reached on across the various wireline and wireless networks" [1]. The service can define rules that allow users to set who can view their presence information (as in skype online, offline, skype me, away, not available, do not disturb, and invisible present set options). Similar features are used by Yahoo IM and other IM systems.

Group List Management

"The group list management service enabler allows users to create and manage network-based group definitions" for use by services deployed in the network [1]. For example, Yahoo IM allows users to define groups based on a user's perception of those they have in their Yahoo contact list. Other group management examples "include: personal buddy lists; 'block' lists; public/private groups (for example, the easy definition of VPN-oriented service packages); access control lists; public or private chat groups" [1]. They are popularly used by online dating sites, chat rooms, and IM.

Routing and Access to Services

Legacy networks are normally user independent and service centric. Prior to the development of the IMS standard, users accessed personal services through one or more service-specific and "user-independent" access points. This included routing in "service-centric" architecture that is often proprietary, and service-specific scalability issues prevailed. The IMS, however, enables a much more user-focused approach for delivering personal services. This change in paradigm means that with IMS, users access personal services through a dynamically associated, "user-centric," "service-independent," and standardized access point—the CSCF. The CSCF is allocated dynamically to the user when a request addressed to the user is received or at log-on. Routing to the server is standardized and service independent, and the service architecture is "highly scalable" and "user centric."

Access to Services

In non-IMS networks, each service often has its own way of authenticating users, which may be standardized or proprietary. The service may not authenticate the user at all but rather rely on lower-level authentication. The operator may also deploy a special single sign-on (SSO) service in order to avoid reauthentication for multiple services [1].

In horizontally integrated networks, the IMS greatly simplifies the sign-on and authentication process for both operators and users. Authentication in IMS networks is handled by the CSCF as the user signs on. Once a user is authenticated through an IMS service, she is able to access all the other IMS services that she is subscribed to use. On receiving a service request, the SIP application server (AS) can verify that the user has been authenticated [1].

The contact list is one of the key interfaces of the IMS. It contains all the contacts, shows their availability for the services and the terminals to use. Therefore, when the user logs on through the terminal (PC software client or mobile phone), the user's new presence state is automatically updated and can be displayed to his or her group members, provided the user has not set to the present state otherwise.

Service Inter-Operator Relation and Interoperability

In current non-IMS networks, when a user desires to access other users' services, routing to the user's service is very specific to the service, and the host operator of the requestor is involved in the process. Furthermore, service-specific network-to-network interface, routing, service access point, and security need to be available, requiring specific inter-operator service agreements as well (Figure 13.8). This is shown as multiple interconnection lines between the services in Figure 13.8.

A key feature of the IMS that changes this is the reuse of inter-operator relations. Rather than develop different and multiple interconnect relations and agreements for each service, the IMS enables a single inter-operator relationship to be

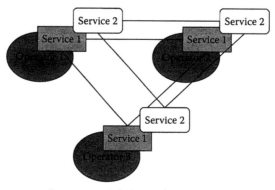

Interconnect Relations and Agreements,
Doubled for Every Service

Figure 13.8   Non-IMS interconnect relations. (From Ericsson, White Paper, 2005. With permission.)

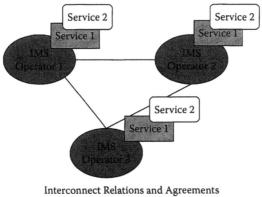

Interconnect Relations and Agreements
Re-used for all IMS Services

Figure 13.9   IMS enables operator relations. (From Ericsson, White Paper, 2006. With permission.)

established and built upon for each service. Once the IMS is in place, access to other users' services is an IMS network issue and this is common to all IMS personal services, as shown in Figure 13.9.

Therefore, the requesting user's operator service setup need not be involved in routing the request. Another advantage is the freedom inherent in not having to establish multiple inter-operator relationships for each service. The single-inter-operator relation in the IMS is shown as a single line linking the operators. The network-to-network relationship between the two operators is therefore established in the IMS, and all elements of the services (such as routing, service network access point, and security) are reused.

Creation of Services in User Terminals

For the user to fully enjoy the benefits of IMS services, the user terminal must contain an IMS/SIP client that includes the service logic, graphical user interface, routing, and service discovery functions. The client is used to communicate with network servers. Thus, the user terminal mirrors the service logic that exists in the network. The IMS/SIP client is structured to reuse the core functions on the same equipment. This reduces the effort needed to deploy new services because components that support their invocation and use already exist in the client in the user terminal. This also allows many applications to be collocated in the same user equipment. The IMS logic in the user terminals also ensures that the IMS architecture is end-to-end.

Interworking with Legacy Networks

Operators do not want to continuously reengineer their legacy networks to support new or existing services that need to work with IMS-enabled networks.

Interworking of IMS-based services with legacy networks is therefore essential. This reduces customer churn, leverages existing investment, and facilitates the uptake of new services. Although the scope and range of interworking between legacy services and IMS-based services will vary and differ according to the services supported in each domain and user terminal, having the ability to do so is a huge incentive for deploying the IMS. This also encourages user quality of experience in the services offered (both old and new). For example, presence in IMS must support interworking between different presence server domains. Therefore, different users can subscribe to the relevant parts of each other's presence services and share them in their contact lists. The IMS also needs to interwork with existing intelligent network (IN) services such as virtual private networks (VPNs). Therefore, existing VPN short numbering can be used by IMS services and allow the authentication server in SIP to interrogate the IN VPN for the full number so as to complete the application.

Enabler of Convergence

One of the major advantages of the IMS is its potential as an enabler for convergence and interworking across fixed and mobile network access—in the control and service layer, the service layer, and the connectivity layer. This standard is also suitable for fixed-to-mobile convergence and is being deployed in that manner by operators.

The application layer in the real sense is oblivious to whether the network is fixed or mobile. That negotiation is done at the control layer. The application and control layer therefore handles both fixed and mobile networks. The common functions and service enablers work in both mobile and fixed networks. Therefore, the presence functions and group list in use either with a mobile phone or PC client are identical for communication so that changes made are reflected in both terminal domains.

The IMS is developed to possess inherent multi-access functionality to be able to execute different services properly. This intelligence is required to understand the requirements for different services for either high bandwidth or low latency or high processing power in the device. The IMS allows the network to be aware of the different characteristics of the access methods. When this is extended with access-aware control and service logic for multimedia services, the IMS offers fixed and mobile operators to finally deliver true fixed-to-mobile convergence. Therefore, the delivered service can be adapted to the characteristics and capabilities of the currently selected device and its network access method.

Device Types versus Networks

Over the previous few years, the size of portable devices has diminished and their capabilities have increased significantly. Thus, with portable devices like "laptops and PDAs, used in conjunction with wireless LANs, the boundary between fixed and mobile communications has become blurred" [1].

The traditional distinction between fixed and mobile calls is that with a mobile call, one calls a person who is free to move with the terminal, whereas with a fixed call one calls a location using the fixed wireline network. The new development using SIP tilts this definition so that with personal SIP addresses, fixed calls can become personal as well, depending on user needs. Progressively, it is becoming clear that the type of device used by the person is becoming more important than the underlying access network and its architecture. This is a result of convergence.

With IMS, operator-controlled services are provided to authenticate subscribers. Hence, secure end-to-end communications services can be built around a number of IMS security and network architectures. The host operator where the communication originates is responsible for the secure communication. For operators to ensure this, "no services are delivered to anonymous or untrustworthy end-users, and no service requests are relayed from anonymous and untrusted operators and enterprises" [1]. The secure communication responsibility is wrapped around the need for IMS authentication, controlled IMS services to authorized users, secure network interconnections, and inter-operator agreements with security responsibilities and mandates. This means that the operator has responsibility to check the transported data for viruses and provides domain-specific access security through SSO and authentication. "Network domain security is provided through site security for hosted solutions, node hardening, virus protection and audit logging. O&M security is supported by management traffic protection and virus protection" [1].

Regulatory Concerns

IP network regulation has not kept pace with the rate of development of telecommunication technology. Hence, it is not uncommon to see the regulation of new technologies such as IP networks being relegated or neglected in many countries. This is due, in part, to the need and attractiveness of providing cheap or free Internet calls for citizens compared to the need to provide lawful intercept and other regulatory functions. When VoIP started to gain ground, there were strong forces in the VoIP community that argued that IP telephony should not be regulated in the same way as the classic telephony network has been. This lack of regulation was matched with operators' understanding of how to charge for bits as opposed to charging for seconds and bandwidth. Since those early years, many countries have undertaken to regulate VoIP and operators, and also better understand how to bill for VoIP services. Right from the beginning, a number of regulatory functions were standardized in the IMS architecture, including lawful intercept. The ability to determine the geographic location of the user for both wireless and wireline networks was scheduled for the next release of the IMS standard.

Evolution Paths to All-IP

The IMS has become the service-driven evolution path to all-IP networks as opposed to hardware- or network-driven evolution. Therefore, it is an attractive means for operators who dream for a future in offering IP multimedia services to start to evolve their networks by deploying the IMS. Technically, as the most prominent SIP-based communication standard, the IMS provides three major advantages—as a standardized, well-structured means of delivering services; as a legacy interworking enabler; and as a fixed-to-mobile convergence enabler. These three advantages position the IMS in front as the technology that simplifies all-IP networking. At the same time, it provides "a future-proof architecture that simplifies and speeds up the service creation and provisioning process" [1].

The vision for an all-IP network is that which enables one core network for multiple access techniques and thus reduces the cost of ownership. That is, different access networks share a common all-IP core network for delivering value-added IP services. There are several evolution paths to the all-IP network.

Wireless Evolution

From a mobile network perspective, the preferred approach is to introduce a common infrastructure, the IMS, for delivering multimedia services and service enablement. The IMS enables VoIP in wireless networks and also push-to-talk over cellular (PoC), and affords the combination of both mobile circuit-switched and packet-switched telephony under a single infrastructure. That is, it is a basis for centralizing the operations of several access techniques, such as GSM (circuit switched), GPRS (circuit and packet switched), and UMTS (packet switched). This enables the delivery of mobile TV and IPTV under a single umbrella technology.

Some of the IP services enabled by the IMS include PoC, which provides feature-rich services for group communication and person-to-person communication with rich settings such as do-not-disturb, transparency and presence management, multi-party conferencing, security, billing, and O&M. It operates entirely in the packet-switched domain.

IMS also enables combinational services for users to interactively share multimedia content in real-time. As such, it enables the sharing of Web content, files, images, and live video feed with the person to whom they are talking. These services can be introduced in parallel with existing voice offerings in an evolutionary manner and with legacy circuit-switched infrastructure as well. Therefore, the traditional telephone call-type of service is enhanced with new value-added multimedia services, thus enriching the circuit-switched telephony market even before the promise of all-IP networking is delivered.

In recent years, the VoIP market has evolved and the uptake of the technology worldwide has increased tremendously. In most countries, VoIP is still offered using legacy voice circuit-switched networks with multiple inter-operator connectivity agreements, thereby introducing major delays in the delivery of voice packets to end users. Effectively, this reduces the user experience and slows the uptake of the technology in many countries. Although VoIP service is capable of providing toll-quality voice most of the time to end users, circuit-switched infrastructure limits its performance. Thus, wireless VoIP offerings with mobile networks and WLANs can become a rich experience with the introduction of IMS.

Wireline Evolution

In wireline communications, there are no roaming problems, essentially no band-width constraints, and the user terminals have adequate processing power. These advantages provide a means for a faster wireline evolution path to an all-IP-based network using IMS than for mobile networks. IMS is the major standard architecture for SIP-based communication in wireline networks adopted by ETSI/TISPAN.

In many countries, wireline operators have huge investments buried under-ground as both copper cables and fiber, and have no interest in losing these investments or adopting completely new wireless technologies as their replace-ment. Therefore, the path through deploying the IMS-based wireline solution looks attractive and provides operators with the opportunity to offer IP telephony, enriched VoIP service, presence management, instant messaging, rich multimedia, and video-conferencing. The IMS will facilitate and reduce the cost of ownership of communication infrastructure and hence the enterprise market will benefit. It allows hosted IP PBX solutions and also a single communication infrastructure for voice, Internet, and data.

QoS in IMS

Key functions of the IMS also include:

- Authentication, Authorization, Accounting (AAA) business and fraud pre-vention functions
- Person-To-Person (P2P), Push-To-Talk (PoC), Instant Messaging (IM), CRBT (Color Ring Back Tones), Conferencing, and Streaming: personal functions (calendar/tones/photo database)
- QoS options: multimedia, video, streaming, content-driven services
- Roaming: business interconnection functions
- Rich Call, also known as MultiMedia Service (MMS): integrated voice/data/video with parallel sharing (live and database) media

Summary

This chapter discussed the IP Multimedia Subsystem (IMS), a generic standard architecture for offering IP multimedia services and also VoIP. As an international standard, the IMS provides the framework and the support for fixed-to-mobile convergence. It aims to put an end to fractured technologies by standardizing the network interfaces. It also provides open network interfaces and open platforms. The IMS provides operators and users with specific advantages. For operators, it provides the basis for a horizontal architecture, thereby taking the concept of layered architecture a step forward. This enables reuse of service enablers and common functions for offering multiple applications. For users, it provides and enables person-to-content and person-to-person communications for voice, text, image, and video content and their hybrids. A three-layered architecture is used to implement an IMS. Two of the most important service enablers in the IMS are *presence* and *group list management.* One of the major advantages of the IMS is its potential as an enabler for convergence and interworking across fixed and mobile network access, in the control and service layer, the service layer, and the connectivity layer. This standard is also suitable for fixed-to-mobile convergence and is being deployed in that manner by operators. The chapter also discussed the evolution path to an all-IP network. The IMS has become the service-driven evolution path to all-IP networks as opposed to hardware- or network-driven evolution. The IMS also supports QoS options, including AAA, P2P, push-to-talk, instant messaging, roaming, and rich calls.

Practice Set: Review Questions

1. The IMS standardizes the:
   a. Network interfaces
   b. Network applications
   c. Multimedia applications
   d. Transport layer
   e. Servers
2. In terms of functionality, what does the IMS offer?
   a. A common means for interoperability and roaming
   b. Better control of bearer
   c. Better control of billing and security
   d. Only (a)
   e. All of (a), (b), and (c)
3. Select the names of the IMS layers.
   a. Connectivity
   b. Control
   c. Service
   d. Conductivity
   e. Cooperation

4. What are CSCFs?
   a. Service control functions
   b. Session control functions
   c. Server control frames
   d. Series control frames
   e. All of the above
5. Vertical and horizontal integrated services are identical with:
   a. Vertical handoffs
   b. Vertical and horizontal handoffs
   c. Only horizontal handoffs
   d. Inter-domain handoffs
   e. None of the above

Exercises

1. List and discuss the roles of the IMS layers.
2. How many CSCFs can be found in a domain? List all the CSCFs and discuss their roles in a typical IMS.
3. What is provisioning, and what are the approaches for provisioning?
4. Define and explain service enablers in IMS. What do they enable?
5. Define and explain what an IP multimedia subsystem (IMS) is.
6. Explain each of the following in detail:
   a. CSCF
   b. P-CSCF
   c. S-CSCF
   d. I-CSCF
7. What are the differences between vertical and horizontal integrated service offerings? What are their limitations and advantages?
8. Discuss presence services enabler.
9. What is a group list management service enabler? Why and when it is necessary to use it?
10. Distinguish among and explain the methods of authentication in IMS and non-IMS networks.
11. List the benefits of IMS in service inter-operator relationships and connectivity.
12. Mention the basic requirements of a user's device for it to be suitable for a user to enjoy the full benefits of an IMS.
13. List reasons why the IMS is considered a service-driven evolution to an all-IP network.
14. What is an all-IP network?
15. List and explain at least five IP services enabled by the IMS.

References

[1] Ericsson, IMS–IP Multimedia Subsystem, White Paper, 2004.

[2] H. Mowafy and Z. Zeb, IP Multimedia Subsystem (IMS) (Industry Status/Expectations/Challenges), Telcordia, Presentation, October 4, 2006.

[3] A. Cuevas, J.I. Moreno, P. Vidales, and H. Einsiedler, The IMS Service Platform: A Solution for Next-Generation Network Operators to Be More than Bit Pipes, *IEEE Communications Magazine,* August 2006, pp. 75–81.

Chapter 14

Real-Time Protocols

Fundamental changes take place by changing the mode of communication from the well-known broadcast technique where the information bearing signal is modulated and sent out into open space and each receiver picks it and demodulates it for display to routing and multicasting. The changes involve not only discretizing the signal, but also assembling the information-carrying bits into packets and assigning headers to packets. Depending on the access techniques, different packets are formed. For example, an IP packet is an IP datagram that carries a TCP datagram. Similarly, an Ethernet packet has an IP datagram inside an Ethernet frame. In general, a datagram is the basic unit of data transfer in an IP environment.

Once the information-carrying bits are assembled into packets, they must be routed to their destinations, and this routing process causes significant delays for processing, in the core and radio networks and when disassembly and reassembly of packets are undertaken. In addition to these delays, packets often take different routes and paths to their destinations and hence arrive late and also out of order. Hence, a real-time process or service at the source is ultimately transformed into a non-real-time data set by the network and seen at the destination as non-real-time. The goal of real-time protocols is to transport, control, and reassemble the information-bearing bits into real-time signals for display at the receiver.

Transporting, controlling, and displaying real-time services are handled by a set of real-time protocols working together to achieve real-time objectives. The real-time protocols include Real-Time Transport Protocol (RTP), Real-Time Control Protocol (RTCP; see Chapter 15), Real-Time Streaming Protocol (RTSP), Resource Reservation Protocol (RSVP), and SIP. These protocols work together and with TCP, UDP, and in some cases SCTP to enable real-time IP services. This chapter discusses only RTP.

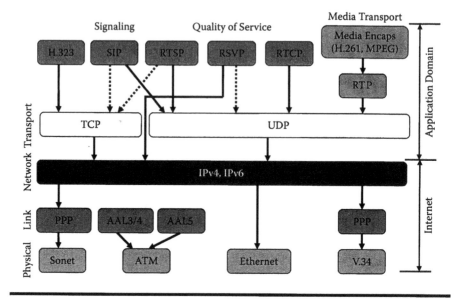

Figure 14.1   Streaming media technology protocols.

Most online music listeners often wonder how the music they download is reconstructed and played back correctly. At all times, reliable downloading is more important than real-time downloading. Usually, one is therefore more interested in receiving the whole music file at the correct rate so that it plays back normally without the annoying stop and start. A real-time transmission ensures that the song is downloaded and listened to in real-time and at the correct playing rate.

Figure 14.1 shows that three signaling protocols can be used: H.323, SIP, and RTSP. H.323 enables legacy PSTN to accommodate streaming media and supports video-conferencing. SIP provides signaling support for IP sessions, and RTSP provides similar support for streaming media. Two protocols are provided for supporting and providing quality-of-service measures: RSVP and RTCP.

RTP: Real-Time Transport Protocol

The RTP is an IETF standard proposed as RFC 1889. The more recent versions are in RFC 3550 and RFC 3551. It acts as an optimal interface between the real-time applications and the transport layer protocols. The RTP does not, however, determine which transport layer protocol is used; it only provides the functions and acts as the enabler for the transport layer protocols to work in real-time environments. It was therefore developed to be scalable and independent of the underlying transport. The scalability feature targets its use for supporting multicast IP. Obviously, therefore, it was designed with an IP packet-switched network in mind. One of the major objectives is for it to provide end-to-end real-time delivery for temporarily sensitive data and support for multicast and unicast delivery. The RTP does not guarantee that the packets will be delivered on time; rather, it guarantees that when they are delivered, they can be reordered correctly for reconstruction and if any packets are missing, they are known.

RTP Functions

The RTP offers the following functions and features:

- *Data packet sequencing to enable correct ordering of packets at the receiver:* Recall that routing de-sequences the IP packets. Hence, by numbering the packets, it is possible to reorder them into the correct sequence for correct display real-time.
- *Identification of data source and type of payload:* This is used by the receiver and intermediate nodes for determining the payload contents.
- *Timing and synchronization:* It is not enough to just order the packets into correct sequence. Many of the real-time services require synchronization. For example, for proper display of a TV signal, the picture and voice must be synchronized.
- *Monitoring of delivery:* For diagnosing, reducing transmission problems, and sending feedback to the sending device of correct delivery and of the quality of data transmission.
- *Integration of traffic sources of different types:* This is used to merge heterogeneous traffic from multiple transmitting sources into a single flow.

The RTP in its functions interacts with different protocols in the transport layer as shown in Figure 14.2. As shown, it does not provide any QoS guarantees and also no data delivery reliability. It only helps, supports, and monitors the control of the flow of information from the transmitter to the receiver. The RTP relies on lower-layer protocols to guarantee reliable delivery and QoS. This is ensured using TCP or UDP.

The RTP is isolated from QoS functions that are undertaken by other protocols but it carries the individual real-time streams with descriptions about their type,

Figure 14.2   Interaction of RTP with other transport layer protocols.

identity, time stamps, and sequencing information. These features or information are used at the receiver in the following manner.

Packets that arrive at their destinations each carry a sequence number and a time stamp. The sequence number is examined and each packet is ordered in its correct sequence position and hence the correct sequence of the data as sent is determined. Lost sequence numbers represent lost packets and hence the percentage of lost packets.

The time stamp is used in the same manner. There are two time stamps: the first time stamp is imprinted by the source and is for when the packet was sent, and the second time stamp is imprinted by the receiver and is for when the packet is received. The difference between the two time stamps gives the information on the length of time this packet has spent in transit from source to receiver and hence on the delay in the network. The inter-packet gaps provide further information on how to regenerate the content at the rate they were encoded and also on jitter. The receiver provides a buffering that the source can use to pace the traffic more correctly and independent of jitter introduced by the network. Synchronization of the source and destination is also facilitated by the time stamps and is necessary to ensure the correct timing of packet events. Source synchronization is also useful for sessions with multiple concurrent types of data (e.g., video sessions and audio-conference sessions). In the case of multiple concurrent sessions, a better approach is to allow each RTP session to carry its unique data content type. On arrival at the receiver, the synchronization fields of the different RTP sessions are examined so as to synchronize audio and video for correct playback. Thus, the RTCP is required for this function, as explained later.

Using RTP

An application that needs to use the RTP creates a session that defines a pair of destination addresses, two ports, and a network address. One of the two ports is assigned to the RTP and the other to the RTCP. The one assigned to the RTCP is used for reporting on the performance and control of the RTP session. The RTP is used in conjunction with other protocols to transport real-time applications over a packet-switched IP network (Figure 14.3). In the example of Figure 14.4, the underlying optical network may use ATM in the data-link layer and IP/UDP in the network layer.

The RTP is used in the transport layer to support audio sessions created from the application layer. Once the audio sessions have been carried by the RTP, the play-back functions are better understood, as in Figure 14.4. In this case, a multimedia (audio) database holds the content that is streamed to the receivers. As the packets arrive at the host receivers, they are sequenced after decoding the payload and the sound contents are reproduced at the receiver. The RTCP provides feedback on the quality of reception to the source. Several types of RTP application can be identified, ranging from audio- and video-conferencing to multicast audio-conferencing.

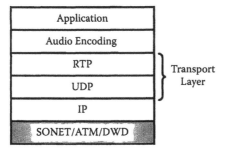

Figure 14.3   Using RTP.

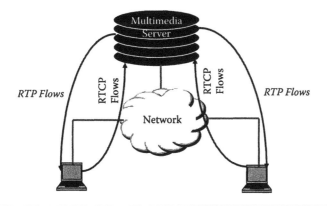

Figure 14.4   Multimedia (video and audio) reception using RTP.

In a simple multicast audio-conference, the initiator of the conference is assumed to have a multicast group address and the pair of ports to use. One port serves as the audio port (RTP) and the other as the control port (RTCP). All the participants have full knowledge of the group and the intended ports to use. They may encrypt both the data and the control packets and for that, too, all group members are aware of the cryptography algorithm to use. For each instance or participant of the audio application in the conference, periodic reports on the quality of reception are issued through the RTCP port. The name of the person issuing the report is also given.

Each instance of the audio application (i.e., each participant) in the conference periodically multicasts a reception report plus the name of its user on the RTCP (control) port. This helps to monitor quality of transmission and also determines who the present participants are.

Mixers and Translators

We have assumed thus far that all the sites use or want to receive the data in the same format. In many instances, users are attached to networks and links of varying

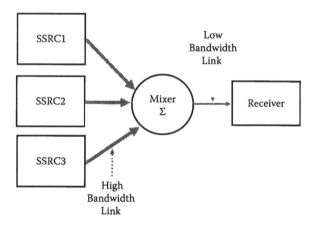

Figure 14.5   RTP mixer.

bandwidth, or they may be working behind firewalls. These users need extra processing and this is done through the use of mixers and translators [3].

When participants use links and networks of lower bandwidth, it might be necessary to degrade the bandwidth offering of better networks so that all of them use a lower bandwidth. This, however, leads to reduced quality due to transcoding or bit rate reduction or modification. The RTP provides a more optimal solution by using an RTP-level relay called a mixer (Figure 14.5). A mixer may be located near the low-bandwidth source or receiver. The mixer resynchronizes incoming audio packets in order to reconstruct the constant speech frames (e.g., 40-ms spacing generated by the sending encoder). It mixes the audio streams into a single stream, transcodes the audio encoding to a lower-bandwidth one (hopefully lossless), and forwards the lower-bandwidth stream across the low-speed link. The mixer may also perform the action of decimating and interpolating images and also composing video streams from several video sources. At the end of the mixing process, the mixer puts its own identification as the source (SSRC) of the packet. It also puts the contributing sources in the CSRC fields.

The Structure of the RTP Packet

The RTP packet has four sections: the IP header, which is used to encapsulate the UDP packet; the UDP header; the RTP header; and the payload.

The payload consists of the actual real-time multimedia data. To this an RTP header is added. For this payload to be transferred over an IP network, it is further encapsulated in a UDP packet and finally as an IP packet with the IP header.

We can view the RTP IP packet structure as a sequence of encapsulations (Figure 14.6). The RTP payload is encapsulated by an RTP Header, the combination is encapsulated in a UDP Header, and finally the three of them are encapsulated

Figure 14.6   RTP IP packet.

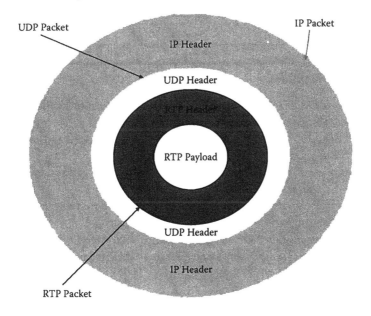

Figure 14.7   Multilayer encapsulation of RTP payload.

by the IP Header. It is like having three layers of envelopes around the RTP Payload so that it can be sent through an IP network (Figure 14.7).

RTP Packet Header

There are two types of RTP packet headers: the uncompressed format and the compressed format. The uncompressed format of the RTP packet header is described and explained in this section and given in Figure 14.8.

■ Two bits are reserved for the Version (V) of the RTP. The Padding (P) is a 1-bit field. When this bit is set to 1, the packet contains one or more additional padding octets at the end that are not part of the payload. The Padding octet contains the count of how many padding octets have been used and they need to be ignored during reconstruction of the data. Padding is used when encryption algorithms are added to the process of constructing the packets, particularly when the encryption uses fixed sizes and there are several RTP packets from a lower-layer protocol datagram.

0 1 2 3 4 5 6 7 8 9 0 1 2 3 4 5 6 7 8 9 0 1 2 3 4 5 6 7 8 9 0 1
V=2	P	X	CC	M	PT	Sequence Number
Time stamp
Synchronization Source (SSRC) Identifier
Contributing Source (CSRC) Identifier
Payload

Figure 14.8 RTP header.

- The extension (X) is a 1-bit field. When set, the fixed header is followed exactly by a one header extension.
- The CSRC count (CC) is a 4-bit field. The count contains the exact number of CSRC identifiers that follow the fixed header.
- The marker (M) is a 1-bit field. It is used by some applications as explained in the application level framing section of this chapter.
- The payload type (PT) is a 7-bit field. These bits are used to identify, for example, the encoding format of the RTP payload.
- The sequence number is a 16-bit field and is incremented by one for each RTP data packet sent. At the receiver, it helps identify the number of packets lost and for restoring the packet sequence. The initial value of the sequence number is selected randomly.
- The timestamp is a 32-bit field. It is an indicator of the sampling instant of the first octet in the RTP packet. It increments monotonically and linearly in practice and helps to compute jitter and allow for synchronization. For example, if the payload is speech sampled at 8 kHz, then at every 125 ms, the time stamp is incremented by one.
- The SSRC is a 32-bit field. It is used to identify the synchronization source that is associated with the data. This identifier is chosen randomly. The number is unique and is chosen to ensure that no two sources within the same RTP session have the same value.
- The CSRC list contains zero (0) to 15 items and each is 32 bits. The CSRC list identifies the contributing sources for the payload in the RTP packet. The header also contains a list of objects that identifies the sources contributing data contained in the packet. The total number of contributors is given in the CC field. If there are more than 15 contributing sources, then only 15 of them may be identified. The CSRC identifiers are normally inserted by mixers using the SSRC identifiers of the contributing sources.

The second type of RTP header is the compressed format. A compressed header is used when only a source is contributing to the payload and there are no ambiguities with respect to the sources or contributors. It is used mostly in non-multicast sessions or unicast sessions. The RTP header is compressed when the first bit of the header or the first bit of the version field is set to zero (0). When the version bit is 0, the header is modified to contain only the fields up to and including the SSRC.

Synchronization

At the receiver, a set of three values is required for synchronization: the synchronization source, ordered packets in order, and the sampling instant of packets that the receiver derives from three header fields [3]. The first value is obtained from the SSRC field. Because the receiver may be receiving data from several sources, it needs to identify the source of each packet.

The second value is the order of the packets. This is obtained from the sequence number. The sequence number increments by one for each RTP data packet sent. The receiver may therefore use them to detect packet losses and also to restore the sequence of received packets. Packets may arrive out of order because they take different paths in the network and are delayed differently on each path.

The third value is obtained from the time stamp. Sometimes, data is not transmitted in the order in which it was generated or created. For example, it is possible that consecutive RTP packets contain time stamps that are not monotonically increasing if they are not transmitted in the order of generation or sampling. This happens in video coding when MPEG video frames are interpolated. Therefore, although the sequence numbers of the packets are still monotonically increasing, the time stamps are not. Hence, the sequence numbers alone are inadequate for synchronization.

For accurate delivery of media data, it is essential to have both the correct order of sequencing and also the sampling instant of the individual packets.

Several consecutive RTP packets may have equal time stamps if they are (logically) generated at once (e.g., belong to the same video frame). Consecutive RTP packets may contain time stamps that are not monotonic if the data is not transmitted in the order in which it was sampled, as in the case of MPEG interpolated video frames. (The sequence numbers of the packets as transmitted will still be monotonic.) So, the sequence number is not enough for synchronization.

Once these values are obtained, reconstruction and synchronization of audio and video for the same scene become possible. The receiver therefore merely performs a matching of the audio data from one channel with the video data from the other channel that corresponds to the same time stamp.

Application Level Framing in RTP

The RTP is intended to be a flexible protocol that provides adequate functionality. This important characteristic of the RTP is referred to as *application level framing*. This characteristic allows each application to specify a profile document that could be completely unique to the application in the way the RTP is used with it. This also means that the application profile could specify or define extensions or modifications to the RTP. This characteristic is also intended to enable users engaged in RTP sessions to agree on a common format for presenting the data, and parts of the header fields can therefore be modified to inform of such manipulations of the RTP. The Marker field (M) of the header is used for this. Its interpretation is a function of the profile. It is intended to facilitate events such as frame boundaries to be shown in the packet stream. To further enhance this feature, a profile could define certain additional marker bits or indicate that there is no marker bit by altering the number of bits in the Payload Type field. Encoding formats are also given by the Payload Type (PT) field of the RTP Packet header. Furthermore, the Extension field may also be set to show that the fixed header is followed by a header extension [3].

Summary

Customers have progressively come to desire real-time services, and the solutions to their desires came in the form of real-time protocols. Using packet switching and the routing of IP packets means that delays are introduced in the radio and core networks and hence a signal acquired in real-time is transformed undesirably in the networks to non-real-time in the receiver. Real-time protocols are used to correct for these delays and the degradation that is introduced during signal processing and IP routing in the network fabric. The goal of real-time protocols is to transport, control, and reassemble the information-bearing bits into real-time signals for display at the receiver. The major real-time protocols are the RTP, RTCP, RTSP, and RSVP. This chapter discussed only the RTP, an IETF standard proposed as RFC 1889. The RTP offers data packet sequencing to enable correct ordering of packets at the receiver because packets can take different routes to their destination. It also provides identification of data source and type of payload, timing and synchronization, monitoring of delivery for diagnosing, reducing transmission problems, and to send feedback to the sending device of correct delivery and of the quality of data transmission. Furthermore, it integrates traffic sources of different types. This is used to merge heterogeneous traffic from multiple transmitting sources into a single flow. Because packets are time stamped at their source and destinations, delays encountered in transit can be estimated and unnecessary jitter can be removed to enable real-time display and enhance QoS. The RTP works hand in hand with RTCP and uses UDP for data transfer.

Practice Set: Review Questions

1. Which of the following are not real-time protocols?
   a. IP
   b. RTP

c. RTCP

d. TCP

e. RSVP

2. RTP operates in the following OSI layer:

a. Session

b. Application

c. Network

d. Transport

e. Data link

3. The RTP acts as:

a. The main interface between non-real-time applications

b. Real-time applications and the transport layer

c. Real-time applications and the network layer

d. Real-time applications and the presentation layer

e. Real-time applications and the application layer

4. The RTP offers at least five functions

a. True

b. False

5. Which one of these are true regarding what you can use RTP packet numbers for?

a. To know the number of hops

b. To for correct sequencing

c. For assessing lost packets

d. For knowing their destinations

e. To know their origin or source

Exercises

1. The main functions of the RTP are to support the transport of real-time service. What are those main functions? Explain them.

2. How do you use the RTP?

3. Discuss the role of mixers in RTP applications.

4. Draw and correctly label the RTP packet.

5. Discuss and explain the fields of the RTP packet.

6. Provide a drawing of the RTP packet header and discuss the fields.

References

[1] IETF RFC 1889.

[2] Introduction to IP with Carrier Extension, Vodafone Australia Training Notes, 2002.

[3] RTP, Geocities, <http://www.geocities.com/intro_to_multimedia/RTP/index.html>.

Chapter 15

RTCP: Real-Time Control Protocol

Real-time data can be transported efficiently by providing a means for reporting the performance of data transfer in the network. The Real Time Control Protocol (RTCP) was proposed for this purpose and is therefore used to report or give feedback on the quality of data distribution. It also is used to provide a persistent transport-level identification for RTP sources. That is, participants in an RTP session are identified with RTCP and may contain more information on participants, such as their e-mail addresses. It provides automatic adjustment of control overhead. Finally, it also provides control information required for maintaining high-quality, efficient real-time sessions.

RTCP

The RTCP is the twin-sister protocol of RTP and is used predominantly for the control of RTP transmissions. Its roles are to expose the states of the client/server to each other so they understand and exchange the parameters of the communication and report on the quality of communication.

The RTCP issues and transmits periodic control packets from participants to all other participants in the session. For that, it uses the same distribution mechanism as for the data packets [2]. Therefore, each one of them is aware of the presence of and the number of participants. The number of participants can be used to calculate the rate at which packets are sent. The greater the number of participants in a session, the less frequently each source can send packets.

RTCP Packets

Two types of RTCP packets are exchanged by the RTCP: the Sender Reports (SRs) and the Receiver Reports (RRs). Fields in the report packets contain descriptions of the state of the session. Compound RTCP packets can be formed by concatenating several report packets and transmitted as one packet. Each SR and RR packet contains the SSRC (synchronization source identifier) of the sender. For RR packets, it contains the SSRC of the first source. The following is a list of the types of packets used by the RTCP:

- SR: Sender report, for transmitting and receiving the statistics from active participants that are senders.
- RR: Receiver report, for receiving the statistics from participants that are not active senders.
- SDES: Source description items, including CNAME.
- BYE: Indicates the end of participation by this participant.
- APP: Application-specific functions.

The SR and RR types of packets contain fields that could be analyzed to determine the percentage of lost packets and hence give an indication of the existing QoS and how to maintain the QoS. The following fields are used to maintain the QoS [2]:

- The fraction of lost packets against the total number of packets sent; this provides or captures how many packets have been lost
- The cumulative number of packets that have been lost since the beginning of the session
- The highest sequence number that was received; this shows the value of the current data in transit that has been received
- Inter-arrival jitter is used to provide how much buffer to provide at the receiver to replay the source efficiently
- Last SR time stamp received to report the time when the last SR was received
- Delay since the last SR time stamp

The structure of the RR that is sent to the sender is given in Figure 15.1.

Determination of QoS Parameters

The RTCP packets obviously report a large amount of information that is highly suitable for determining the QoS but is also subject to varying and different interpretations. For example:

0 1	2	3 4 5 6 7	8 9 0 1 2 3 4 5 6 7	8 9 0 1 2 3 4 5 6 7 8 9 0 1
V=2	P	RR Count	Packet Type	Message Length
SSRC of Sender's Report			
NTP Time stamp (two 32 bit blocks)			
RTP Time stamp			
Sender's Cumulative Packet Count			
Sender's Cumulative Byte Count			
Receiver Report Block 1			
Receiver Report Block 2			
⋮			

Figure 15.1   The structure of the RTCP reception report.

- Throughput can be estimated, on average, by comparing the number of packets and bytes transmitted since the last report.
- The level of packet losses is determined by the fraction of lost packets since the previous RR or SR packet sent or the total number of the RTP data packets lost.
- The round-trip time (RTT) between the source and receiver is given by subtracting the arrival time from the sum of the time stamp of the last report and the delay since that last report. This is useful also for determining the state of the path or route because increased RTT indicates delayed packets and could be a result of longer paths being taken or congestion in the current path.
- The temporal state of the nodes and network can be determined for multicast sessions by analyzing the RTCP packets.
- Inter-packet arrival times or jitter is obtained by determining the difference between the time stamp at the source and the time stamp at the receiver for a sequence of packets. Jitter can be used as an indicator of transient congestion, and packet loss values are an indicator of persistent congestion.
- The number of participants experiencing congestion is also given by comparing lost packet values from different sources.
- In unicast situations, the measured jitter from the differences of the time stamps at the source and receiver for each packet is a reasonable indicator of congestion in the network.

Therefore, the following are some of the simple expressions that can be used to estimate these parameters [1].

Ratio of Lost Packets (η)

The number of packets lost (ρ) is actually the difference between the number of packets expected (μ)and the number of packets received (λ), so that:

$$\rho - \mu - \lambda$$

The number of packets expected by the receiver is defined as the difference between highest sequence number received and the initial sequence number, which is the value:

$$\mu = \theta_b - \theta_i$$

The ratio of lost packets is determined from the ratio of actual lost packets to the number of packets expected by the receiver:

$$\eta = \frac{no.\,of\ packets\ lost}{no.\,of\ expected\ packets} = \frac{\rho}{\mu}$$

$$= \frac{\mu - \lambda}{\mu} = 1 - \frac{\lambda}{\mu}$$

We define the RTCP transmission efficiency (α) as a ratio of the number of packets received and the ratio of packets expected, or:

$$\alpha = \frac{\lambda}{\mu}$$

The larger the transmission efficiency of the protocol (and hence the network), the better the quality of the received signal.

Jitter Introduced by the Network

We can also estimate the amount of jitter that is introduced by the network. This is also estimated in terms of the inter-arrival times or the statistical variation of the RTP data packet inter-arrival times. The instantaneous jitter (τ) is given by the expression:

$$\tau = \left| \left(r_j - r_{j-1} \right) - \left(s_j - s_{j-1} \right) \right|; \quad 0 < j$$

Jitter can be computed in a recursive manner whereby the value of the current jitter is obtained based on the previous estimate of the jitter with the expression:

$$\tau_k = \tau_{k-1} + \frac{\tau - \tau_{k-1}}{16} = old \ jitter + \frac{instantaneous \ jitter - old \ jitter}{16}$$

Round-Trip-Time

The round-trip-time (RTT) is computed using the last sender's report time stamp received and the delay seen since receiving the last sender's report fields in the RR block. This is used to estimate the RTT. Let

$$t_{RR} = the \ time \ a \ source \ receives \ this \ RR$$

$$t_{SR} = the \ last \ SR \ timestamp \ received \ field$$

$$\delta = delay \sin ce \ last \ SR \ report \ field$$

Then, the RTT is:

$$RTT = t_{RR} - t_{SR} - \delta$$

Structure of RTCP Messages

The structure of an RTCP message is shown in Figure 15.2.

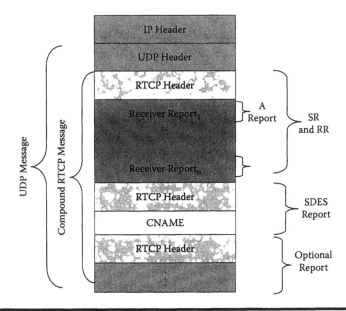

Figure 15.2 **RTCP message encapsulation.**

RTCP Overhead

The RTCP attempts to keep its overhead traffic very low, to about 5 percent of the session bandwidth. Thus, if the sender is sending an IPTV stream at a rate of 4 Mbps, then the RTCP overhead traffic will be limited to a maximum of 200 kbps. RTCP shares this rate among the senders and receivers in the ratio of receivers (75 percent) and senders (25 percent). Therefore in this example, 150 kbps will be reserved for all the receivers and the remaining 50 kbps for the senders. Thus, if there are R = 25 receivers in the multicast session, each one will be allocated 150/R (or 6) kbps for reporting. The greater the number of receivers in an RTCP session, the less the bandwidth reserved for reporting—and thus the number of receivers is critical for sending reports. This can have detrimental effects in commercial real-time services when a large number of subscribers are acting as receivers. Hence, it is a good idea to combine multiple receiver messages before sending them as a single compound RTCP message.

Summary

RTCP also works hand in hand with RTP, as discussed in Chapter 14. Its function is to provide a means for reporting the performance of data transfer in the network. Its roles are to expose the states of the client/server to each other so that they understand and exchange the parameters of the communication and report on the quality of communication. RTCP issues and transmits periodic control packets from participants to all other participants in a session. Two types of RTCP packets are exchanged by the RTCP: the Sender Reports (SRs) and the Receiver Reports (RRs). Fields in the report packets contain descriptions of the state of the session. Compound RTCP packets can be formed by concatenating several report packets and transmitting as one packet. Each SR and RR packet contains the SSRC of the sender. For the RR packet, it contains the SSRC of the first source. QoS can be determined using the parameters in the reports provided by the RTCP. The RTCP reports on the number of packets sent and received since the last report (throughput) and hence the number of lost packets. Because packets are time stamped, the RTCP also provides the round-trip delay, the state of the paths, and of course, the jitter associated with the packets. This chapter has also provided expressions for how to determine these QoS variables from the RTCP reports.

Practice Set: Review Questions

1. The RTCP and RTP have nothing in common.
   a. True
   b. False

2. The role of the RTCP is to provide a means for reporting the transport of real-time data. Which of the following are true?
   a. Expose the states of clients to each other.
   b. Expose the roles of servers to each other.
   c. Expose the states of the client/server to each other.
   d. Expose the state of the link.
   e. Expose the locations of the servers to each other.
3. The role of the RTCP is performed using what?
   a. SIP messages
   b. Issuing and transmitting periodic control packets
   c. Issuing and transmitting TCP packets
   d. Issuing and transmitting periodic frame relay packets
   e. Issuing and transmitting TCP frame packets
4. RTCP is always agnostic to the quality of data distribution.
   a. False
   b. True
5. The RTCP packets are of how many types?
   a. One
   b. Two
   c. Three
   d. Four
   e. Five
6. What can you use the RTCP sender and receivers' reports for?
   a. Percentage of lost packets
   b. To maintain QoS
   c. To assess the level of QoS
   d. All of the above
   e. None of the above

Exercises

1. Discuss the types of RTCP packets.
2. What fields in the RTCP sender and receivers' reports enable the assessment of QoS?
3. Elaborate on how QoS parameters are determined through the use of RTCP reports.
4. A real-time device expected to receive 2000 packets per second but received 1500, of which 500 have identical packet numbers. What is the packet error rate? Explain your answer. What is the efficiency of the RTCP system?
5. What is the ratio of lost packets in Exercise 15.10?
6. Explain how you can use the RTCP report to compute the round-trip time.

References

[1] K. Jeffay, The Multimedia Control Protocol RTCP, Lecture Notes, University of North Carolina at Chapel Hill.

[2] Introduction to IP with Carrier Extension, Vodafone Australia, Training Notes, 2002.

[3] RTP, Geocities, <http://www.geocities.com/intro_to_multimedia/RTP/index.html>.

Chapter 16

RTSP: Real-Time Streaming Protocol

The RTP and RTCP provide mechanisms for transporting real-time traffic through the IP network and reporting on the reliability and quality of transmission, respectively. They, however, do not provide the means for requesting real-time content.

Real-Time Streaming Protocol

The Real Time Streaming Protocol (RTSP) RFC 2326 has been developed to work in the application layer of the OSI to supply the mechanism for making requests for delivery of real-time content. Unfortunately, the RTSP is the least well known of all the real-time protocols and yet its role is by no means trivial. The role of the RTSP is to provide the means for notifying the network of the bandwidth for the upcoming transmission and the means for requesting content from the multimedia server.

Real-time content can exist in three distinct forms, and the RTSP has the capability to request all of them for transmission:

1. Real-time live content: Real-time live content supplied, for example, from television or radio stations. The content is converted into digital form and streamed to the user live as it is produced. Any delays inherent in the process are a result of local conversions of the data or editing (as in broadcasting).
2. Real-time interactive media clip: Includes video-conferencing stream contents created by each participant, pre-recorded multimedia content, digitally stored on a server or database.

3. Non-real-time stored clip: In its traditional form, it includes content that is typically transmitted via MIME (e-mail) or HTTP (the Web).

The RTSP is built and operates around the client/server approach. The RTSP client generates requests or messages that are transmitted to the content server and in doing so provides the signaling functions and control protocol to access content in a random manner. It does not by itself carry the real-time content. In this aspect, its name is a misnomer because it tends to suggest to the engineer that, in fact, it is responsible for streaming the content. In some sense, it has that responsibility without carrying the actual content. It is purely an information bearer with the intent that real-time content is required and how to supply it. The function of carrying the content to the receiver is reserved for RTP. The RTSP provides the following functionalities:

1. A means of requesting real-time content
2. A means of control of content playback (start, stop, pause)
3. A means of random access to the content and for starting the playback at any point in the content stream (which means it can request a particular track on a CD or a playback from a specific frame or point in a digital video stream)
4. A means of requesting specific information about the content that allows selection of the transport layer protocol that matches the content

The RTSP uses TCP for exchanging its control messages with the content server or between the client and the server. Its choice of TCP instead of UDP is apparent. TCP permits retransmission in case of errors while UDP does not. The RTSP is designed to operate in both unicast and multicast modes and environments.

Operation of RTSP

In terms of syntax and operation, the RTSP is similar to HTTP/1.1. They are, however, different in many respects. For the RTSP, both client and server can make requests when they are interacting. In HTTP, it is the client that always makes the requests for documents. Furthermore, the RTSP always maintains a state. This is important for streaming multimedia. HTTP, on the other hand, is stateless. HTTP in the real sense is unable to retain the memory of the identity of each client that connects to a Web site. This means that it treats every request as unique and independent, irrespective of previous connections from the same client. Of course, we understand that Web browsers now fulfill these roles for HTTP but it needs to be understood that it is not the protocol itself that does that.

Streaming is playing back a multimedia content while it is still downloading. Therefore, there is no need to store it completely at the playing device first or *a priori*. Thus, it is not necessary to wait until the entire file is received before hearing or seeing the content. The RTSP plays the central role of supporting streaming media and framework. The client first registers with the server. The real-time server therefore knows it and responds to requests or messages from the client. After it receives the RTSP request for transmission of content, it uses the RTP for transporting the content to the client and the RTCP for reporting on the quality of transmission and reception. In many practical situations, the content contains voice, video, and data that will necessitate the use of multiple streams between the client and server. Each stream carries a unique type of content (voice only or video only) to the client. The RTSP functions by exchanging signaling messages between the client and server using a request/response mode. Three types of control messages are used:

1. Connection control is used to establish, maintain, and terminate individual data streams.
2. Global control is used to control all sessions between the server and client.
3. Custom control is used to report on the exchange of messages, the so-called exception messaging.

Global Control Messages

Global control messages form the first set of messages that must be exchanged between the client and the server. The following global control messages are available from the RTSP:

HELLO message is a request for registration with the server.
IDENTIFY message is the server's request for the client to identify itself for authentication.
IDENTIFY REPLY is the client's reply for the request for authentication.
REDIRECT is a response from the server pointing the client to other content servers.
OPTIONS message is used to indicate miscellaneous functionalities.
GOODBYE message is a request to terminate the global session.

The HELLO message is used to begin the registration process. This message is issued by the client to the server. The server's reply is a similar HELLO message, and the body of the message has the server's version number of the RTSP. This version number guides the client and server to choose which one has the lower version number of the RTSP between them, and they use the lower one for communication with the assumption that the entity with the higher version number is backward compatible with the lower version. If the server does not have the contents required by the client and if it knows the location of the content on another server, the server REDIRECTS the client with a REDIRECT message, which also contains the identity (URL) of the new location for the client to go and access the content.

Next, either the server or the client will request authentication by generating identification that requests the other party to authenticate themselves or prove their identity. The method of authentication is known by the client and the server and may be a plain password or a more complex challenge/response mechanism. If the authentication process fails, the TCP connection is closed immediately.

A GOODBYE message is used to terminate a global session, and clients use it to close their TCP session gracefully. The OPTIONS message is provided for use in improving the RTSP with time—such as improved functionality or new versions.

The RTSP URL is of the form:

```
rtsp://media.example.com:554/brenda
```

A track within Brenda's CD may be requested for playback as:

```
rtsp://media.example.com:554/brenda/nonono
```

Content Request

The RTSP content request comes after the global messages have been exchanged and the client/server have authenticated themselves. Next is the content request. The following methods are available for requesting content:

FETCH
PLAY RANGE
STREAM HEADER
SET SPEED
SET TRANSPORT
SET BLOCK SIZE
STREAM SYNC
SEND REPORT
STOP
RESUME
RESEND

Request

The process for requesting media is straightforward because the client specifies the object of its interest and what to use during transmission. For example:

```
la request (client _ server or server _ client)
```

To be specific,

```
PLAY rtsp://video.example.com/Yvonne/video RTSP/1.0
```

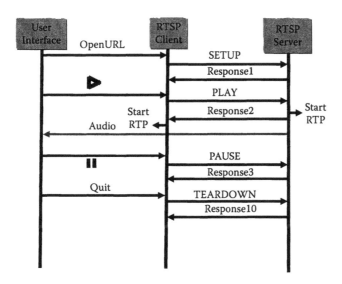

Figure 16.1   Client/server signaling sequence.

Here, the method to use is "PLAY," then the URL. The RTSP version in this case is 1.

```
CSeq: 2 (sequence number for request/response pair)
Session: 12345678 (session identifier)
Range: smpte=0:15:00- (play from this offset for an
        undefined duration)
```

The response to the request is:

```
RTSP /1.0 200 OK
CSeq: 2 (sequence number for request/response pair)
Session: 12345678 (session identifier)
Range: smpte=0:15:00-0:30:00 (play from this offset for
        the duration shown)
RTP-Info:url=rtsp://video.example.com/Yvonne/video;
Seq=13314232; rtptime=797412821
```

RTSP Methods

The following are the RTSP methods and their descriptions, and the sequence of client/server behavior and signaling is shown in Figure 16.1.

Method	Description
ANNOUNCE	Posts the description of the stream

Method	Description
DESCRIBE	Retrieval of the media description
GET_PARAMETER	Retrieval of the value of a parameter
OPTIONS	Querying of the available methods
PLAY	Starts sending data
RECORD	Starts receiving data
PAUSE	Temporal halting of delivery of streams
SETUP	Specifies the transport mechanism
REDIRECT	Informs to connect another server location
SET_PARAMETER	Request to set the value of a parameter
TEARDOWN	Pulls down the stream delivery and frees resources

The RTSP has the following three states—INIT, READY, PLAYING—as shown in Figure 16.2.

- SETUP: the server allocates resources for a client/server session.
- PLAY: the server delivers a stream of media to a client session.
- PAUSE: the server suspends delivery of a stream to the client.
- TEARDOWN: the server pulls down the connection and releases the resources allocated for the session.

The time the session remains in these states is both a function of the implementation and the extent of the network. The SETUP delay is a function of the time the server takes to respond to the request by the client and sets up the resources for the session. The PLAY state is also a function of the duration of the stream and how long it takes to deliver the signaling to make the receiver ready to start playing

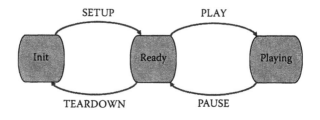

Figure 16.2   RTSP states.

the stream of media. With respect to the RTSP, the following parameters define the QoS:

■ Delay: The time that elapses between when a request is first submitted to when the desired result is produced.
■ Throughput: The total amount of packets delivered during a specific time interval.
■ Jitter: The variable delays that occur during playback of a stream.
■ Reliability: How errors are handled during transmission and processing of the media.

Summary

The Real Time Streaming Protocol provides a means for requesting data or packets. This service is conspicuously absent in both the RTP and RTCP. The RTSP developed to work in the application layer of the OSI to supply the mechanism for making requests for delivery of real-time content. Real-time content can exist in three distinct forms: real-time live content, real-time interactive media clips, and non-real-time stored clips. The RTSP uses TCP for exchanging its control messages with the content server or between the client and the server. Its choice of TCP instead of UDP is apparent. TCP permits retransmission in case of errors while UDP does not. The RTSP is designed to operate in both unicast and multicast modes and environments. It supports all the major functions of a plug-and-play control (such as fast forward, rewind, pause, etc.).

Practice Set: Review Questions

1. The RTSP operates in which layer of the OSI?
   a. Service
   b. Transport
   c. Application
   d. Data link
   e. Network
2. The role of RTSP is:
   a. Service support
   b. To support the transport layer
   c. To stream real-time applications
   d. To serve as an interface to the RTCP
   e. To serve as an interface to the RTP
3. The RTSP provides the mechanism for:
   a. Requesting real-time content

    b. Reporting on real-time content

    c. Adding real-time content into servers

    d. Loading real-time content into devices

    e. All of the above

4. The RTSP methods for playback does not include:

    a. Pause

    b. Options

    c. Fast forward

    d. Record

    e. Setup

5. Which one of the following are RTSP control messages?

    a. Connection

    b. Global

    c. Connect

    d. Custom

    e. Disconnect

Exercises

1. Discuss the various types of real-time content.
2. List and discuss all the RTSP methods.
3. Name the three types of RTSP control messages. Discuss each one in detail.
4. How does an RTSP device request real-time content?

References

[1] <http://www.javvin.com/protocolRTSP.html>.

[2] <http://www.cisco.com/en/US/docs/ios/12_0t/12_0t7/feature/guide/fw_rtsp.html>;
<http://www.cisco.com/univercd/cc/td/doc/product/software/ios120/120newft/
120t/120t7/fw_rtsp.htm>.

Chapter 17

VoIP: Voice over Internet Protocol

Globally, telephony infrastructure is rapidly evolving from circuit switching to IP-based packet switching. Therefore, no link is dedicated to a voice call. This is a huge advantage and frees organizations from the costs of leased lines that are wholly dedicated to voice calls and another for data. Voice over Internet Protocol (VoIP) is therefore becoming more common and popular. IP telephony thus is emerging as one of the fastest-growing IP network services.

How to transmit voice over a packet network was developed decades ago. Voice is typically sampled at 8 bits per sample (8 kHz), 8000 samples per second (64 kbps), and transmitted in pulse code modulation (PCM) format. Therefore, personal computers and laptops can be used routinely for playing back voice and hence telephony.

Fundamentals of VoIP Technology

VoIP is a digital packet-based technique for sending real-time, full-duplex voice communication over the Internet or intranet. It is also known by several names, including Internet telephony, digital telephony, and IP telephony.

The objectives for developing VoIP are to provide the efficiency of a packet-switched network and at the same time rival the voice quality of a circuit-switched network (toll quality). Achieving toll-quality voice means the voice of the person on the other end of the call can be clearly recognized as his or hers. In general, the intention is to provide real-time voice calls through VoIP. Unfortunately, voice signals are intolerant to lengthy delays, jitter, packet losses, and packets-out-of-order at the receiving end. These unwanted signal processing and recombination problems conspire to gravely degrade the quality of the voice signal at the receiver. These problems are further exacerbated in wireless networks because of problems in the links, higher packet dropouts or error rates, greater latency or delays, and more jitter.

VoIP can be implemented in many ways. A PSTN can be used for VoIP communications with the Internet in between the two telephones. VoIP can also be provided without using the PSTN and this chapter examines those other modalities for providing voice over IP.

VoIP is primarily a software feature. In VoIP, the digital signal processor (DSP) segments the voice signal into frames, processes them, and stores them in voice packets. In the original version, these voice packets were transported using IP in compliance with the ITU-T specification H.323, the specification for transmitting multimedia (voice, video, and data) across a network. Voice is a delay-sensitive application; therefore, a well-engineered end-to-end network is required to successfully use VoIP. Traffic shaping considerations must also be taken into account to ensure the reliability of the voice connection.

Benefits

There are several factors influencing the adoption of VoIP technology. The major ones include [1]:

1. *Integrated infrastructure:* Corporations and individuals are interested in eliminating the cost of two phone lines, one for data and the other for voice. VoIP allows simultaneous Internet access and voice traffic over a single phone line, which potentially eliminates the need for two phone lines.

2. *Cost savings:* VoIP permits more efficient use of bandwidth and fewer long-distance trunks between switches. Therefore, billing is no longer dependent on distance (as in STD calls) but on the amount of data carried. The cost of a packet-switched network for VoIP could be as much as half that of a traditional circuit-switched network (such as the PSTN) for voice transmission. For Third-World countries without an existing or efficient telephone infrastructure, VoIP is a saving grace to reduce the unnecessarily huge bills in international calls and to reduce the costs to less than what is required for a local call.

3. *Improved network utilization:* When voice calls are made in traditional circuit-switched networks, a full-duplex 64-kbps channel is dedicated for each of the calls for the duration of the call. However, with VoIP networks, the bandwidth is used only on demand when something must be transmitted. Therefore, by using the bandwidth more efficiently, more calls are carried over a single link without requiring the installation of new lines to further augment network capacity. In fact, a single line to the home can be used simultaneously to make a voice call, access the Internet, and send a fax.

4. *Opportunities for value-added services:* VoIP offers other features, including caller ID and call forwarding, that can be added to VoIP networks at minimal extra cost.

5. *Progressive deployment:* IP telephony is added onto existing communications networks because it can be easily integrated with existing PSTN infrastructure and networks.

6. *Mobility:* Free Internet telephony generally applies to the case where a call is made from one computer to another computer (basic Skype). This requires both parties to be simultaneously logged on to the Internet and running compatible IP telephony applications. The fact that it requires both parties to be present is an inconvenience. The real Internet telephony exploits the use of gateways. A gateway is a node that interconnects data networks of different types (PSTN, LAN, Internet, WAN). A gateway performs lossless and seamless transcoding of data from one coding format to another and the conversion from a packet data stream that is digital to an analog voice signal, and vice versa. Therefore, the gateway allows users to call each other using IP from analog telephones.

Coding

Voice coding is a well-known technology and voice over IP implementations are based on established and standardized codecs (coder/decoder). This section provides an overview of the fundamentals of voice coding with specific emphasis on voice sampling, compression, and packetization. The characteristics of several codecs used for VoIP implementations are given.

Figure 17.1 provides the building blocks of a typical VoIP implementation. The sender's side has a microphone, an analog-to-digital converter (ADC), and a speech compression algorithm. Analog voice is traditionally sampled at 8 kHz, with each sample represented as 8 bits. Therefore, there are 64,000 bits being produced by the ADC per second. This is the linear PCM rate, an uncompressed audio format or compressed u-law (G.711). Traditionally, the acquired voice bits are then

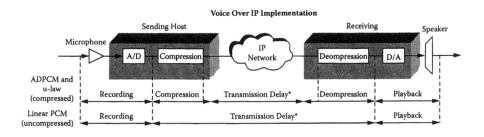

Figure 17.1 VoIP implementation details. (From S. Zeadally, F. Siddiqui, and P. Kubher, *IEE Proc.-Commun.*, 151, 3, June 2004, 263–269. With permission.)

Table 17.1   End-to-End Delay in PCM and ADPCM Codecs

Delay Component	Linear PCM (ms)	ADPCM (ms)	u-law (ms)
Recording	12.5	12.5	12.5
Compression	0 (0%)	0.2 (1.2%)	12.4 (42.5%)
Transmission	3.99	4.04	4.05
Decompression	0 (0%)	0.13 (0.76%)	0.08 (0.3%)
Playback	0.13	0.13	0.13
End-to-end Delay	16.62	17.0	29.6

transmitted in UDP packets over the network. At the receiver, speech packets are decompressed (in the case of compressed samples) and put in the play-out buffer of a sound reproduction system (e.g., a sound card), which then plays back the decoded speech either to a headset or a loudspeaker.

Over the years, to reduce bandwidth utilization for transmission of speech, many speech coders have been developed. One of the oldest ones is the adaptive differential PCM (ADPCM, G.726) that compresses speech from 64 kbps to 32 kbps. Table 17.1 lists the expected end-to-end delay when compressing speech using the ADPCM.

The values in parentheses in Table 17.1 are the percentages of the delay overhead of the total end-to-end delay due to compression and decompression. The transmission delay in this table includes the delay involved in transferring the packet between the voice application and the physical network (at sender and receiver) and the network delay. If the speech is sampled at 8 kHz, for real-time, it is expected that the time available for the transmission of an 8-bit sample on a PCM channel (64 kbps) is 125 μs. The time required to place the same 8 bits in an OC-3 channel (155 Mbps) is 0.05 μs.

Codecs

The ADPCM speech coder is an ITU initiative. It is used in PSTN. Since its creation, other faster speech coders have been developed. AT&T spearheaded the development of code-excited linear predictor (CELP) coders and various variants exist. G.728, also known as low-delay CELP (LD-CELP), provides 16-kbps toll-quality speech compression. This is equivalent to representing one speech sample with 4 bits, although its implementation does not translate to that directly.

There are three main types of codecs which are waveform codecs, source coders, and hybrid coders. PCM and ADPCM are traditionally "waveform" codecs. Waveform codecs use the redundant parts of voice waveforms to compress it. Source codecs compress speech by extracting and sending only simplified

Table 17.2   Bit Rates and MOS of Typical VoIP Codecs

Codec	Bit Rate (kbps)	Frame Length (ms)	MOS	Compression Delay (ms)
G.711 (PCM)	64	1	4.1	0.75
G.723.1 (MP-MLQ	6.3	30 (240)	3.9	30
G.723.1 (ACELP)	5.3	30 (240)	3.65	30
G.726 (G.721, ADPCM)	32	0.125 (1)	3.85	1
G.728 (LD-CELP)	16	30 (240)	3.61	3–5
G.729 (CS-ACELP)	8	10 (80)	3.92	10
G.729A (CS-CELP)	8	10 (80)	3.7	10
G.729 x 2 Encodings	8	10 (80)	3.27	10
G.729 x 3 Encodings	8	10 (80)	2.68	10

parametric information about voice transmission and traditionally lead to better compression or bandwidth. Source codecs include linear predictive coding (LPC), code-excited linear prediction (CELP), conjugate structure-code-excited linear prediction (CS-CELP), conjugate structure-algebraic code-excited linear prediction (CS-ACELP), and multi-pulse-multilevel quantization (MP-MLQ). Hybrid coders include CELP, VSELP, and RPE.

Table 17.2 is a summary of the bit rates and mean opinion scores (MOS) achievable with the most popular VoIP codecs. The frame length is the number of speech samples compressed together at once and is shown in parentheses in Table 17.2. In Table 17.2, tandem (multiple) encodings rather than enhancing speech quality truly degrades it. Therefore, while it is useful to compress speech, it is destructive to speech quality to use tandem encoding. This is also true for tandem transcoding.

The ITU has standardized speech coders in its G-series of recommendations. For example, G.711 is the 64-kbps PCM voice coding technique. In G.711, encoded voice is already in the correct format for digital voice delivery for fixed telephones. G.729 describes the CELP. A code from a code table is used for exciting or as input to a linear predictor algorithm. G.729 CELP compresses voice into 8 kbps streams. There are two variants of this standard (G.729 and G.729 Annex A). They differ mainly in computational complexity and both provide speech quality similar to ADPCM at 32 kbps.

Mean Opinion Score (MOS) Rating of Speech Coders

There are several approaches for evaluating the performance of speech coders in terms of how close their compressed outputs are to toll-quality speech. The term "toll-quality speech" is used to refer to the quality of speech received from a fixed-line telephone handset. Usually, it describes the fact that the output of the telephone handset can be recognized as belonging to the speaker without confusion. It indicates the extent to which the voice of the speaker remains unchanged at the receiving end of the telephone line. Each codec is therefore rated in terms of how close it represents the speaker's voice or how much it has altered the naturally sounding voice of the person. The most accepted method of measuring the quality of speech from speech coders is obtained from a subjective response of the listener. To estimate the MOS, a wide range of listeners are used to judge the quality of compressed speech from the coder and they are asked to rate it on a scale of 1 (bad), 2 (poor), 3 (fair), 4 (good), and 5 (excellent). The scores are averaged to provide the MOS for that codec. To make the results acceptable and believable, the test room of choice is normally an anechoic chamber, and various types of speech files are played back to a listener who then rates them. Speech files are acquired under different noisy and noise-free scenarios.

Delay

One of the most important design considerations in implementing voice is minimizing one-way, end-to-end delay. Voice traffic is real-time traffic; if there is too long a delay in voice packet delivery, speech will be unrecognizable. Delay is inherent in voice networking and is caused by a number of different factors. An acceptable delay is less than 200 ms.

There are basically two kinds of delay inherent in today's telephony networks: propagation delay and processing or handling delay. Propagation delay is caused by the characteristics of the speed of light traveling via a copper-based medium or through fiber-optic. Handling delay is caused by the devices that handle voice information. Handling delays have a significant impact on voice quality in a packet network.

Packetization delay is another processing delay. It is the time it takes to generate a voice packet. In VoIP, the DSP generates frames every 10 ms. Two frames are then placed within one voice packet, thus making the packet delay 20 ms.

Most receivers will choose to use either the G.723 or G.729A codecs because their maximum bit rate is much lower than that of G.711. The lower bit-rate codecs perform better when errors are present and bandwidth is limited.

Signaling

Although the Signaling System Number 7 (SS7) protocol is popular in fixed-line telephony, the popular signaling protocol for VoIP is H.323. H.323 is an umbrella of protocol recommendations from the ITU that sets standards for multimedia communications over IP-based networks that do not provide a guaranteed QoS (e.g., the Internet). H.323 uses the RTP as a foundation to counteract the effects of network latency and lack of guaranteed QoS. H.323 uses the RTP as an application layer protocol to provide a degree of QoS over IP-based networks.

The SIP developed by the IETF has become the *de facto* signaling protocol for IP services [9]. It has also been adopted by the 3GPP as the signaling protocol for future generations of 3G and 4G networks. SIP has gained ground as the signaling protocol for Internet telephony. It specifies user location (i.e., directory lookup), handshake to set up a session, and how the invitee responds to an invite message.

When VoIP is used with the SIP, IP telephony gives users a media-agnostic' signal that various devices can use to set up sessions and negotiate what capabilities those sessions will have, whether it's video-conferencing or application sharing. What device the other party is using really does not matter.

Challenges

IP telephony faces many technical challenges, including packet loss, delay, and jitter. This section looks at these technical challenges and some of the solutions for IP telephony.

Packet Loss

All packet-switching networks, including IP networks, exhibit packet loss to varying degrees. Unlike circuit-switched networks (e.g., PSTN), no end-to-end physical circuits are established in IP networks. IP packets from many sources are queued for transmission over an outgoing link in a router and are transmitted one by one from the head of the queue.

Once the queue is full, an arriving packet is lost in the network because there is no space left in the queue. As more and more people use the Internet, routers often become congested, resulting in packet loss. Packet loss can cause severe damage to voice quality for IP telephony. Each IP packet contains a 40- to 80-ms frame of speech information (depending on the type of codec in use). The frame sizes match the duration of critical units of speech (called phonemes).

Therefore, when a packet is lost, a phoneme is lost in the continuous speech. Normally, the human brain is capable of reconstructing a few lost phonemes in speech. However, a large packet loss makes a voice unintelligible.

There are several approaches for dealing with packet losses in IP telephony and they all rely on the objective of trying to reduce packet losses and/or repairing the damage caused by packet losses—provided the loss is manageable.

Upgrading the IP Network

Packet loss in an IP router is a direct result of insufficient link bandwidth and/or the speed with which the router processes packets. Therefore, the first approach is to deal with the capabilities of the routers (network upgrading). Upgrading the IP network infrastructure, the links, and the routers is a direct engineering solution to the packet loss problem. High-speed transmission technologies such as ATM for megabit-per-second, Synchronous Optical Network (SONET) for gigabit-per-second, and wavelength division multiplexing (WDM) for terabit-per-second line speeds are some of the candidates to consider in upgrading the transmission capacity of the IP backbone lines and routers. In recent years, new technologies in router design have provided the means for complementing the phenomenal increase in link capacity. High-speed, switching-based router technologies have emerged with the capability to process packets on the order of millions of packets per second.

Network upgrade is a long-term viable solution at very high cost to the operator. It will normally help reduce packet losses but will not solve the problem forever. Less costly technical methods focus on repairing the damage done by packet loss to voice quality. Several techniques are used, including silence substitution, noise substitution, packet repetition, packet interpolation, frame interleaving, and forward error correction [2].

Silence Substitution

When a packet is lost in a network, the content of the packet cannot be played out and appears as a silence interval in the playback device. On arrival at their destinations, the contents of the packets are played back to reconstruct the original voice. Some VoIP solutions substitute silence (comfort silence) in place of a missing packet. This allows the playback device at the destination to continue to play the voice without any disruption. Experience has shown that silence substitution causes voice clipping, which deteriorates the quality of the voice significantly. Therefore, the toll quality of the received voice will decrease. This is particularly true for high loss rates and large packets. For speech sampled at 8 kHz, studies on silence substitution revealed that it achieves adequate performance only for packet sizes smaller than 16 ms (128 samples) at loss rates up to 1 percent [3].

Noise Substitution

A better approach than silence substitution is noise substitution. White background noise is substituted for the lost packets. The better performance is due to the ability of the human brain to repair a received message if there is some background noise—a technique known as phonemic restoration [3, 4].

Repetition of Packets

For this method, the last correctly received packet is replayed in place of the lost packet. In a GSM system, to ensure improved quality, it is recommended that the repeated signal be damped or faded.

Interpolation of Packets

To account for a lost packet, the interpolation-based method uses either the pitch of the voice signal or the time scale on samples of speech in the neighborhood of the lost packet to repair and produce new speech sample as replacement. This ensures that the replacement follows the changing characteristics of the whole voice stream. This method is found to outperform both silence substitution and packet repetition [5].

Frame Interleaving

Interleaving of voice frames across different packets can reduce the effects of packet loss. It involves rearranging the original frames so that previously consecutive frames are separated at transmission. At the destination, the received packets are rearranged back to their original sequence. The loss of a packet therefore will not appear to be concentrated at one spot but is spread across several frames. Therefore, the loss of a packet after interleaving only results in multiple short gaps in different areas of the received data and the receiver is therefore better equipped to tolerate the short gaps rather than long gaps of lost packets. Unfortunately, frame interleaving has the disadvantage of increasing delay because packets must be reassembled as a result of destroying their original consecutive order. The objective should therefore be to perform frame interleaving within the delay budget of the system.

Forward Error Correction

Redundancy in transmission can result in a reduced effect of packet loss. An approach for doing this uses forward error correction (FEC) so that packets are redundantly transmitted. Therefore, if an original packet is lost, it can be reconstructed from subsequent packets. The redundancy may either be independent of the data stream, or use the stream characteristics to enhance the repair process. Because the RTP is the protocol used to support IP telephony over the Internet, it is appropriate to have mechanisms within the RTP to carry redundant voice packets. An RTP payload format to carry redundant voice packets is discussed in [2, 6].

Delays

Apart from packet loss, delay is perhaps the worst enemy of VoIP systems. Data transmission and voice transmission in a packet-based network have significant differences in requirements. Data is delay tolerant but loss sensitive. Voice, on the other hand, is delay sensitive and loss tolerant. For these reasons, it is easier to account for packet losses for VoIP, as seen above. Therefore, VoIP systems in the transport layer of the protocol stack use the UDP to carry voice instead of the traditional TCP. While TCP is used to carry the signaling message and teardown connections, the RTP is run on top of UDP for end-to-end delivery of real-time voice data.

Timing is a critical characteristic of voice. Two syllables of a word are normally uttered with an interval that is as much a part of the voice as the uttered syllable. If extra delay is inserted between syllables, the rhythm of voice is lost [2]. "Too much delay can impair voice in several ways. First, long delays cause two speakers to enter a half-duplex communications mode, where one speaks, and the other listens and pauses to make sure the other is done. If the pausing is ill timed, the speakers end up 'stepping' on each other's speech. Second, long delay exacerbates echo because the reflected signal comes back to the sender after the sender finishes transmission" [2].

How much delay can a VoIP system tolerate before it becomes annoying to the listener or destroys the intelligibility of the conversation? Normally, delays below 150 ms are acceptable for most applications. When delays exceed 150 ms, users start to run into echo and step on each others' speech. However, delays below 400 ms are still acceptable for long-distance communications. Delays on the order of 500 ms or more are unacceptable and cause echoes to be heard coming from the channel. Delays up to 500 ms are usually due to unavoidable signal propagation delay. Voice quality deteriorates significantly above 400 ms delay and is unacceptable in most real-time cases.

Delay varies with the type of network being used. In circuit-switched networks such as PSTN or ISDN, delay is not a very serious issue for short-distance voice communications. However, in long-distance communications (for example, between Sydney and New York), delay becomes a prominent issue. The primary source of delay in PSTN is propagation delay (the time it takes for the microwave signal to reach the receiver), which is directly a function of distance (delay = distance/speed of light). In a duplex system, this delay is doubled.

There are many contributing sources to the delay sum in a VoIP system. Consider the case of VoIP over PSTN. The delay budget is due to queuing delay, propagation delay, serialization delay, encoding and decoding delay, and jitter (Figure 17.2). It is therefore very easy for the total delay to exceed the 400-ms mark. Of these, queuing delay can build up very fast, particularly if the network consists of many intermediate nodes at which packets must wait for transmission.

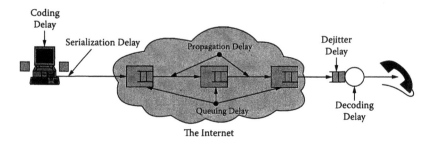

Figure 17.2 Delay points in VoIP. (From M. Hassan, A. Nayandoro, and M. Atiquzzaman, *IEEE Communications Magazine*, April 2000, 96–103. With permission.)

Coding Delay

The first requirement for transmitting voice as packets is coding. The function is to convert analog voice to digital data (coding), and the device or software used to do this is called the coder. The coder also performs voice compression. Traditionally, if voice is sampled at 8 kHz, and each bit is represented by 8 bits, then the bandwidth of the channel to carry the sampled speech without compression must be 64 kilobits per second. Hence, one voice channel has capacity:

$$C = f_s * Q$$

where f_s is the sampling frequency and must be at least twice the highest frequency content of voice (taken to be about 3.4 kHz), and Q is the quantization bits (number of bits per sample). Transmitting voice at this bandwidth is very inefficient because only a few customers can be supported per channel per second—hence the need for compression. Coders therefore perform voice compression to reduce the bandwidth requirement of voice transmission over digital networks. A coder consists of an analog-to-digital conversion (ADC) at its front end and a voice compression algorithm that is used to reduce the bit rate to below 64 kbps. The system therefore introduces delays in the coder, that is, the amount of time required by the ADC and the compression, the algorithm, and putting the bits into packets (packetization). Higher compression is achieved at the price of longer delays.

Two factors significantly contribute to the total encoding (input coding) delay and they are frame processing delay and look-ahead delay. Frame processing delay is the delay in processing a single voice frame (the number of samples to be packed into one packet). The lengths of voice frames (number of voice sample chunks processed together per encoding step) are chosen depending on the type of algorithm. Typical sizes range from 64 samples to about 256 samples. Look-ahead delay is the delay in the processing part of the next frame to exploit any correlation in successive voice frames. This is done as a predictive process to ensure that the present frame contains part of the next sets of speech samples so that in times of packet loss, it is easier to recover. Normally, the decoder is almost a replica of the encoder, and is used to reverse the encoding process. Decoding delay at the receiver is typically as low as half the encoding delay at the sender, or more. Table 17.3 is a list of encoding and decoding delays for several voice coding standards used mainly in IP telephony. The lower bit-rate codecs perform much better when errors are present in the system and bandwidth is limited.

Serialization Delay

Serialization delay is the time it takes to place a packet on the transmission line, and is determined by the speed (capacity) of the line. With higher line speeds, serialization delay can be greatly reduced. For example, it takes 125 μs (this is the same as

Table 17.3   Coding Delays in VoIP Coders

Coding Standard	Compression Algorithm	Bit Rate (kbps)	Frame Processing Delay (ms)	Lookahead Delay (ms)	Total Encoding Delay (ms)	Typical Decoding Delay (ms)
G.711	PCM	64	0	0	0	0
G.729	CS-ACELP	8	10	5	15	7.5
G.723.1	ACELP	5.3/6.4	30	7.5	37.5	8.75

Source: From M. Hassan, A. Nayandoro, and M. Atiquzzaman, *IEEE Communications Magazine*, April 2000, 96–103. With permission.

the sample interval for voice sampled at 8 kHz) to place 1 byte of information on a 64-kbps line, while it takes only 0.05 μs to place the same amount of information on an OC-3 line of capacity 155 Mbps. Note that serialization delay also depends on the frame size used by a coder. Longer frames result in higher delays in transmitting the packet.

Queuing Delay

Delay at the transmission points such as switches, routers, and gateways is called queuing delay. There, voice packets are queued behind each other to be transmitted over the same outgoing link. The number of packets waiting in the queue is a function of the arrival process and in many cases is approximated as a Poisson process. If a user has control of the infrastructure, then queuing delays can be reduced. The RSVP and DiffServ are two protocols that can be used to reduce queuing delays by prioritizing packets from delay-sensitive applications such as voice.

Propagation Delay

Propagation delay occurs at the links. It is the time required by signals to travel from one point in space to another. This time is fixed and is a function of the distance separating the source and receiver and the speed of light (in the medium used as link).

The worst propagation delays normally occur over satellite links, satellite telephony, or architectures based on satellites because the calls are routed over a satellite link, especially geostationary earth orbit (GEO), which is around 170 ms (one way), or low earth-orbiting (LEO) satellites, less than for GEO satellites. Propagation delay for long-distance satellite calls is a problem for traditional long-distance intercontinental telephony. Propagation delays are reduced when LEO satellites like Iridium are in use. However, extra delay is involved in the inter-satellite handover from source to destination. This gives rise to variable delay paths because LEO satellites move with respect to the ground station and also involve buffering at the nodes during connection handover.

Other Sources of Delay

"Some delay sources are specific to certain implementations of VoIP systems. In dialup networks, there are delays caused by modems. Such delays can be avoided by using digital lines. Packet voice systems using multimedia PCs also incur delays due to operating system inefficiencies and sound card delays. Such problems can be addressed using gateway cards that use fast, specialized DSPs" [2].

Inter-arrival packet delay: Inter-arrival delay is defined as the difference in the arrival times of two consecutive packets at the receiver. Because the packets have time stamps at source and destination, the differences time can be measured. The difference between two consecutive time stamps of two consecutive packets gives the inter-arrival packet delay given by the expression:

$$D_k = T_k - T_{k-1}$$

Inter-arrival packet delay is used to assess jitter in the network, as discussed in the next subsection. The variation in mean inter-arrival packet delays is proportional to the packet sizes. Linear PCM is known to yield, on average, the lowest inter-arrival delay and u-law the highest. Unfortunately, because of its large bandwidth requirement, it is not used for VoIP. The inter-arrival packet delays for ADPCM, on average, tend to be slightly higher than those packets that use PCM. On average, the inter-arrival time increases with distance. The end-to-end delay in a VoIP network is therefore given by the expression:

$$\tau_k = \tau_{enc} + \tau_{pack} + \tau_{dec} + \tau_{play} + \sum_{j \in hops} \left(q^i_j + d_j + \frac{s_k}{C_j} \right)$$

where k is the packet number, q is the queuing delay, d is the propagation delay, s is the packet size, and C is the capacity of the link (the maximum transmission speed at which it can forward packets).

Influence of packet size: End-to-end delay increases with packet size in a linear manner. This is caused by the transmission delay, which increases with increasing packet size, leading to an increase in the overall end-to-end delay.

It has been observed that the end-to-end delays with u-law for almost all packet sizes are about twice those obtained with PCM and ADPCM. This is mainly due to the higher compression delay incurred compared to PCM or ADPCM.

Table 17.4   One-Way Transmission Delay from Wayne State University, Michigan, USA, to Various Destinations (Packet Size = 200 bytes)

Destination		One-Way Transmission Delay (ms)
North America	Concordia University, Montreal, Canada	40
Europe	University of Cambridge, England	72
South America	University of Buenos, Argentina	114
Asia	University of Sydney, Australia	113
Australia	Vidyasagar University, West Bengal, India	115
Africa	Cairo University, Egypt	130

Jitter

Jitter is defined as the difference between the average of the inter-arrival delay and the individual inter-arrival delay, D_k. It is caused by the dynamic nature of the network.

Jitter or the inter-arrival time between two packets is therefore calculated as:

$$J_k = t_k - t_{k-1}$$

The variability of queuing delays at the network nodes and the fact that packets take different paths from source to destination makes it hard to eliminate jitter. Jitter is potentially more disruptive for IP telephony than the other delays discussed above. Network jitter can be significant even for low or average network delays. Legacy equipment and variations in interconnectivity and routes taken by packets from network to network and between countries can cause significant jitter. If an IP packet is unnecessarily delayed, it will not arrive in time at the receiver and will be considered lost.

The IETF in RFC 1889 provides a recursive expression for computing jitter as:

$$J_k = J_{k-1} + \frac{\left(\left|D\left(k-1,k\right)\right| - J_{k-1}\right)}{16}$$

To combat variable packet arrival times and still achieve a steady stream of packets, the receiver keeps the first packet in a jitter buffer for a time proportional to the size of the jitter buffer before playing it out. Therefore, a 60-ms hold time means 60 ms jitter buffer, and the jitter buffer delay time adds to the overall delay. Delay arising from jitter buffer can significantly alter the balance of the overall perceived delay. For example, if the voice packet experiences a large jitter delay of about 80 ms when the delay in the network is as low as 30 ms, the perceived overall delay becomes 110 ms, which is high. It is therefore essential to make adequate decisions as to the levels of acceptable jitter buffer delays, or to set the size of the jitter buffer to the level of acceptable overall delay in the VoIP application. Setting the jitter buffer very small may cause some packet loss while setting the buffer too large will lead to large jitter buffer delay. In practice, the jitter buffer size is dynamic and changes in proportion to the ratio of late packets to those that arrive in time. A reasonable choice of jitter buffer size should be between 50 and 100 ms.

For the experiment conducted in [1], Table 17.5 provides the jitter characteristics obtained. The quality of VoIP sessions can be quantified by the network delay, packet loss, and packet jitters. Delays over this value adversely affect the quality of the conversation. The mean jitter was found to lie in the range of 7 to 9 ms and is uncorrelated with distance. The mean jitter for u-law is constant around 7.5 ms and increases slightly for the ADPCM codec.

Table 17.5   Delays Encountered when Using Various Coders

Destination	Codec	Minimum (ms)	Maximum (ms)	Mean (ms)
Network Lab	PCM	0.08	20.3	3.4
	ADPCM	2.9	32.7	8.2
	u-law	4.7	17.4	7.5
College of Nursing	PCM	4.4	18.5	8.1
	ADPCM	4.5	12.4	9.1
	u-law	1.2	17.8	7.5
College of Pharmacy	PCM	1.2	27.0	9.2
	ADPCM	1.7	62.9	9.1
	u-law	0.6	20.8	7.5
Eastern Michigan University	PCM	2.1	35.3	8.1
	ADPCM	6.7	27.9	14.0
	u-law	0.2	25.5	7.6

Echo

Echo is hearing your own voice reflected back to you in the telephone receiver while you are talking. If the duration of echo exceeds approximately 25 ms, it can be distracting and cause breaks in the conversation. Echo in a traditional telephony network is caused by impedance mismatch from the four-wire network switch conversion to the two-wire local loop and can be controlled by echo cancellers. Echo cancellers are set by design by the total amount of time they will wait to receive the reflected speech. This is known as an echo trail and is normally about 32 ms. The ITU-T echo cancellation standards are G.165 and G.168.

Implementation

Architectures are based on the signaling type adopted. This section considers three based on SS7, H.323, SIP, and VPNs.

Each section describes the methods of deployment of VoIP in terms of the signaling protocol in use, including those using SS7, H.323, and SIP as the signaling protocol.

Integration of IP and PSTN

There are several architectures for end users of VoIP. One is a PC-PC architecture using SS7 as the signaling protocol. In this example, users are equipped with multimedia-capable computers directly connected to the Internet using network

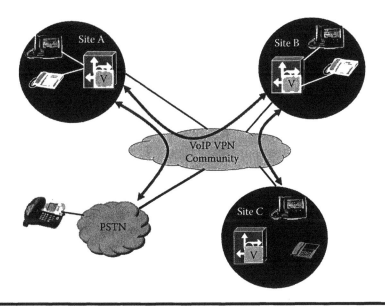

Figure 17.3   VoIP VPN architecture.

interface cards in the case of a LAN or via modem/cable modem when the connection is through an Internet service provider. At both ends of the call, microphones and PC-based speakers are used for reproduction of the sound. This means that unless a user chooses to use an earphone, conversations can be easily heard by people nearby. The signal processing functions required for the VoIP to work, such as sampling, compression, coding, and decoding, happen at the computers. This places enormous load on the CPU unless a hardware card is used to carry out the above functions. Calls between users are established using IP addresses. In this architecture, the IP network and PSTN continue to operate independently. PC-phone behaves as an alternative IP telephony architecture that allows a PC user to establish a call with a conventional fixed phone user. This architecture highlights the issue involved in the integration of IP networks and PSTN.

Phone–Internet–Phone Architecture

The phone–Internet–phone architecture is an extension of the PC–phone architecture. In this architecture, the Internet is used in the background to reduce telephone costs for traditional fixed phone users. This is the second most popular architecture as of today and is supported by Skype. In a phone–Internet–phone architecture, conventional telephone sets are used. A user intending to call another user calls a particular number, which is the gateway between the PSTN and the Internet, and then enters the desired telephone number. All sampling and coding take place at the gateway. Voice packets are then carried over the Internet to a gateway close

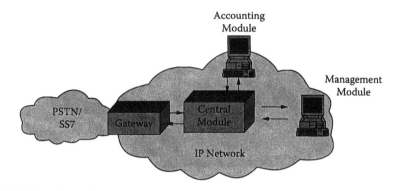

Figure 17.4 Carrier-class IP telephony. (From M. Hassan, A. Nayandoro, and M. Atiquzzaman, *IEEE Communications Magazine,* April 2000, 96–103. With permission.)

to the second user. The second gateway does all the decoding and conversion to an analog signal, which is then carried over the PSTN to the second user. This architecture therefore places most of the signal processing functions at the gateway (Figure 17.4).

These two types of architectures are likely going to continue to be the popular as they permit the best use of home PCs for normal voice calls. A variant of this architecture replaces the fixed phone with an IP phone so that the signal processing functions take place in the IP phone.

VoIP over Mobile Radio Using H.323

"In keeping with the goal of being cost-effective, this system is being designed using mostly off-the-shelf components. The hardware used to host the voice conference is a server-class computer. The server uses First Virtual Communications' (FVC) Conference Server software based multipoint control unit (MCU)" [7]. Conference Server enables the hosting of multiple simultaneous conferences with audio and video mixing. Conference Server supports the H.323, SIP, and T.120 conferencing standards.

This solution (Figure 17.5) is most suitable for security and emergency services. It involves establishing a wide area VoIP network consisting of a single central server and a client located at each of the participating locations. The central server uses H.232 to host a voice conference with standard audio, video, and data communications across the Internet. A radio that is compatible with the individual locations' mobile radio units is interfaced to the desktop computer at the client. When a mobile unit desires to communicate with a mobile unit from a different agency, instead of doing so directly, the transmission traverses the established network. A voice conference is first created on the server if one is not present. All clients have the ability to perform conference creation or deletion. Transmissions begin at the mobile unit and travel to its base station. At its base station, the transmission is

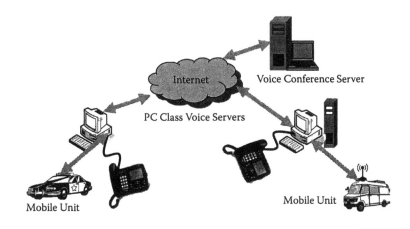

Figure 17.5 Voice over IP solution for mobile radio interoperability. (From J. H. Mock and W.T. Miller, *IEEE 56th Vehicular Technology Conference Proceedings,* **2002, 1338–1341. Wth permission.)**

processed by the client and sent to the conference on the server. The server then distributes the transmission to whoever has joined the conference. Each client computer at the different agencies then sends out the transmission using a radio that is compatible with its mobile units.

Performance Measures

Summary of Perceptual QoS Assessment Determinants and Aspects of Quality in VoIP Systems

Figure 17.6 shows the various aspects of the perceptual QoS of a VoIP system and how these are determined. In conventional telephony until the 1980s, where the signal bandwidth was fixed between 0.3 and 3.4 kHz, the impairment factors were transmission loss, frequency distortion, stationary circuit noise, and, in digital systems, signal-correlated quantization noise associated with PCM. These properties are usually described in terms of an SNR. However, VoIP systems are based on new coding technologies and a new transmission technology; the primary determinants of the perceptual QoS of a VoIP service are distortion caused by speech coding and packet loss, loudness, delay (network and terminal delay), and echo in a link. Echo is caused by inadequate circuit termination, which results in sound signal reflection back to the source.

Perceptual QoS Design and Management

The prime criterion for the quality of audio and video communications services is subjective quality (quality as heard by everyone of us), the users' perceptions of service quality. This differs from person to person and is to some extent a function of our listening patterns. This can be measured through subjective quality assessment.

The most widely used metric is the mean opinion score (MOS). However, while subjective quality assessment is the most reliable method, it is also time-consuming and expensive. It requires first having a collection of speech signals under different listening conditions such as in an office, market, crowd, and in places with adequate sound proofing (e.g., music or TV studios). Second, it requires the availability of an anechoic chamber. An anechoic chamber is a sound-proof room specifically designed to absorb all possible sound sources outside of the speech files. Third, it requires using many experts to listen to the speech files and then receive from them their opinions on each of the voice signals played to them. The arithmetic mean of all the opinion scores collected is the MOS.

Methods for estimating subjective quality from physical quality parameters are desirable. This process is called objective quality assessment.

The subjective quality factors are mapped to network and terminal quality parameters as shown in Figure 17.6. Because service providers use quality assessment technologies in order to design and manage QoS in a way that takes users' perceptions into account, they need to further map these quality parameters to parameters that are designed and/or managed.

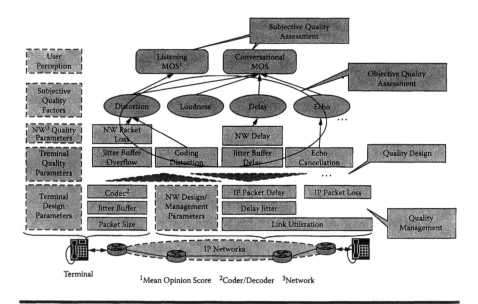

Figure 17.6 Factors that determine the quality of a VoIP service. (From A. Takahashi, H. Yoshino, and Nobuhiko Kitawaki, *IEEE Communications Magazine,* July 2004, 28–34. With permission.)

Subjective Quality Assessment or Opinion Rating

The most widely used subjective quality assessment method is opinion rating, which is defined in ITU-T Recommendation P.800. The performance of the system under test is rated directly (absolute category rating, ACR) or relative to the subjective quality of a reference system (degradation category rating, DCR). The ITU-T opinion scale used in an ACR test is the most frequently used, with the following scoring patterns: excellent (5), good (4), fair (3), poor (2), and bad (1).

In a DCR test, the subjects are instructed to rate the conditions according to the following five-point degradation category scale: degradation is inaudible (5), audible but not annoying (4), slightly annoying (3), annoying (2), and very annoying (1). The mean value of the results is called the degradation mean opinion score (DMOS). Some of the most used degradation introduced to the speech files includes babble noise, street noise (cars, horns, and people), background noise such as air conditioning noise, and large audible noise.

Subjective quality is categorized into listening quality and conversational quality, as shown in Figure 17.6. Conversational quality assessment involves two-way communication, while in listening quality assessment, subjects are simply provided with recorded speech material played back to them. In practice, the overall quality of VoIP must be discussed in terms of conversational quality. However, listening quality assessment is also quite helpful in diagnosing the effects of individual quality factors such as speech coding and packet loss.

Opinion Equivalent-Q Method

The MOS is experiment dependent due to differences in the testing date and the mix of quality levels in the experiment. It can vary slightly as per time conducted. For example, if one employs a lot of good-quality conditions, the MOS for a certain condition becomes lower than that obtained if one uses fewer good-quality conditions. Therefore, these effects need to be removed from the MOS. An approach for doing this is the so-called opinion equivalent-Q method, in which the modulated noise reference unit (MNRU) is used. This approach is standardized as ITU-T Recommendation P.810. "MNRU is a reference system that outputs a speech signal and speech-amplitude-correlated noise with a flat spectrum. The ratio of signal to speech correlated noise in dB is called the Q value." The opinion equivalent-Q is defined as the Q-value of MNRU speech with quality equivalent to that of the speech under evaluation.

Objective Quality Assessment

Subjective quality assessment is time-consuming and expensive. A test conducted by this author in 1998 for measuring the MOS of LD-CELP required more than 200 listeners of mixed backgrounds, and many different speech files of at least more than several seconds duration. It lasted about 3 months before a convergent MOS rating was obtained. These expensive and time-consuming requirements mean that we need a method for estimating subjective quality by measuring the physical characteristics of the terminals and networks.

We can categorize the objective quality assessment methodologies into several groups from the viewpoints of the aim, measurement procedure, input information, and MOS for estimation. These are opinion models, speech layer objective models, and packet layer objective models.

Opinion Models

Three objective opinion models are defined in this subsection. The objective opinion models exploit network and terminal quality parameters, as shown in Figure 17.6, and results in estimates of conversational MOS. The speech layer objective models require speech signals as inputs and produce estimates of listening MOS. The third category, which exploits IP packet characteristics to produce estimates of listening MOS, are called packet layer objective models. Both the speech- and packet layer models estimate listening quality. They are, however, used in different scenarios. When it is impossible or difficult to obtain actual speech samples using in-service quality monitoring, packet layer objective models should be used. When it is difficult to capture necessary packet information or when one needs to obtain quality estimates that are as accurate as possible, the speech layer objective models should be used.

The ITU-T has long led the study of opinion models. It was impossible to settle on a single algorithm as the international standard and hence an informative document was created with four different models introduced. The E-model was proposed by the ITU-T for standardization. In it, the factors responsible for quality degradation are summed on a psychological scale.

The E-model and the TR model from AT&T produce similar outputs: a score on a psychological scale is produced as an index of overall quality. In the E-model, the quality degradation introduced by speech coding, bit error, and packet loss is treated collectively as an equipment impairment factor. The ITU-T standardized the E-model as Recommendation G.107 in 1998. It was also adopted by the European Telecommunications Standards Institute (ETSI) and the Telecommunications Industry Association (TIA) as a network planning tool [2, 3] and has become the most widely used opinion model in the world.

The E-model has 20 input parameters that represent the terminal, network, and environmental quality factors. Its output is called the R-value, which is a function of the 20 input parameters. First, the degrees of quality degradation due to individual quality factors such as loudness, echo, delay, and distortion are calculated on the same psychological scale. Then these values are subtracted from the reference value.

Recommendation G.107 provides a set of default values that can be used when network planners assume that terminals and the usage environment are normal.

An index of overall quality should thus represent the perceptions of a user using a normal terminal under normal circumstances. Because a basic assumption for the E-model is telephone band (300 to 3400 Hz) handset communications, it is inapplicable to the evaluation of hands-free or wideband (e.g., 150 to 7000 Hz) communications. Taking into account the quality evaluation of future speech and multimedia communications services, it is quite important to expand the scope of the E-model.

Although the R-values produced by the E-model have some correlation with subjective conversational MOS and are useful in network planning, they are not necessarily accurate as estimators of subjective quality. In particular, the validity of the additive property assumed in the E-model is sometimes questionable [4, 5]. This is discussed in detail later. The accuracy of E-model prediction is being thoroughly studied by the ITU-T. This might lead to a revision of the existing Recommendation (G.107). We can also expect the integration of opinion models with speech layer and/or packet layer objective models, which are introduced in the next subsection. One such example is ITU-T Recommendation P.834, which provides a means of converting the results of a speech layer objective model into a quantity that can be incorporated in Recommendation G.107; the E-model's scope is currently limited to the evaluation of error-free speech.

Speech Layer Objective Models

The study of speech layer objective models started with the use of the signal-to-noise ratio (SNR) as a means of evaluating PCM-coded speech.

Strategies

There are three strategies being adopted in the cable industry for VoIP. The basic strategy adopted by many that offer VoIP in the first place is mainly to win a new customer away from the incumbents. Basically, the strategy is to displace the incumbent and gain a new customer for other services beyond VoIP.

Hence, cable operators bundle voice over IP with other services such as cable television and high-speed data. The following models are used to deliver the strategy to home users.

Outsourcing

A third party is used by the cable TV operator to deliver the service, and the outsourcing company receives a cut of the revenue in proportion to the number of lines sold. In the second model, the cable TV company owns the last mile of cables to the houses and outsources the directory services and interconnections with the incumbent exchange carrier. The third approach is for the cable TV company to deliver the whole service by itself.

Billing

Spending on OSS and customer care by companies wishing to offer VoIP will increase substantially at the initial stage to support VoIP services. Thereafter, the main challenges will be IP activation, QoS, and self-care and self-provisioning. At the initial stage if the state of the network is poor, customer churn will become a major key issue. This will therefore drive further investments in managing QoS. Network management will thus become prominent as the company must track a customer's usage, and watch for latency and echo. This will further increase investment in echo cancellations but this expenditure will be modest. There will also be a need for network diagnostics and expenditure on diagnostic tools.

Validating that the VoIP service is working correctly after installation is essential. Therefore, the need to recruit skilled staff in voice processing and implementation will come to the fore. These new skilled staff will be used to verify the QoS and troubleshoot if a service is working correctly. Their main functions will include providing information on jitter, delays, and packet loss to customer service representatives, so they can manage complaints and service quality.

Billing for VoIP is a natural extension of billing for and managing high-speed data. Because the service no longer depends on the network, the OSS needs to be media neutral. Previously, OSS was wrapped around the service. In today's terms, they must support any service or application irrespective of the network carrying it.

It is essential that in offering VoIP to offices and homes, the wiring from the offices to the incumbents' equipment must be bypassed instead of being completely disconnected because by disconnecting them, in times of severing ties from a provider, there should not be the need for elaborate technical work to be done by the technician sent to the office or homes. Because VoIP does not need the houses or offices to be rewired, bypassing the telephone connections to the premises is all that is required. In Africa, where most homes do not have telephone wiring in the first place, it may be necessary for wiring to be added or, cheaper still, the last mile should be a wireless feed to the house from a nearby base station or wireless device.

For customers who are used to receiving rich fixed phone services, there must be incentives to switch to VoIP, such as reduced tariffs, and companies also need to differentiate themselves with new features, such as the ability of the user to establish multiple voice sessions over one line, distinctive ringing tones by phone numbers, and voicemail that can be read directly like an e-mail on a PC or home TV. Therefore, VoIP could be supported and complemented with text messaging, instant messaging, and speech-to-text. Speech-to-text service on VoIP is particularly attractive to a deaf customer, just as text-to-speech is to a blind customer.

Unlike traditional telephony, which is designed to supply power to the phone for some time after power outages occur so it can ring, VoIP requires electricity at the customers' premises for the customer to obtain a dial tone. In Third-World countries prone to power failures, it will be necessary to have battery- or solar-powered services instead of attempting to build complex systems that supply power over the line.

Summary

This chapter discussed a major innovation in voice communication—that of VoIP. VoIP is the result of many voice technologies working together, including voice compression and encoding, IP PBX, and voice transcoders. VoIP is a digital packet-based technique for sending real-time, full-duplex voice communication over the Internet or intranet. It is also known by several other names, including Internet telephony, digital telephony, and IP telephony. VoIP provides the efficiency of a packet-switched network and at the same time rivals the voice quality of a circuit-switched network (toll quality). It leads to cost savings to both the end user and the telecommunication operator. It enables carrying of voice over IP networks and hence aids network integration. The chapter fully discussed the technology and the associated delays introduced by the radio and core networks.

Practice Set: Review Questions

1. A voice over IP source signal is sampled at an assumed Nyquist rate of 8 kHz. What is the assumed bandwidth of the voice source?
   a. 4 kHz
   b. 3.4 kHz
   c. 6 kHz
   d. 12 kHz
   e. 24 kHz
2. A voice over IP source signal is sampled at an assumed Nyquist rate of 12 kHz. Each sample is represented by 8 bits. What is the data rate from the source?
   a. 20 kbps
   b. 96 kbps
   c. 1.5 kbps
   d. 16 kbps
   e. 64 kbps
3. A voice over IP source signal is sampled at an assumed Nyquist rate of 8 kHz. Each sample is represented by 8 bits and then compressed using LD-CELP. What is the compression ratio?
   a. 2
   b. 9
   c. 3
   d. 4
   d. 7

4. A voice over IP source signal is sampled at an assumed Nyquist rate of 8 kHz. Each sample is represented by 8 bits and then compressed using LD-CELP. What is this VoIP bit rate?
   a. 20 kbps
   b. 96 kbps
   c. 1.5 kbps
   d. 8 kbps
   e. 16 kbps

5. The bit rate given by Adaptive CELP (ACELP) is:
   a. 5.3 kbps
   b. 9.6 kbps
   c. 1.5 kbps
   d. 8.5 kbps
   e. 6.4 kbps

6. What is the MOS of ADPCM?
   a. 3.88 kbps
   b. 4.89 kbps
   c. 2.65 kbps
   d. 3.85 kbps
   e. 4.99 kbps

7. The frame length of CS-ACELP is:
   a. 35 ms
   b. 40 ms
   c. 25 ms
   d. 45 ms
   e. 30 ms

8. Voice, in general, is delay what?
   a. Sensitive
   b. Insensitive
   c. Cumbersome
   d. Assisted
   e. Pronounced

Exercises

1. What is the meaning of the term "mean opinion scores?"
2. Assume that the one-way transit distance between Sydney and Lagos is about 16,000 km. How would you go about computing the expected round-trip delay if a point-to-point voice communication is possible between the two cities directly?

3. If voice over IP data is sent as ATM cells of 53 bytes as opposed to using bigger frame relay packets, which version of VoIP transmission could lead to more voice degradation in times of lost cells or lost frame relay packets?
4. Define jitter. Ten packets were received with delays of 200, 210, 190, 205, 220, 200, 185, 220, 230, and 198 ms. Compute the average packet delay. Compute the amount of jitter experienced by each packet.
5. What are silence and comfort noise substitutions? What advantages do they provide in digital voice communications?
6. Distinguish between queuing and serialization delays.

References

[1] S. Zeadally, F. Siddiqui, and P. Kubher, Voice over IP in Intranet and Internet Environments, *IEE Proceedings Communications,* 151(3), June 2004, 263–269.
[2] M Hassan, A. Nayandoro, and M. Atiquzzaman, Internet Telephony: Services, Technical Challenges, and Products, *IEEE Communications Magazine,* April 2000, pp. 96–103.
[3] V. Hardman et al., Reliable Audio for Use over the Internet, *Proceedings INET '95,* Hawaii, 1995.
[4] C. Perkins, O. Hodson, and V. Hardman, A Survey of Packet-Loss Recovery for Streaming Audio, *IEEE Network,* 12(5), 1988, 40–48.
[5] D. Goodman, O. Lockart, and W. Wong, Waveform Substitution Techniques for Recovering Missing Speech Segments in Packet Voice Communications, *IEEE Transactions Acoustics, Speech and Signal Processing,* 34(6), 1986, 1440–1448.
[6] C. Perkins et al., RTP Payload for Redundant Audio Data, IETF RFC 2198, September 1997.
[7] J.H. Mock and W.T. Miller, A Voice over IP Solution for Mobile Radio interoperability, *IEEE 56th Vehicular Technology Conference Proceedings,* 2002, pp. 1338–1341.
[8] A. Takahashi, H. Yoshino, and N. Kitawaki, Perceptual QoS Assessment Technologies for VoIP, *IEEE Communications Magazine,* July 2004, pp. 28–34.
[9] E. L. Tivoli and J. I. Agbinya, "Communication Cost of SIP Signalling in Wireless Networks and Services," *Proc. ICT 2005—12th International Conference on Telecommunications,* Cape Town, South Africa, 3–6 May 2005.

Chapter 18

The VoIP Market

The global Internet infrastructure removes centralized dominant controls for voice calls for the first time from the hands of incumbents and hands them over to mobile operators. No single entity will eventually have full control of national and international voice telephony. This change of paradigm cannot be anything else other than good for the consumer. This chapter discusses the VoIP (voice over Internet Protocol; voice over IP) market, and the challenges and prospects it offers.

There are several motivating factors why VoIP is popular among voice callers. It provides an alternative voice service to high-cost mobile phone access and reduces the cost of communications for the consumer.

The High Cost of Mobile Phone Access

Traditional voice service is rapidly changing from being the primary revenue generator for traditional telecommunications operator worldwide. As VoIP uptake grows, traditional voice calling will gradually be replaced by voice over Internet Protocol. This analog-to-digital substitution, or voice-to-data substitution, is a result of the Internet becoming the major communication infrastructure of the future.

Mobile phones currently cost organizations a great deal of money. In part, this is because calls are more expensive, particularly if you are roaming and on plans. That is not a problem, however, when people are working away from the organizations' offices—the ability to keep in touch delivers benefits that offset the additional costs. But increasingly, employees are using their mobile phones as their principal means of making and receiving calls, regardless of whether or not they are in an office where a lower-cost alternative is available.

Why VoIP Was Not Popular with Telecommunication Operators

Traditional telecommunications operators want to maintain a healthy revenue flow and also a hold on voice calls. However, VoIP will lead to a rapid decrease in traditional voice revenue. This rapid decrease in PSTN revenue will be caused by the increasing use of VoIP. VoIP, however, needs efficient connectivity for toll-quality voice reproduction with reduced delay and echo, and therefore telecommunication companies that have built more modern networks can offer better VoIP.

VoIP Reduces Cost of Telecommunications

VoIP is an essential technology for messaging that telecom administrators need to understand because it can reduce the cost of telecommunications and provide a number of other technical and productivity benefits compared to circuit-switched telephony [7]. VoIP is attractive for toll bypass and elimination of monthly line rentals from telcos. The motivating factors for migrating to VoIP therefore include:

- Lower telephony costs, improved end-user productivity, and improved communications with remote sites are key motivators.
- Convergence of voice and data networks leads to lower infrastructure costs.
- Voice fidelity, adherence to industry standards, software-based VoIP client support, and the ability to integrate VoIP with legacy PBX systems are perceived as important or very important attributes for VoIP.
- The ability to encrypt VoIP communications for more secure voice communications is a major factor in organizations and individuals who decide to deploy the technology.
- Ease of use for VoIP handsets and soft phone peripheral selection.
- Better end-user features.
- Overall network efficiency.
- Security fears for VoWLAN (voice over wireless LAN) have been resolved with the ratification of the IEEE 802.11i standard. Therefore, the increase in WLAN deployments also provides a customer base to tap into with customers beginning to appreciate the cost-effective benefits of using voice applications over WLANs. With better handset capabilities and enhanced WLAN infrastructure, VoWLAN has the potential to amalgamate both voice and data services into an attractive package, thereby compelling both mobile and fixed-line carriers to consider this converged technology as part of their overall strategic solutions.

The not-so-good concerns of VoIP for users include security, poor QoS, and limited emergency response systems. The technology is more complex compared with the vanilla PSTN. Therefore, retraining of personnel is a detraction and a necessity.

Line rentals are only good for the incumbent and no one else. Line rentals are seen by consumers as that "never-ending penance imposed by incumbents and service providers." Consumers need a pipe dream—and they may have just found it in VoIP.

VoIP Sectors

Apart from the traditional telecommunication providers, the following sectors have walked with their money to VoIP:

- Government departments and agencies (in Australia, the local government councils)
- Small- and medium-sized businesses (SMBs)
- Universities and colleges

Migration to VoIP

This section derives content from Kiehn [8]. When an enterprise accepts the challenge of internally hosting VoIP services instead of outsourcing IP PBX and enhanced feature servers to telecommunications service providers, there are critical issues to which it must pay attention. The issues include managing VoIP hardware and software implementations themselves, rolling out sophisticated VoIP services to headquarters and branch offices across the enterprise, and maintaining hands-on development of enhanced VoIP features.

For this to succeed, the enterprise must acquire reliable, secure, WAN local access and backbone transport to interconnect remote enterprise locations as well as provide connectivity to the PSTN.

Technology Selection to Maximize Profit

The IT managers need to consider several technology criteria when selecting WAN services, including the network's ability to provide security for enterprise communications as well as provide cost-efficient converged services for all voice, data, and video traffic. Consideration for the overall WAN network's QoS capabilities and ability to back them up with meaningful service level agreements is essential. Finally, enterprises should consider and, where possible, insist on partnering with a WAN provider that fully leverages the latest standards to ensure interoperability, improve flexibility, and reduce costs [8].

Integrating Voice

One of the many reasons for enterprising IT managers to pursue "self-hosted" VoIP services is to eliminate the administrative and capital costs of maintaining local PBXs at several offices, including the costs of local procurement and local staffing. They are also consolidating applications over a single service provider's MPLS backbone network to reduce overall network costs by eliminating expensive point-to-point links and increasing the enterprise's utilization of available network bandwidth.

"However, until recently, voice communication has been conspicuously absent from the converged WAN services model, limiting the cost benefits of a multiservice network. With current VoIP implementations, enterprises typically run site-to-site traffic over the WAN but access the PSTN via separate local voice connections from each site. This means IT managers have had to continue to manage numerous local voice circuits and multiple off-network vendors as well as wasting available WAN bandwidth" [8].

"Two key barriers have stood in the way of integrating voice with other applications in a converged network: signaling incompatibility between enterprise and service provider VoIP networks, and the lack of an ability to guarantee voice QoS and security on corporate WANs. These challenges have forced enterprises to continue to use TDM connections separate from their converged WANs to access the PSTN" [8].

Now these barriers are beginning to fall, thanks to efforts by enterprises, service providers, and equipment vendors to deploy IP-enabled PBXs and adopt common signaling protocols, and by service provider efforts to transform wholesale PSTN gateway services into integrated components of retail converged WAN access and transport services. Leading WAN service providers can now offer cost-efficient converged services access for all VoIP, data, and video traffic over a single multiservice access infrastructure.

Parameters for Selecting a VoIP Provider

Conformance to SIP Signaling Protocol

One of the most critical criteria for selecting a VoIP service provider is its conformance to the SIP, a VoIP signaling protocol. It permits the establishment of mobility and real-time voice sessions between endpoints. At the enterprise level, virtually all major IP PBX suppliers now offer SIP-based solutions. Enterprises are adopting SIP in large part for its compatibility with common Internet standards such as TCP/IP, DNS, and DHCP, allowing voice to fit seamlessly into existing routing and address schemes, including IPv4 and IPv6. Because SIP is an application layer protocol, it presents a common real-time session control mechanism not only for VoIP, but also for instant messaging, presence, and a multitude of additional IP multimedia applications.

Furthermore, because SIP is adopted by both the IETF and 3GPP (UMTS or 3G), the service provider community also is now largely committed to SIP as the signaling protocol of choice—not only for VoIP, but also for IP Multimedia Subsystem (IMS) fixed/mobile convergence architectures. Traditional telephone companies have begun to deploy SIP at the core of their networks, using border elements such as signaling and media gateways to provide interworking with other protocols, including the PSTN's SS7. SIP phones are also being sold and shipped as replacements for the traditional SS7-based non-IP phones.

"Such widespread adoption of IP PBXs and SIP, both inside and outside the enterprise, is making seamless peering between enterprise and service provider VoIP networks possible. Using SIP-compliant IP PBXs, businesses can extend the reach and functionality of a single IP PBX across an entire geographically distributed enterprise via MPLS-based IP VPNs, without an expensive PSTN gateway card in every branch office router. The enterprise can migrate its internal voice communication to IP by routing intra-company voice traffic over a private WAN or managed MPLS network, allowing it to eliminate intra-company toll and tariff charges" [8].

As the SIP standard is widely adopted, it overcomes variances in interoperability between carrier and enterprise voice equipment as well as easing the task of troubleshooting problems.

QoS

Many types of different communications traffic now run over a converged network. Therefore, select a WAN service provider that supports hard, deterministic QoS for each type with strict priority queuing. This also places the onus on the DIY integrator to understand QoS requirements and how to measure them to be satisfied with the offered service level agreements (SLA). This is in recognition that different types of enterprise communications traffic will have different QoS requirements.

Voice and video applications require rigorous timing control and performance metrics. Sensitivity to delays and echo places strict priority considerations on the VoIP service provider. Priority data includes mission-critical business applications with lower delay sensitivity than voice/video applications, such as surveillance video and applications with flow-control-capable transport layers. Standard data includes sporadic LAN-to-LAN traffic that need not be given high priority. Internet-class applications such as e-mail and Web browsing have the lowest QoS requirements and can be transported as best-effort traffic.

The enterprise should only purchase the bandwidth required, thus maximizing flexibility and cost effectiveness.

Routing of VoIP Calls

Like traditional IP packets, VoIP also can use all the routing methods on the Internet, including packet forwarding in MPLS. The following IP routing methods are therefore at the disposal of the network administrator:

■ *Dynamic least cost routing:* Dynamic least cost routing (LCR) is a simple, powerful feature that ensures that wholesale routing is always optimized for maximum gross profit.

  – LCR may be configured globally or selectively for specific customers or routes.
  – Automated linking between billing system rate plans and route provisioning.
  – Find and eliminate low-margin routes.

■ *Quality-of-service routing:* Use a traffic analyzer to monitor call detail records and provide traffic analysis statistics such as answer seizure ratio, average call duration, and post-dial delay for all routes. Use this intelligence to dynamically shift traffic from routes with a low quality of service to routes with a higher quality of service.

■ *Optimized routing:* Use optimized route provisioning to simplify the creation and management of hundreds of routes.

■ *Product code routing:* Route calls based on product codes or type of service. In addition to routing flexibility based on customer, source IP address, time of day, day of week, calling number, or called number, calls can also be routed based on product codes. Product code routing enables the operator to route calls based on the type of service.

■ *Eliminate revenue leaks:* Maximize profits from interconnect services and eliminate revenue leaks using:

  – Secure inter-domain access control to eliminate fraudulent inter-carrier calls.
  – Collect CDRs in real-time from both source and destination networks to eliminate settlement disputes.
  – Use wholesale prepaid billing to eliminate bad debt risk from inter-carrier traffic.
  – Use least cost routing alarms for any routes that do not meet minimum gross margin thresholds.

■ *Retail PSTN gateway services "in the cloud":* "Given broad adoption of SIP in the IP PBX realm, service providers are well positioned to turn wholesale PSTN gateway services into retail gateway services in support of SIP-based enterprise WANs.

"On the service provider's side, on-net calls are sent over the enterprise's private IP or MPLS backbone, while off-net SIP calls to the PSTN ride the carrier IP network to a service provider owned SIP gateway that converts VoIP to TDM for calls to PSTN parties. The service providers' economies of scale have replaced a cost structure based on TDM gateways at every enterprise location with a handful of regional gateways interconnected by low-cost, high-quality IP circuits.

Such gateway services in the cloud are all the more efficient if the service provider operates both a national IP/MPLS network and a national TDM voice network complete with robust circuit switch support of E911, local number portability, and other vital telephony services. This SIP-enabled scenario enables transport of all data and voice traffic over a single converged WAN connection and the centralization of all PSTN calling through a single VoIP connection to a national carrier.

In terms of capital and operating costs, on-net and off-net SIP voice connections remove the need for on-premise PSTN gateway equipment. The converged WAN connection removes the need to contract with a local phone company in each market for local PRI lines to the PSTN, while also reducing toll charges to the PSTN for on-net long distance and local access calls that are now carried over the WAN. On top of this, nationwide pooling of VoIP ports allows enterprises to aggregate PSTN capacity on large circuits rather than manage channels on many different individual trunks, resulting in greater oversubscription, higher port utilization, and lower costs.

Additionally, data WAN access options, such as per-application QoS offerings, now also apply to VoIP as well as to any other WAN data application, thereby moving the enterprise toward a usage-based access and transport capacity model for voice. As a critical bonus, the SIP-based WAN positions the enterprise to exploit the coming integration of voice with IM, presence, multimedia conferencing, IP video and other SIP-based IP multimedia applications for enhanced enterprise mobility, productivity, and business agility" [8].

A WAN to Match Your Enterprise-Hosted VoIP Network

"Given the large investment enterprise IT managers are making in implementing Hosted VoIP networks, it is important to take the next step and select a WAN service provider [that] can offer the best match in security, converged services, QoS, and network protocols. Only then will the enterprise gain the full benefits of IP communications technology" [8].

VoIP-Based Services

Hosted IP PBX [5]

VoIP gives small companies the capabilities of a PBX solution for just a small, up-front investment without any of the management headaches. At some point in time, every home and every business, irrespective of the size of the business, has the responsibility to install a phone line or the challenge to purchase a phone system. Even if a new phone system is not the objective, businesses often outgrow their communication infrastructure and must upgrade or expand. The questions facing small and medium business (SMB) owners are the same questions a homeowner must answer at different scales. They must ask themselves what features they want and need, how much they are willing to spend, and if and how fast they expect to grow. For a homeowner, it may be necessary to invite an expert to install the system; and for a business, it might be necessary to outsource the tasks or hire a specialist to deploy and manage the solution in-house.

For a business, choosing between Hosted PBX technology and deploying a PBX at the customer location often comes down to a strategic decision between deploying technology in-house, which gives it all the control and security that it implies, and letting its provider host services in their network, while the business focuses on its core business competencies. In a Hosted PBX service, the service provider provides all the necessary functions of housing the company's PBX and managing it. In a nutshell, each SMB must decide against the levels of external technical distractions it is willing to accept. For many SMBs, the choice for communication is likely to be that of Hosted PBX services.

The advantages are many. The solution allows the SMB to do what it does best, while allowing the service providers to handle their communications worries.

The goal of a Hosted PBX solution is to simplify communications for the SMB while providing some of the must-have features of the online world, such as e-mail integration and Web-based management.

What are the pros and cons of a hosted versus owned PBX? Because traditionally, small business owners do not have as many choices and do not have any influence to sway manufacturers of phone systems to provide features and cost structures geared toward the SMB market, resorting to a Hosted PBX technology therefore enables small businesses to purchase connectivity with the features that they need "at a price that is within their budget" [5]. In appearance, a Hosted PBX makes an SMB look like a much larger enterprise. "One way to describe this is: "Big business benefits on a small business budget"[5]. Among the cost considerations, deploying a Hosted PBX solution means:

- There is no significant up-front commitment to installation and maintenance.
- There is no commitment to a steep learning curve on how to deploy and use the system.
- This translates to lower capital outlays on staffing and support infrastructure.
- If a problem occurs, service providers can begin to address it long before the customer realizes that there is a problem.
- A fixed monthly cost, with service and software upgrades rolled in, makes it an easier sell to those who hold the purse strings.
- The lower monthly costs of making phone calls will also go a long way to help make the CFO smile.

Benefits of a complete Hosted IP-based system include:

- "IP-based solutions deliver increased efficiencies, with easier integration of voice into other existing business processes, such as integration with Microsoft Outlook for click to dial directly from the corporate directory or integrating your communications with back end customer databases for increased customer service or contact center functionality" [5].

■ Moves, adds, and changes, administration of new employees and employees changing physical locations cost a fortune in time and money become a simple matter of a mouse click or two.

■ By taking advantage of a network-based Hosted PBX solution, whole remote offices can be brought online and reconfigured in a matter of hours by a single remote user sitting at a computer (integration, configuration, and reconfiguration of corporate communication services from a single desk).

■ This eliminates the process of porting existing telephone numbers from the LEC, a process that can take weeks; but in the case of new numbers, it is a much faster process. "This is especially beneficial in the case of a business that has the need to set up temporary campaign-type situations or other transient work locations, or multiple locations (without skilled IT personnel on site)" [5].

■ Hosted PBX facilitates disaster recovery. In today's dangerous terms, "the need to prepare a business' communications systems for unexpected disasters such as terrorist acts and natural disasters has become table stakes; In a traditional, premises equipment-based scenario, the costs of disaster preparedness are high, when one factors in the necessary levels of redundancy, fail-over, multiple site networking, etc. A network-based Hosted PBX solution affords a much less expensive alternative. If a major disaster were to hit, you could simply have all your calls redirected to a temporary telephone number (such as a cell phone), send your employees home or relocate your equipment to another IP-enabled facility. Since all the call processing and features are tied to servers that sit in the network, all the features are tied to the 'cloud' ensuring availability in the event of a disaster" [5].

A hosted PBX can be used to increase employee productivity when they are mobile. Taking advantage of a softphone on a laptop computer or PDA, for example, "an employee can log in from anywhere in the world, provided they have a broadband connection, and most of the features and functionality of the phone that sits on their desk back at headquarters is transferred to the softphone regardless of location. This makes the employees more responsive to calls, and the ability to transfer calls or bridge third parties into a conference from anywhere in the world speaks volumes regarding increased productivity and a professional-looking response to customer needs. And again, it gives the small business an outward appearance of a large, competent, professional organization" [5].

■ A "hosted solution is not necessarily tied to a specific phone vendor, the SMB is free to choose phones based on their needs, be they cost, features, vendor preference, or the like. The increasing array of available endpoints means that SMBs are free to choose what works best for them" [5].

■ "If the SMB contracts for Hosted PBX services from a service provider, they would do well to consider a carrier that provides the broadband 'pipe' as well as the Hosted PBX functionality. By choosing a vendor that provides the full package of connectivity and services, SMBs enjoy other benefits such as Quality of Service guarantees (and one provider to point the finger at if those levels are not met), a secure and reliable network that provides voice traffic priority, access to constant solution upgrades, and more" [5].

As mentioned earlier, the decision to embrace a Hosted PBX solution often comes down to a religious debate. But as research and analysis from nearly every major analyst firm reveal, enterprises of all sizes—particularly the small and medium-sized business segment—are increasingly setting their sights on Hosted PBX as the communications solution of choice. The benefits are clear.

Hosted VoIP

By simplifying and consolidating resources, companies can "focus on the business that really matters: their own." For companies with offices and employees spread across cities, provinces, and states, VoIP is a preferred approach for integrating data and voice services.

The system should streamline phone management, let the organization outsource the headaches, and keep control of the administration. A Hosted VoIP solution with a Web tool that simplifies moves and also includes automated call routing, four-digit dialing between all sites, and unlimited long-distance calls is desirable [4]. A Hosted VoIP solution can enable employees to stay connected as if they are in the office. When mobile employees use unified messaging, which delivers voice messages to their e-mail accounts, and a "find me/follow me" function, which ensures that calls always reach employees, VoIP should be considered. Hosted VoIP offers the potential for more telephone lines at less cost to a company. It offers predictable monthly bills and no nasty surprises. It is estimated that savings of up to 20 percent in IT time each month are possible. Savings in long-distance charges, phone system connectivity costs, and conference calling representing up to 25 percent to 27 percent of a company's communications costs can be expected.

It is apparent that more and more Web hosting companies will undertake Hosted VoIP to protect their Web hosting services and lock in their customers to a single point of communication reference.

Managed VoIP Services

In Managed VoIP Services, a VoIP service provider manages its clients' VoIP. This takes part of the fixed voice telephony away from incumbents.

Advantages of Hosted VoIP include:

■ It is cheaper for the customer.
■ It is easier to manage and for billing.
■ It provides control at manageable cost.

Key attributes include:

- No equipment leases and maintenance agreements are required.
- Administration is Web-based, which greatly reduces IT time.
- Businesses enjoy nationwide access between their offices, without dedicated circuits.
- Unlimited long-distance calling is possible.

There are three basic models for companies making the switch to VoIP:

- Model 1: Ownership
  - The company owns all the equipment.
  - The company has total control, but higher capital expenditures.
  - The company shoulders all the management headaches.
- Model 2: Hosted
  - Focuses on outsourcing responsibilities to experts
  - Costs are less
  - Lower level of control

The days of Internet hosting ushered in the era of dedicated Web developers who charge their customers on the basis of the quality of hosting and the number and quality of the Web pages designed and developed. In hosting VoIP, maintaining QoS is as essential as maintaining a functioning network. Hence, QoS maintenance may be integrated into the billing regime. The industry is perhaps at the point that it should consider offering paid QoS as part of the regular VoIP product offering to customers. Hybrid approaches are possible.

Best option parameters include:

- Installation, maintenance, and configuration management costs
- Chance for customized application development; thus, the benefits of hosted solutions may outweigh the negatives

VoIP Peering

VoIP peering uses a set of interconnection facilities to offer voice over IP. The Voice Peering Fabric (VPF) is a layer 2 Ethernet exchange allowing VoIP peering to take place. These are session boarder control operations and do transcoding as well match disparate data rates. VoIP peering might be a useful venture for interconnection between differing types of African international networks.

Architectures

Architectures are based on the signaling type adopted. This section considers two based on H.323, SIP, and VPNs.

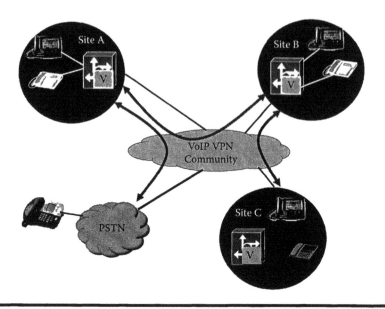

Figure 18.1   VoIP VPN architecture.

VPN

It is both crucial and important for organizations to connect many corporate sites to a single network and yet preserve the isolation of each site or group communicating on the shared infrastructure. VoIP VPN can provide this functionality as well.

Each group is able to use optimized private dialing plans and rely on the VoIP VPN to manage the communication between the sites in a manner similar to Figure 18.1. The VoIP VPN is an application layer overlay on any IP connectivity technology and can be deployed with or without a data-level VPN. This is facilitated with protocols such as IPSec.

The VoIP VPN should not require that significant changes be made to the VoIP network architecture, as it is an exact replacement of the PSTN VPN. Therefore, replacing the leased lines between the PBXs or the equivalent intelligent network closed User Group service should not lead to major alterations at the customer premises.

The initial investment is limited to one CCS (call control server) softswitch that is capable of VoIP VPN operation and installation of dedicated VoIP gateways to the PBXs.

VoIP VPN should facilitate VoIP value-added services, including advanced number routing and call center services. The VoIP VPN service provider therefore requires the following infrastructure: VoIP gateways for installation in multiple points of presence (POPs).

Basic call control infrastructure for routing phone-to-phone calls providing prepaid and post-paid telephony. A VPN feature should facilitate the aggregation of many different VoIP telephony services on the same backbone, including hosted Internet Protocol PBX (IPBX), residential gateways, IP phones, and corporate PBXs. Calling capabilities therefore include:

- *On-net to on-net (calls over the corporate intranet):* This can help simplify dialing plans that span several different companies with intranets.
- *On-net to off-net:* The architecture should permit a set of least-cost routing features so that the best routes and terminations can be selected according to the load conditions and prices; the VoIP VPN may then use the home DSL, cable connection, or broadband wireless to extend the VoIP VPN reach with single-step dialing;
- *Off-net to on-net:* This feature allows voice VPN to be called from PSTN using direct extension. Virtual local phone numbers can be allocated for some VoIP extensions. Therefore, while travelling in a foreign country, a local number in the foreign country can be called to reach long-distance numbers in the home country or office.

Integration of IP and PSTN

This subsection describes the methods of deployment of VoIP in terms of the signaling protocol in use and includes those using SS7, H.323, and SIP as the signaling protocol.

There are several architectures for end users of VoIP. The first one is a PC-PC architecture using SS7 as the signaling protocol. In this example, users are equipped with multimedia-capable computers directly connected to the Internet using network interface cards in the case of a LAN or via modem/cable modem when the connection is through an Internet service provider. At both ends of the call, microphones and PC-based speakers are used for reproduction of the sound. This means that unless a user chooses to use an earphone, conversations can be heard easily by people nearby. The signal processing functions required for the VoIP to work, such as sampling, compression, coding, and decoding, happen at the computers. This places enormous load on the CPU unless a hardware card is used to carry out the above functions. Calls between users are established using IP addresses. In this architecture, the IP network and PSTN continue to operate independently. PC–phone behaves as an alternative IP telephony architecture that allows a PC user to establish a call with a conventional fixed phone user. This architecture highlights the issue involved in the integration of IP networks and PSTN.

Phone–Internet–Phone Architecture

The phone-Internet-phone architecture is an extension of the PC–phone architecture. In this architecture, the Internet is used in the background to reduce telephone costs for traditional fixed phone users. This is the second most popular architecture

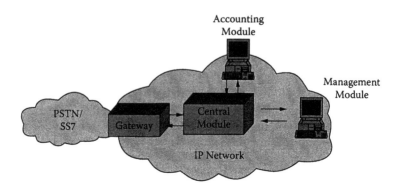

Figure 18.2 Carrier-class IP telephony. (From G.Audin, "Universities Grade VoIP," Friday, May 19, 2006. With permission.)

as of today and is supported by Skype. In a phone–Internet–phone architecture, conventional telephone sets are used. A user intending to call another user calls a particular number, which is the gateway between the PSTN and the Internet, and then enters the desired telephone number. All sampling and coding take place at the gateway. Voice packets are then carried over the Internet to a gateway close to the second user. The second gateway does all the decoding and conversion to an analog signal, which is then carried over the PSTN to the second user. This architecture therefore places most of the signal processing functions at the gateway.

These two types of architectures are likely going to continue to be the popular as they permit the best use of home PCs for normal voice calls. A variant of this architecture replaces the fixed phone with an IP phone so that the signal processing functions take place in the IP phone.

VoIP over Mobile Radio Using H.323

"In keeping with the goal of being cost-effective, this system is being designed using mostly off-the-shelf components. The hardware used to host the voice conference is a server-class computer. The server uses First Virtual Communications' (FVC) Conference Server software based multipoint control unit (MCU)" [8]. Conference Server enables the hosting of multiple simultaneous conferences with audio and video mixing. Conference Server supports the H.323, SIP, and T.120 conferencing standards.

This solution is most suitable for security and emergency services. It involves establishing a wide area VoIP network consisting of a single central server and a client located at each of the participating locations. The central server uses H.232 to host a voice conference with standard for audio, video, and data communications across the Internet. A radio that is compatible with the individual locations' mobile radio units is interfaced to the desktop computer at the client. When a mobile unit

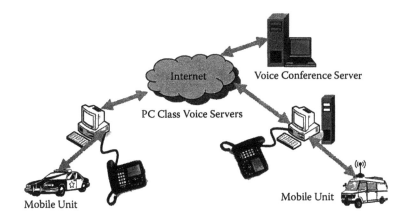

Figure 18.3 VoIP solution for mobile radio interoperability. (From M. Saadi, "Metric: VoIP to steal $100B in PSTN revenues," www.fiercevoip.com, May 18, 2006. With permission.)

desires to communicate with a mobile unit from a different agency, instead of doing so directly, the transmission traverses the established network. A voice conference is first created on the server if one is not present. All clients have the ability to perform conference creation or deletion. Transmissions begin at the mobile unit and travel to its base station. At its base station, the transmission is processed by the client and sent to the conference on the server. The server then distributes the transmission to whoever has joined the conference. Each client computer at the different agencies then sends out the transmission using a radio that is compatible with its mobile units (Figure 18.3).

VoIP Deployment Issues and Tests

There are SIX VoIP deployment issues:

1. Reliability
2. Hidden costs
3. Security
4. Sound quality
5. Handset issues
6. Network provisioning uncertainties

None of these issues is strong enough to derail the effectiveness of VoIP but smart carriers will have answers to the problems that smart customers will raise.

Tests are revealing that most VoIP equipment is interoperable; the bad news is that "most" is not the same thing as "all."

Testing

Testers found that turning on the router's QoS feature improved voice quality dramatically, even over non-busy networks. However, the NAT (network address translation) feature that many companies use as a security feature on their routers breaks VoIP entirely. In a recent article by Tim Green [1], volunteers set up five model enterprise networks fitted with VoIP equipment, network firewalls, application firewalls, Wi-Fi access points, and VPNs; they ran VoIP calls through them, using a variety of VoIP phones—softphones, hardphones, Wi-Fi handsets, and PDAs. The tests involved equipment from two dozen vendors. The calls ran over a combination of the public phone network and the Internet, using a service provider that supports SIP signaling, and then testers tried to disrupt the calls and measure the results.

Results

NAT, the masking of private IP addresses from public view, can break VoIP by making it impossible to set up SIP-based calls over the Internet to devices with private IP addresses. The best option the labs found was to get rid of NAT if possible. If not, get a SIP proxy server that can ignore the public addresses on VoIP packets and find the actual addresses within.

An alternative is to install a server outside the NAT device—usually a firewall—that keeps track of where packets come from and shepherds them through the NAT.

As for the use of QoS on Wi-Fi networks, while the labs did not quantify the difference, testers say that the improvement in quality jumped dramatically when QoS was turned on and was noticeable even on non-busy networks—it cut delays.

VPNs do not disrupt VoIP. This came as a surprise to testers, who expected that encapsulating real-time UDP voice packets inside TCP packets would cause delay, but that was not the case. With IPSec and SSL VPNs, there was no significant degradation. It is observed that the quality over SSL VPNs is better.

Billing Options [13]

As an IP service, VoIP lends itself to micro-billing and macro-billing processes. It could be billed per bit consumed, per time spent online, or based on a reserved bandwidth and quality of service. When IPBX hosting is part of the equation, the billing method will have to accommodate the service level agreements promised. Many billing options are at the discretion of the operator and they include:

- Real-time billing of customer's and partner carrier's traffic
- Flexible billing engine, which allows a personalized customer approach
- Affiliate system with retailer commissioning and self-care
- Multiple business models and service management

- Statistics and monitoring
- Support access of an unlimited number of rating plans
- Flat rate or usage-based rating
- Billing:
  - By time of the day
  - By day of the week
  - Based on destinations
  - By originating gateway
  - By DNIS
  - By traffic
- Full prepaid/post-paid convergence
- Concurrent calls and time increment management
- Hierarchical structure of the accounts
- Yearly, quarterly, monthly, weekly, biweekly, daily, and anniversary billing cycles
- Receiving and accounting payments
- Customer self-care module module
- Invoices available on the Web or sent to customers by e-mail or fax
- Easy customizable invoices by Crystal Reports
- Carrier settlement
- Credit card payment via the Web
- Powerful reporting on accounts, calls, carriers, traffic, usage, profit, loss, service quality, and payments
- Own RADIUS server
- Group access permissions
- 100 Percent Web based
- Traffic analysis
- Fraud detection

Sources of Competition in VoIP

- Infrastructure managers
- Traditional service telecommunication companies
- Cable TV operators (have advantages over telcos of possessing broadband access and cable customers in the home)
- IT services
- Managed desktop service providers
- Dedicated VoIP services providers such as Skype, Vonage, and Packet8
- Grey market

New costs to be considered in deploying VoIP include, but are not limited to the following:

- The costs of softswitch, where used

- PBX and voice codecs
- Added security features (some Internet security features can be used on VoIP packets)
- Training of staff to service new VoIP offerings
- Improvement in the core networks
- Platforms to support VoIP
- Management of VoIP services
- New business support systems (BSS)

Security

Security is a major consideration in an IP network. IP voice is prone to security breaches as e-mail and Internet data. Enterprises should look for service providers with private VoIP WANs not directly exposed to the Internet. The VoIP WAN should have access to it controlled via session border controllers, essentially purpose-built VoIP firewalls, to ensure that only authorized signaling and media packets reach the core VoIP network. This network security arrangement helps to secure enterprise communications end-to-end from eavesdropping, denial-of-service, phishing, and other hacker attacks. VoIP exposes communication networks to a wide range of Internet security problems and in this way, the following security issues become prominent:

1. Calls can be recorded.
2. Eavesdropping is more likely to happen in business than residential VoIP implementations.
3. Legal interceptions will be easier.
4. SPIT: VoIP spam may also emerge as a big problem.
5. Call centers could face more challenging security problems.
6. VoIP malware, phishing, viruses, and hacking could be rife—the old TDM world did not experience these.
7. The use of VoIP means that an eavesdropper no longer needs to have physical access to the line, as he or she does in a TDM environment.
8. Having both voice and data on the same network means that all the organization's eggs are in the same basket. If the data network goes down, the company's ability to communicate at all is compromised.
9. Using VoIP with SIP adds the security problems of SIP to the equation.

Security of Enterprise VoIP Networks

"Enterprises today are a heterogeneous mix of IP-based applications, and a wide variety of tools are necessary to ensure security. This approach includes initiatives aimed at network infrastructure, call processing systems, the end-points, and the applications. Keys at the network infrastructure level are the switches, routers, and connecting links that carry the IP data and voice and video traffic" [8]. The call processing systems cover servers and associated call management, control, and accounting systems. The end-points are the IP phones, soft phones, video terminals, and similar devices. Applications range from unified messaging, to conferencing, customer contact, and custom tools.

VoIP SPAM (SPIT)

VoIP, like e-mail, is a packet-based service and is subject to the same security problems as expected for other IP services. Will it happen in the future that if you log on to your voicemail and it announces that you have 30 new messages—and that 25 of them were unsolicited commercial broadcasting calls? Or you are out to dinner with your children, your phone rings, and it's a pornographic voice spam message. The word SPIT has been coined to represent "SPam over Internet Telephony." It appears that SPIT will be more deadly than its e-mail cousin. Why? Although e-mail spam will degrade service and clog bandwidth and mailboxes or be delayed by a few minutes or hours due to spam, the problems caused may not be too severe. However, with VoIP spam, the gateway is hit directly, causing congestion and degradation of voice quality directly. Therefore, end users will notice the effect of SPIT in IP voice communications. VoIP so far is open to attack by spammers. The expected spams will include audio commercials sent to customers' VoIP mailboxes. Because VoIP packets are unencrypted and require no authentication, it becomes very vulnerable to anyone just calling your VoIP voicebox. Caller IDs can be hacked as well. So far in most countries, steps have not been taken to regulate this aspect of VoIP and therefore the level of protection that used to be in place in analog voice calls will be greatly eroded. Any open IP-phone system (SIPPhones, free online calling programs, etc.) is therefore a target for spitters.

VoIP services that operate over closed systems are more immune to spitting attacks. In real terms, any network architecture is vulnerable to hacker attack and VoIP users could be subjected to unsolicited voice broadcast messages.

Despite this negative aspect of VoIP service, the ability to send IP broadcast to phones has benefits in times of emergencies, alerts, and disaster. This enables a large population to be reached fairly easily.

Spitting can be reduced at the network level by tagging all foreign packets and performance measuring and monitoring of VoIP lines.

Summary

The VoIP market emerged in response to the introduction of new technology and posed threats to network operators and provided competition from ISP, cable networks, and broadcasting companies. As a growing popular medium for communication, it has found use in instant messaging services, including Yahoo, Skype, and other similar services. This chapter is thus only an enumeration of the VoIP services, the market, the competition, and the benefits that it offers to end users.

Practice Set: Review Questions

1. Which of the following advantages does voice over IP provide end users?
   a. Flexible network access
   b. Cheaper call rates
   c. Freedom from operators
   d. New handsets
   e. New networks
2. Most operators are leery of voice over IP because:
   a. It is a new source of revenue.
   b. It represents fresh competition.
   c. It will lead to drops in voice revenue.
   d. They are unable to bill for bits.
   e. They hate new technologies.

Exercises

1. List and discuss at least seven reasons why operators should embrace voice over IP.
2. Discuss a possible path to migrate a network to VoIP by incumbents.
3. Selecting an inefficient VoIP provider can cost an organization both voice quality and productivity. How then should one select the best provider?
4. Discuss at least three methods of routing voice over IP calls.
5. How will interconnectivity affect VoIP in long-distance communications?
6. What VoIP services can you recommend to a new ISP trying to enter into the voice market, and why?
7. What is VoIP peering and VPN?
8. Is SPIT a problem? Why?
9. What are the key security concerns in VoIP services?

References

[1] T. Greene Framingham, Interoperable VoIP: It Works, But Needs Tweaking, *ComputerWorld: The Voice of the ICT Community*, Monday, May 15, 2006.
[2] G. Audin, Universities Grade VoIP, Friday, May 19, 2006.
[3] G. Galitzine, Editorial Director of *Internet Telephony Magazine*, 2006.
[4] B. Chatterley, Hosted VoIP, *Internet Telephony Magazine*, 2005.

[5] G. Galitzine, Why SMBs Need to Consider Hosted PBX, Editorial Director of *Internet Telephony Magazine*, 2006.

[6] C. Champion, VoIP: Challenges & Rewards; Convergys Corporation's Information Management Group, <http://www.convergys.com>.

[7] Osterman Research, November 15–18, 2005.

[8] T. Kiehn, Select Carefully to Get the Most from Converged Services Networks, Broadwing, <http://www.broadwing.com>.

[9] J. Hall, "Viral about Voice", May 22, 2006.

[10] D. Haskin, VoIP Eats Into Traditional Voice Services Revenue, TechWeb. com, May 16, 2006, <http://www.informationweek.com/story/showArticle. jhtml?articleID=187203672>.

[11] M. Saadi, Metric: VoIP to steal $100B in PSTN revenues, www.fiercevoip.com, May 18, 2006.

[12] J.H. Mock and W.T. Miller, A Voice over IP Solution for Mobile Radio Interoperability, *Proceedings of the Vehicular Technology Conference*, 2002, Vol. 3, pp. 1338–1341.

[13] Alitel VoIP Billing, <http://www.alitel.com/contacts/index.html>.

Chapter 19

IPTV: Internet Protocol Television

With the emergence of IP services, real-time video on demand (VOD), broadband TV, and video-over-IP networks have come to be known popularly as IPTV. IPTV is a suite of services consisting of digital TV, VOD, VoIP, and Web/e-mail all in one. Although it resembles cable TV, the two are not identical. Cable TV picture quality is still better than what is offered by IPTV. However, IPTV offers an interactive experience and comes with multimedia streaming features such as play, pause, fast-forward, and rewind. These features are absent in traditional cable TV offerings. IPTV, however, requires broadband access and needs a bandwidth of between 2 Mbps/channel and 8 Mbps/channel. At 8 Mbps/channel, its quality is comparable to low-end HDTV. HDTV typically requires about 20 Mbps/channel and in times of channel changes, it requires about 40 Mbps/channel for two channel streams. Hence, a typical household requires at least 40 Mbps bandwidth, which in today's terms is best provided with optical networks. A household with two or three TV sets therefore needs about 100 Mbps for them to concurrently run IPTV services.

IPTV

IPTV represents the convergence of television broadcasting, telecommunications, multimedia content distribution, and computing. It is offered and carried primarily as multicast IP packets by first coding the picture frames. One of the common standards for embedding MPEG-2 TS (transport stream) video into IP frames to be carried as video over IP is given by the PRO-MPEG organization [1, 2]. It is expected that the market for IPTV will increase tenfold over the next five years, making IPTV a target IP service for telecommunication operators, TV broadcasters, and Internet service providers (ISPs). A typical IPTV stream consists of video, audio, and information on the channel being used, with the dominant part being the compressed video content. IPTV is a system where a digital television service is delivered to subscribers using the Internet protocol over a broadband connection (e.g., ADSL2+ (20 Mbps) or VDSL2 (50 Mbps), WLAN, and fiber-to-the-X (FTTX, 100 Mbps)). IPTV is of no real use without broadband access. The playback system for IPTV is either a personal computer or a set-top box (STB) connected to a television. The video content is typically Moving Picture Experts Group2 Transport Stream (MPEG2-TS). The compressed video is delivered with IP multicast, a method in which information is sent to multiple subscribers at the same time. The underlying protocol to achieve this is IGMP (Internet Group Multicast Protocol) version 2. It is used for channel change (zapping) and signaling for live TV.

IPTV Services offer great opportunities for service providers such as mobile phone organizations to diversify their revenue stream and move aggressively into the potentially lucrative broadcast TV and emerging interactive video markets. IPTV could emerge as the most sought-after "killer application" but it will consume a great deal of bandwidth. Hence, changes must be made in the physical infrastructure that brings triple-play services to the home. Triple-play services bundle video, voice, and Internet access as one offering. The last 100 m to the home could pose serious bottlenecks for many operators seeking to offer IPTV to the home. Hence, IPTV could see a rise in the uptake of ADSL and more WLAN (802.11b/g) deployments and widespread deployment of the WiMAX backbone. Dependence on traditional phone lines in many countries could strangle efforts to offer IPTV.

Over the past 15 years, video compression technology has progressed to a level that makes it possible to send high-quality video signals over narrowband channels. As a result, the past few years have witnessed rapid increases in transporting video over IP networks such as the Internet. This increase, to a great extent, is led by deployment of optic fiber infrastructure for fixed networks and large uptake of WLANs, 802.11a/b/g. Transporting video over IP networks requires many protocols to work together to ensure efficient real-time delivery. These protocols are used at various layers of the OSI reference model. Traditionally, video over IP is carried at the presentation and application layers as the MPEG-2 Transport Stream Protocol and either MPEG-2 or MPEG-4 is used as the compression standard.

At the transport layer, the UDP is preferred over TCP. The choice of UDP over TCP is due to the fact that while TCP requires acknowledgment for every packet received, no such requirement is needed for UDP. Hence, delays in transmission are reduced, thereby enhancing real-time transmissions. Acknowledgment of received packets is also not compatible with multicast services. Hence, TCP is not that great for real-time transmissions.

To support real-time transmission of video over IP, several protocols must work together. The additional protocol used with UDP is the RTP. RTP is vitally

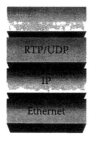

Figure 19.1   IPTV protocol stack.

importance in transporting and streaming of video. In the first place, it allows one to number sent packets so that when packets are received, they can be ordered to align with the sent order. RTP also permits the use of time stamps to ensure that received packets can be played at the receiver side with correct sequence and timing. Furthermore, RTP permits the use of forward error correction (FEC) schemes, which are added when IP networks are used to carry video information. Figure 19.1 is therefore the summary of the protocol stack required for carrying video over IP or IPTV [3].

This protocol stack specifies Codes of Practice (COPs). COP 3 and 4 are the most important. While COP 3 deals with MPEG-2 TS video compression, COP 4 focuses on uncompressed video at 270 Mbps and higher. The Ethernet layer limits the size of the output IP packet to the maximum transmission unit (MTU) of 1500 bytes. The end device (e.g., a laptop, a PDA) is normally connected to the Ethernet and is required to support anything from 1 to 7 MPEG-TS packets per IP packet. The number is determined by the ability of the end device, with the minimum size being 1, 4, or 7 transport stream packets. This size is kept constant during a send-receive session.

There are, however, several problems that must be thoroughly addressed by research. They are admission control mechanisms, multicast admission control, admission controls for Ethernet, congestion control, WLAN (wireless IPTV) and DSL, security, and standardization and communications among admission control schemes.

IPTV System Architecture

IPTV represents the convergence of TV broadcasting and telecommunications. It integrates video content with voice and data into a single service. Figure 19.2 outlines a typical IPTV system. The IPTV system consists of three major parts: the head-end video server (VS), the client's (C) home apparatus, and the network routers (R) with connecting infrastructure. A low-latency transport network is required for efficient transmission of IPTV, and the performance of the system is also impacted by the capability of the video server and the client's equipment. Hence, efficient solutions of these units can greatly enhance the service offering.

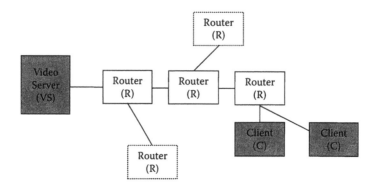

Figure 19.2   IPTV system architecture.

Compression and IP packet formation takes place in the VS. The quality of the video stream depends on both the quality of the input source (camera) and the efficiency of compression. This quality can further be degraded by the effects of the network, such as lost packets and jitter. Decompression of received video IP packets takes place at the client device. The client device has a corresponding decoder (decompressor) that is used to recover the sent video stream.

IPTV System Description

The source of the IPTV pictures from Figure 19.3 is either a live digital or analog camera feed, and/or a video source derived from videotapes, DVD recorders, other

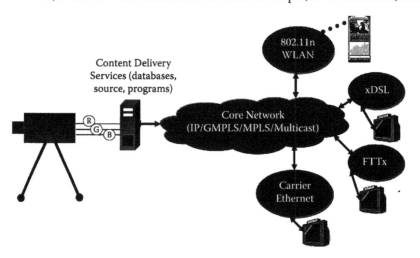

Figure 19.3   IPTV delivery mechanism. (From Y. Kukhmay, K.Glasman, A.Peregudov, and A.Logunov, "Video Over IP Networks: Subjective Assessment of Packet Loss," 2006. With permission.)

analog video sources, and a database of video. If the source is an analog video source, then analog-to-digital conversion of the signal is required to transform the source video frames into pixels. Whether the source is digital or analog, each video pixel is 24 bits long, consisting of R (red), G (green), and B (blue) components. The R, G, and B components are individually 8 bits (or 256 resolution levels) stored in such a manner that the bits 0 to 7 form the red component, bits 8 to 15 form the green component, and bits 16 to 23 form the blue component of the colored pixels. Therefore, the storage in the processing unit is a set of picture frames (25 frames per second if the source is PAL European/African and 30 frames per second for NTSC formats, United States and North America). This raw format requires huge amounts of storage and a very large transmission bandwidth to carry it. Hence, each video frame must be compressed.

Example 19.1

How many bits are in a second of a PAL video sequence of 25 frames, each of size 1024×1024?
Solution: $25 \times 1024 \times 1024 \times 3 \times 8 = 629,145,600$ bits = 629 Mbits.

Example 19.2

Using the data Example 1, what is the required bandwidth for 1 minute of this uncompressed video sequence?
Solution: The required bandwidth is equal to $60 \times 629,145,600$ bits = 37.75 Gbits (in 1 minute!)

From these examples, we observe that transmitting raw video without compression is ill-advised and hence over the past 30 years enormous research has been conducted into how to compress images and video to reduce the bit rate required for transmission and also retain significant video quality. For IPTV, two video compression standards are used for encoding and decoding. These are the Motion Picture Expert Group (MPEG) compression standards: MPEG-2 and MPEG-4. Both MPEG-2 and MPEG-4 are lossy compression standards. That is, the exact replica of the input video frame is not obtained at the decoder due to compression. Rather, an excellent reproduction of the input is desired. Although in practice the compression does not affect the physical dimensions of the video frames, it reduces the number of bits used to represent one picture element (pixel), that is, it is lossy. This reduction fits the video frame onto the transmission bandwidth so that it can be carried in near-real-time or in real-time. Often, more is done to reduce the bit rate further from the raw size to fit the transmission channel. In many instances, a reduced number of frames (below 25 frames per second) are carried and at the decoder the missing frames are reconstructed using interpolation techniques [9–11].

Core Network Requirements

From Figure 19.3, the input to the core network is a series of compressed video as IP packets. This is wherefrom IPTV derives its name. The TV frames are prepared as IP packets and transmitted through the core IP network. Where an IP core network is not available, a media gateway is used to re-cast the IP packets into the form that the core network is able to carry (for example, into ATM cells and carried by an ATM-based network). At the receiving end, they are re-cast back into IP packets and reconstructed for viewing.

The high-speed core Internetwork functions as a multicast network so it is able to serve many customers simultaneously, providing picture quality with guaranteed QoS. Usually, the core network is a high-speed optical network backbone.

In the core network, the GMPLS and MPLS switching protocols between layers 2 and 3 may be used. MPLS provides superior traffic engineering compared to IP, such that it enables connectionless IP to behave more like connection oriented. The path between the source and destination is predetermined and labeled. This is similar to the ATM protocol. MPLS is also similar to DiffServ because it labels traffic at the ingress and removes the labels at the egress points. MPLS is used to label the end-to-end paths and these paths are called label-switched paths (LSPs). MPLS adds labels to its headers and forwards the labeled packets using switching instead of routing in the corresponding paths. Unlike IP headers, the MPLS headers do not identify the type of service carried in the paths. Therefore, the service provider does not have the means to prioritize different data streams based on the type of data. Generalized MPLS (GMPLS) is an extension of MPLS by adding a routing control plane and signaling specifically for devices in the packet domain, and wavelength, time, and fiber domains. Thus, it provides an end-to-end QoS and provisioning of connections. GMPLS therefore services broadband applications better and hence is well suited to IPTV and controls all the service layers. In the optical plane, it can therefore prove very useful in labeling the optical paths and separating the data planes.

Access Networks Requirements

The core network deals predominantly with the requirements of the network backbone and not the last mile or the home- or office-front. Beyond the core network in Figure 19.3, high-speed access networks in the form of ADSL, ADSL2, VDSL, WLAN, carrier-grade Ethernet, fiber-to-the-curb (FTTC) and DSL, and fiber-to-the-home (FTTH) carry the IPTV to homes or commercial premises.

The IPTV receiver devices are either digital TV or HDTV receivers. The digital TV receivers decode the IP packets and use the MPEG decoder in the TV set to decode the compressed video frames for normal display for viewing. Apart from incorporating MPEG decoders, the receivers also incorporate local storage devices so that viewers can store and watch their programs later.

The deployment and assurance of IPTV service involves enormous challenges. "To assure IPTV service, the network must provide sufficient bandwidth to assure delivery without too much loss or delay. Service assurance thus requires bandwidth provisioning in the access network and capacity management in the core network." QoS must be ensured and guaranteed for all components of the service (video, voice, and data). For example, a standard definition TV requires a bandwidth of between 1 and 4 Mbps, and an HDTV requires 4 to 12 Mbps. At the homefront or at the customer premises, different access technologies can be used. Figure 19.3 shows several types of access networks that can be used for IPTV services within the last meters. In this section we discuss them in more detail. We discuss DSL technologies, carrier-grade Ethernet, FTTH, and FTTC. We have discussed the IEEE 802.11n in its various forms in a previous chapter.

DSL and Carrier-Grade Ethernet

Digital subscriber line (DSL) technologies provide high-speed digital data transmissions over standard telephone copper from the operators' end officers to users (home and commercial precincts). Depending on the provider, the downlink speeds are typically in the range of 128 kbps to 24 Mbps. Over a distance of about 2 km, ADSL can provide about 8 Mbps, and 24 Mbps for ADSL2+ (depending on the distance between the user and service point). ADSL is, however, the more widely used or deployed version (Figure 19.4).

Multiple computers can be connected to a DSL modem via Ethernet or the IEEE 802.11 WLAN. Much more can be achieved using a very high bit-rate (VDSL) standard, which has a theoretical bit rate of 52 Mbps downstream and 12 Mbps upstream

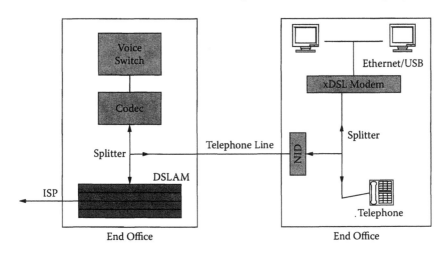

Figure 19.4   DSL services. (From Y. Xiao, X. Du, J. Zhang, F. Hu, and S. Guizani, *IEEE Communications Magazine,* **November 2007, 126–134. With permission.)**

from a user to the operator's equipment. VDSL can use up to two frequency bands for upstream and two frequency bands for downstream with quadrature amplitude modulation (QAM) or discrete multitone (DMT) modulation techniques. Better still, the very high bit rate digital subscriber line 2 described as the ITU-T G.993.2 standard provides full duplex total data rates up to 200 Mbps at a bandwidth of up to 30 MHz. For many countries with very little fiber underground, DSL technologies are the preferred access network techniques for delivering IPTV.

Ethernet has remained the dominant access technology for LANs. It can support up to 10 Gbps access speed and eight classes of service (CSs). It also supports both unicast and multicast as well as broadcast modes of operations over virtual LANs. In the optical (single mode) domain, the IEEE 802.3ae is a full duplex 10 Gbps Ethernet standard. It is a viable access technology for IPTV services that can provide high data rates at reasonable QoS.

Fiber-To-The-Last Meter

To date, optic fiber provides the largest access bandwidth at the home and business front. Within homes and offices, fiber is deployed in various forms as FTTH or FTTP (fiber-to-the-premises), FTTC, FTTN (fiber-to-the-node), and FTTB (fiber-to-the-business). These are generally referred to as fiber-to-the-X (FTTX). This section covers FTTH(P) and FTTC. For FTTN, fiber runs to the node and then the remaining distance to the house or premises is covered by copper as ADSL2+ and VDSL with speeds up to 3 to 6 Mbps (symmetric) and 14 to 25 Mbps asymmetric at distances of between 1 and 1.5 km.

Until a few years ago, copper reigned supreme as the major industry cable for carrying narrowband and broadband applications. For that reason, most networks deploy copper all through the network backbone right to the homefront (Figure 19.5). This setup has been used and supports ADSL and its other flavors.

Fiber was introduced to replace copper, due in part to the bandwidth limitation of copper and also because it is highly prone to environmental effects and noise. This structure is predominant in most developing countries. As an evolution, initially fiber was deployed to not too far away from the office and homefront, and the homefront was reserved for copper (Figure 19.6). This setup still exists in many countries.

The architectures in Figures 19.5 and 19.6 are optimized for voice. The last mile supports bit rates of about 28 kbps to 1.5 Mbps and still exists in many countries.

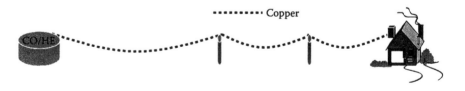

Figure 19.5   Copper-to-the-home and -office.

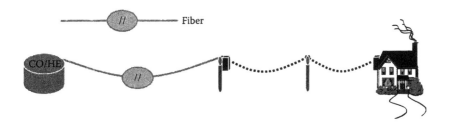

Figure 19.6   Fiber-in-the-core network only.

Many countries have, however, deployed fiber to the homefront. Optic fiber provides several advantages. It is immune to electromagnetic interference; is strong, reliable, and flexible; a large bandwidth can be achieved over long distances; and it is easily deployed and upgraded. Furthermore, it is secure and not attractive for use in making ornaments and rings, as is experienced in many developing countries with copper. For IPTV, the major driver is to achieve a large bandwidth (broadband) right from the backbone to the home or the premises, and also to maintain the optimum low bit error rate quality supported by fiber. FTTH therefore bridges the last meter to the home with fiber, as shown in Figure 19.7, at a large bit rate of about 20 Mbps to 1 Gbps and above.

Optical networks are optimized for voice, video, and data. Between the central office (CO) and the home environment (HE), fiber is used exclusively. The optical access network (OAN) architecture also includes two network units, the optical line terminator (OLT) and the optical network unit shown in Figure 19.7 at the customer premises. The optical fiber network therefore provides enormous capacity throughout the backbone and to the homefront, and allows full symmetric transmission. Hence, the cost of maintenance and using an optic fiber network is reduced compared to copper installations. In general, compared with other access techniques, up to 1 Gbps can be achieved with single-mode optic fiber over a distance of more than 100 km. This cannot be achieved with copper.

Countries that have the financial capacity to lay fiber within metro areas, Internet download speeds up to 40 Mbps, and upstream speed of about 80 Mbps

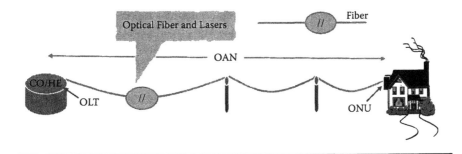

Figure 19.7   Fiber-to-the-Home (FTTH).

Figure 19.8 Connections using optic fiber.

can be achieved. The advantages in using fiber within metro areas continue to outstrip what copper can offer. Copper can support about six phone calls per strand while a single fiber strand can carry about 2.5 million simultaneous voice calls [7] (at 64 channels at 2.5 Gbps)! Optic fiber can be more easily installed for beauty and ease. An optic fiber with the same information-carrying capacity as copper is less than 1 percent the size and weight of the copper cable required. Fiber has a very low thermal expansion coefficient and hence is not as highly subject to thermal noise as copper. This makes fiber highly suited to hot and humid tropical climates.

FTTH and FTTC

FTTH supports both Ethernet and ATM transport. When ATM cells are used, 5 bytes of the 53-byte ATM cells are used for the overhead and the remainder is payload. ATM is connection oriented, and the established connections remain on for the duration of the communication (Figure 19.8). The color bands in Figure 19.8 represent whose data is being sent down a particular fiber.

Ethernet, on the other hand, is connectionless oriented. The Ethernet packet has 22 bytes overhead and the payload can be up to 1500 bytes (Figure 19.9 and Figure 19.10).

There are three types of FTTH architectures: passive, active, and hybrid optical network. A passive optical network (PON) shares fiber strands for a portion of the optical distribution network (Figure 19.11). Optical splitters are then used to split and aggregate the optical signals. Hence, power is only required at the endpoints. By using passive splitters, the need to build equipment cabinets to support PON applications is avoided. PONs perform both layer 2 and 3 functions at the operators' central offices instead of dedicated equipment as in the active modes.

The line splitter can, for example, split the light signal in 1×16 or to groups as (1×2 and 1×8) or, alternatively, for a 1×32 into (1×4 and 1×8). In the active network architecture, the equipment is built to support about 400 to 500 customers and performs layer 2 and 3 functions. Based on the IEEE 802.3,a full duplex data rate of about 100 Mbps can be supported.

In the active optical network (Figure 19.12), each user has a dedicated fiber-optic strand to serve him or her. The nodes are active or powered and are used to

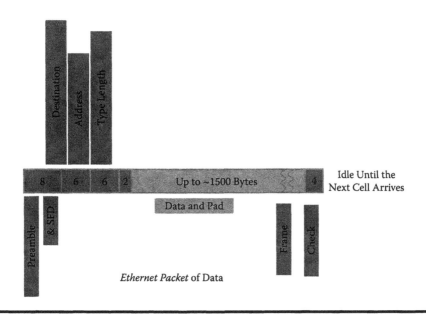

Figure 19.9 Ethernet packet structure.

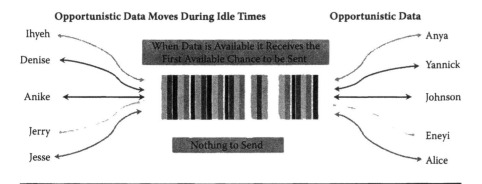

Figure 19.10 Ethernet on fiber.

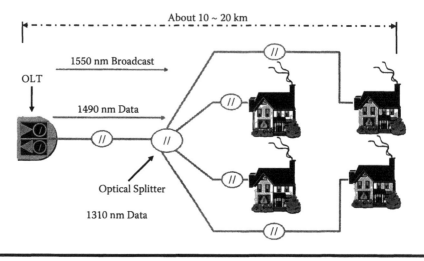

Figure 19.11   Passive optical network architecture.

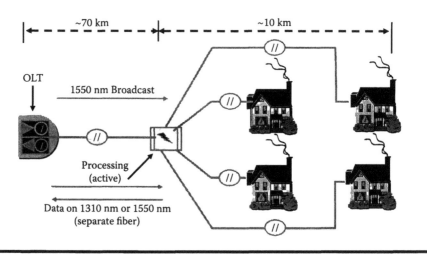

Figure 19.12   Active optical network architecture.

manage the signal distribution. The hybrid optical networks combine the features of passive and active optical networks.

In the active mode, 1550 nm fiber is used for broadcasting and either 1310 nm or 1550 nm fiber is used depending on distance and may be used to carry data.

The hybrid optical network architecture is shown in Figure 19.13. The light signal is carried to very close to the premises in the active mode and then split close to the premises, and the passive mode network is then used from then on to the homes. Between the optical splitter and the active processing unit, a single-strand optic fiber is used at 1550 nm for broadcast and at 1310 nm for bi-directional data.

In all the three architectures, during the planning and implementation phases it is essential to take into consideration the bandwidth required per home (to support voice, video, or IPTV, data or broadcast mode) in deciding the type of fiber and the splitting point of the signal. Other issues to consider are the required QoS and the security levels required for each subscriber to avoid data meant for one house being snooped upon or being directed to the wrong premises. Therefore, efficiency in sharing the channels to protect the customers' data and still offer the required SLA is required. These requirements translate to considering offering competitive bandwidth compared to ADSL or cable modems (at ~0.5 to ~1.5 Mbps) in an asymmetrical manner. For example, FTTH can provide several 100 Mbps in a shared manner (symmetrical) or about 10 to 30 Mbps non-shared. Different services require different data rates. SDTV video requires about 2 to 4 Mbps at the IP level and HDTV consumes five times the capacity reserved for SDTV. When audio is streamed, about 380 kbps is required. Depending on the SLAs with different customers, data for some homes will need higher priority in terms of transmission and security.

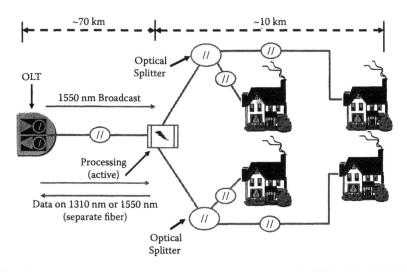

Figure 19.13   Hybrid optical network architecture.

Table 19.1   Analog Video Link Budgets

Split	Nominal Splitting Loss (dB)	Available Fiber Loss (dB)	Nominal Distance (km)
4	7	11	44
8	10.5	7.5	30
16	14	4	16
32	17.5	2.5	10
64	21	−1	−4

Source: From J. Farmer and L. Ray, "Fiber-to-the-Home Overview and Technical Tutorial", PowerPoint presentation, Fiber-to-the-Home Council, 2003. With permission.

Every time the optical signal is split into two components in the PON and hybrid architectures, each channel receives half of the signal and there is a theoretical loss of 3 dB in the signal energy (about 3.5 dB to 4 dB in practice or a cost of 16 km of transmitted distance). It is estimated that every two-way split of the light costs about 10 km in transmission distance. This is in addition to the losses in the optic fiber itself, which for the 1550 nm fiber is about 0.25 dB/km and about 0.4 dB/km for shorter wavelength fibers (260 to 1360 nm). For the 1550 nm fiber, the result is that the light energy can be transmitted for about 80 km without amplification. Hence, the theoretical distance over which a particular fiber can be used depends on the amount of energy lost per split and the losses in the fiber. Once the energy has light signal energy has reduced to below the receiver threshold, no more splitting or transfer is possible. Therefore, the network is limited by the number of times an operator can split the light energy before reaching its customers' premises. A typical link budget for an optic fiber network therefore can be summarized as in Table 19.1 [7].

The values in Table 19.1 are based on nominal splitter and fiber loss, rather than the worst case [7]. The achievable distances in practice are always less than what is estimated in Table 19.1 and will include about 2 dB for connection. These values are based on a 1550 nm externally modulated transmitter. The losses are nominal, and in practice will include a 2 dB loss for connection. These values are based on a 1310 nm DFB fiber.

Standards

The ITU-T has developed several standards for FTTH, and Table 19.3 provides a listing of such standards.

In general therefore, the picture in terms of how to use either full optic fiber or a combination of fiber and copper to offer IPTV looks like that in Figure 19.14. This architecture does not include broadband wireless IP access to the home. In

Table 19.2   Gigabit EPON Link Budget

Split	Nominal Splitting Loss (dB)	Distance (km) 40 km GBIC	Distance (km) 70 km GBIC
4	7	14	44
8	10.5	4	34
16*	14	—	24
32	17.5	—	14
64	21	—	4

Source: From J. Farmer and L. Ray, "Fiber-to-the-Home Overview and Technical Tutorial", PowerPoint presentation, Fiber-to-the-Home Council, 2003. With permission.

Table 19.3   FTTH Standards

Year Published	ITU-T G.98x.y Standard and IEEE 802.3x,z	Access Type	Bit Rate Support	Service
1998	G.983.1	Basic ATM PON		
1998	G.983.2	ONT management and control interface		
1998	G.983.3	WDM		Enhanced services (analog video)
1998	G.984.1	GPON (hybrid ATM)		
1998	G.984.2	GPON (hybrid Ethernet)		
1998	G.985	Ethernet (point-to-point)	100 Mbps	
2004	IEEE 802.3a,h	Ethernet-in-the-first-mile (EFM)	1 Gbps	Point-to-multipoint PON

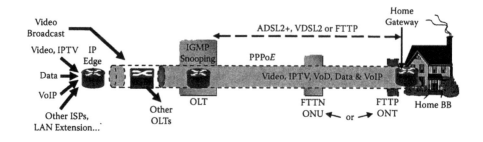

Figure 19.14 Integrated IP services on optical networks.

certain situations, both broadband access techniques will coexist at the customer and homefronts.

For economic reasons, FTTC can be used up to a few tens of meters to the house or business, and the remainder of the distance is covered by xDSL, as in Figure 19.14. The rest of the distance to the home is cable or carrier-grade Ethernet.

Multicasting of IPTV

IPTV is clearly a service that requires a lot of bandwidth and is delivered to many users through multiple paths. This makes it difficult to offer if broadcasting is not used because the bandwidth requirement is enormous. Delivery of IPTV is done using IP multicast. In a commercially available service offered to many customers, the probability that only one person is watching or receiving the service is low, and this is used to save bandwidth in the core and access networks. Because there is a high probability that many subscribers are watching the program, it can be delivered as a single send in the core network and redistributed or multicast close to the service homefront or business. The combination of IGMP, RTSP, RTP, RTCP, and TCP (UDP) makes IPTV multicasting feasible.

IP multicast is a method of sending a message to multiple receiving nodes simultaneously. For this multicast group, addresses ranging from 224.x.x.x to 239.x.x.x are used. Multicasting delivers the same information to many end nodes simultaneously. The end nodes receive a combination of video, voice, and data using RTP/TCP (or UDP)/IP protocols. Each multicasting sink has a unique IP address. Clearly, because of the large bandwidth of the delivery backbone, scalability is a problem in multicasting. Nowhere is this more apparent than in bandwidth-hungry IPTV.

There are three types of multicasting: source specific, dense multicast, and sparse multicast. The source-specific multicast mechanism originates from a known specific source to multiple addresses that request the data. Hence, the list of sites to receive the data is trimmed by the number of requests.

In dense multicast, a tree is constructed for delivering data to the multicast users. In this case, the source node broadcasts the information to all the routers and all nodes, and they in turn send pruned packets if they have no interest in the multicast data. This prevents routers and nodes from sending corresponding packets to them. To prevent loops from forming in the network, reverse-path forwarding is used. Dense multicasting leads to large network overhead as many messages and packets are sent. It is therefore very expensive in terms of bandwidth utilization.

Sparse multicasting constructs a tree for sending packets to multicast nodes. A node that wishes to join the multicast group sends a join/prune message using the Internet Group Management Protocol (IGMP) to a router near it, and the router forwards the information to the multicast group. The join/prune messages are sent in a periodic manner to a group-designated rendezvous point (RP) by a designated for each known active group. In the designated router, a route entry state is maintained representing the distribution tree for all the group members. Reverse-path forwarding is used by the routers to ensure that loops are not formed with packets or messages circulating.

When unicast is used to deliver IPTV, the service is basically video on demand (VOD) for a single customer.

IPTV over Wireless

Thus far we have discussed offering IPTV over cable (copper and fiber) and not yet delved into how IPTV can be delivered with wireless networks. Delivering IPTV over wireless media is not a far-fetched thought because many wireless network standards offer speeds that are suitable for IPTV. The bandwidths of all the 802.11xx standards can support IPTV and so can HSDPA (high-speed downlink packet access) and WiMAX. In this section we discuss recent advances in how to deliver IPTV using wireless network access.

IPTV over WiMAX

One of the major limitations in delivering IPTV over optical networks and xDSLs is that they fail to support mobility and line-of-sight communications. WiMAX, apart from providing the required speed that is suited to IPTV, also supports mobility and line-of-sight communications that rival single-length optical drops. WiMAX can therefore be used to provide IPTV anywhere and anytime and more so in rural areas of the world. WiMAX or 802.16 (2004) and 802.16e (2005) standards provide for fixed and mobile wireless access in metro areas or the so-called metropolitan area networks (MANs). Supporting data rates of up 70 Mbps over a range of about 30 km, WiMAX is indeed a technology that is suited for IPTV service. It therefore is the preferred wireless technology for delivering wireless IPTV. The WiMAX multicast capability is a significant incentive for its selection. With it, a single WiMAX base station sends video streams to many receivers in a wide geographical coverage and they could all be mobile subscribers as well! There are, however, problems that must be overcome. First, many broadband wireless video streams need to be beamed to the receiver from the provider—each stream for a unique channel. Video streams from both channels must be buffered when changing channels and this requires very large buffers to cover the time. Changing channels requires a minimum delay and that delay could be anywhere from 5 s to 15 s before the video on the channel is seen on the TV screen. It might therefore require buffers to be provided for all the envisaged IPTV channels so that at all times, the buffers have the most current video stream from the corresponding channel. This limits the delay time for channel change, but consumes bandwidth. Fortunately, the price of memory has fallen dramatically over the past 5 years.

WiMAX, however, is subject to jitter- and mobility-related network QoS degradation effects (e.g., delays, congestion, and packets arriving out of sequence).

Peer-to-Peer (P2P) IPTV

The description of IPTV so far has assumed that there is an infrastructure network that exists to serve in delivering the IPTV service. Thus, IPTV is within the complete control of the operators. IPTV can, however, be delivered without an established and centralized communication network that is managed by an operator. By taking advantage of ad-hoc networking and emerging mesh networking technology, a new kind of IPTV, in which each IPTV user is potentially an IPTV server node and multicasting received content to other IPTV users [8], is practically possible. This new method of offering IPTV is called P2P IPTV (P2P = peer-to-peer). In a P2P IPTV system, subscribers serve as peers and participate in video data sharing [8].

Packet Loss in IP Networks

All practical networks are imperfect, and network errors and delays are bound to occur and often degrade the service quality. For IPTV, the quality of the received video stream must be high and the delays incurred in transit must be very small. Hence, the bit error rates and number of lost packets must be very small for the service to be acceptable. The QoS metrics for IPTV therefore must take into consideration video jitter, packet loss probability, delay, out-of-sequence packets, multicast join time, and network fault probability. These metrics are within the administrative and network management concerns of the IPTV provider. For voice, the metrics include jitter, mean opinion scores for the speech coder in use, delay, echo, and voice packet losses. For the overall IPTV service, the QoS metrics include channel availability, channel start time, channel change time, channel change failure rate, and the quality of the service itself. In the next sections, some of these metrics are discussed further.

The ITU-T over the years has set some benchmarks for evaluating the performance of video streaming networks. This performance measure is based on the

Table 19.4   ITU-T J.241 Standard (Performance Classifications Used for TV Services)

Packet Loss Rate (PLR)	QoS
$\leq 10^{-5}$	Excellent service quality (ESQ)
$10^{-5} < PLR \leq 2 \times 10^{-4}$	Intermediate service quality (ISQ)
$2 \times 10^{-4} < PLR < PLR\_out = 0.01$	Poor service quality (PSQ)
$PLR\_out = 0.01 < PLR$	IP end-to-end service not available

packet loss ratio (PLR). For high-quality video streaming service, the PLR should be in the range $10^{-7} < PLR < 10^4$ and better. The required delay (a few hundred milliseconds) and jitter should also be a few tens of milliseconds. These requirements are captured in the ITU-T Standard J.241 and listed in Table 19.4.

A means for evaluating the performance of the network and the quality of received video is the so-called objective quality index—MDI or Media Delivery Index—first proposed by IneoQuest Technologies and Cisco Systems, the so-called IETF Network Working Group Internet-Draft (draft-welch-mdi-03.txt) the RFC 4445 published in April 2006 [http://www.ietf.org/rfc/rfc4445.txt]. The MDI is represented mathematically by two numbers. One of the numbers is the Delay Factor (DF) or network jitter, and the other number is the Media Loss Rate (MLR). The MLR states the number of Transport Stream (TS) packets lost over a period of time. These two numbers, however, do not account for the subscribers' perceived subjective quality of the received video or their quality of experience (QoE).

IP networks are highly heterogeneous. Therefore, the features of the information transmitted have a tendency to change permanently, causing occasional network errors. Normally, the IP network layer and the Ethernet physical and link layers go hand in hand and IP packets are dropped due to single bit errors in transmission. Hence, the transmitted packets are either received correctly or in error. Three main factors lead to packet losses:

1. Bit errors caused by noise or equipment failures
2. Buffer overflow as a result of queuing packets and delay caused by congestion and traffic flow jitter in the network
3. Rerouting of packets to avoid bottlenecks in the network

Due to the nature of the network, packets will arrive with varying delays (jitter) and some will arrive late. This is due to the routes taken by some packets and the traffic load on the network. This is shown in Figure 19.15.

All receivers have limited buffer for storage of arriving packets. Therefore, packets that arrive too late could be construed as lost, thus requiring retransmission in TCP domains.

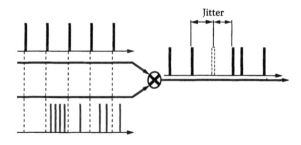

Figure 19.15   Network jitter.

The so-called MDI is one of the methods for assessing the level of jitter in the IP network for streaming media such as MPEG video. It is a measure of the cumulative jitter induced by the network and thus helps in selecting the level of buffer required and is an indicator of lost packets. The MDI is a performance number and two forms are used. On the one hand, it represents the DF and in the other hand it also represents the MLR. The DF is the maximum time difference observed at the end of a media stream between the drain of the media data and the arrival of the media data. For constant bit rate traffic, the drain rate of the media is the normal rate of reception of the traffic. For variable bit rate traffic, it is the computed piece-wise traffic rate. Hence, it can be used to estimate the buffer size at the receiver. Because it also indicates the amount of jitter in the data traffic received, it provides a good indication of the delay in the network and how long data should be buffered.

The MLR is also used as an objective measure of the received picture quality and is derived from the expression:

$$\psi = MLR = \frac{DF}{MDI}$$

It is obtained from the count of lost or out-of-order flow packets over a selected time of usually about a second. It should be observed, however, that the MLR does not provide a measure of the quality of the picture at the source of transmission (e.g., in the database, camera, or at the output of the encoder).

Despite these performance measures, from the user point of view, the perception measure is preferable. The perceptual performance of IPTV requires expert evaluators to measure distortions in the video segment. The ITU-T BT.500-7 standard provides five grades of impairment measuring scales to be used for estimating perception:

5: imperceptible
4: perceptible, but not annoying
3: slightly annoying
2: annoying
1: very annoying

These values represent opinion scores and are similar to the mean opinion scores used for evaluating voice coding standards.

Summary

IPTV is another service supported by IP networks. It is a logical addition to the communication trend of giving each person a telephone number, a handset, a camera, a radio, a microphone, the Internet, and logically a TV in the handset. This adds to the growing attributes of gadgets that we all carry around in our daily lives. IPTV represents the convergence of television broadcasting, telecommunications, multimedia content distribution, and computing. It is offered and carried primarily as multicast IP packets by first coding the picture frames. IPTV can be offered to your handset or directly to a set-top box at home. The chapter provided the architecture and the modes of delivery of IPTV to homes through the support of optic fiber technologies. IPTV is clearly a service that requires a lot of bandwidth and is delivered to many users through multiple paths. This makes it difficult to offer if broadcasting is not used because the bandwidth requirement is enormous. Delivery of IPTV is done using IP multicast. IPTV can also be offered over wireless networks, WiMAX, and peer-to-peer.

Practice Set: Review Question

1. IPTV is:
   a. Stored TV beamed to a receiver
   b. TV delivered using Internet Protocol
   c. The same as MMS
   d. The same as SMS
   e. All of the above
2. IPTV is coded with:
   a. MPEG-2
   b. MPEG-4
   c. MPEG-7
   d. JPEG
   e. JPEG2000
3. IPTV may be delivered to users with which of the following access methods?
   a. Optical networks
   b. Wireless LAN
   c. DSL
   d. GSM
   e. UMTS

4. Why is UDP preferred over TCP for delivering IPTV over the transport layer?
   a. TCP is too old to be used.
   b. No real benefits with UDP.
   c. Higher data rates.
   d. Reduced error rates.
   e. Reduced delay with no acknowledgment required.

Exercises

1. What is IPTV?
2. List some modes of IPTV delivery.
3. A full HDTV frame of size 1920×1080 pixels was acquired by a source at 50 frames per second (RGB). What is the bit rate?
4. Discuss the system architecture of IPTV. Illustrate your discussion with diagrams.
5. Compare and contrast the access methods that could be used for delivering IPTV.
6. Explain the differences among the following:
   a. FTTN
   b. FTTH
   c. FTTC
7. Distinguish between passive and active optical networks. Discuss each one with diagrams and why certain wavelengths are preferred.
8. In what ways is the hybrid optical network a compromise between active and passive networks?
9. Discuss the merits of transporting IPTV over WiMAX. In doing so, consider the link budget involved in both fixed and mobile WiMAX.
10. Discuss the following concepts in relation to IPTV:
    a. Media loss ratio (MLR)
    b. Media delivery index
    c. Delay factor
11. Explain the ITU-T BT.500-7 standard method of measuring impairment in IPTV service.

References

[1] Pro-MPEG Code of Practice #3 release 2. Transmission of Professional MPEG-2 Transport Streams over IP Networks.
[2] Pro-MPEG Code of Practice #4 release 1. Transmission of High Bit Rate Studio Streams over IP Networks.

[3] Y. Kukhmay, K. Glasman, A. Peregudov, and A. Logunov, Video over IP Networks: Subjective Assessment of Packet Loss, 2006.

[4] Recommendation ITU-R BT.500-7. Methodology for the Subjective Assessment of the Quality of Television Pictures.

[5] J. Welch and J. Clark, A Proposed Media Delivery Index, RFC4445 August 2005.

[6] Y. Xiao, X, Du, J. Zhang, F. Hu, and S. Guizani, Internet Protocol Television (IPTV): The Killer Application for the Next-Generation Internet, *IEEE Communication Magazine*, November 2007, pp. 126–134.

[7] J. Farmer and L. Ray, Fiber-to-the-Home Overview and Technical Tutorial, PowerPoint presentation, Fiber-to-the-Home Council, 2003.

[8] X. Hei, C. Liang, J. Liang, Y. Liu, and K.W. Ross, Insight into PPLive: Measurement Study of a Large-Scale P2P IPTV System, *Proceedings of the IPTV Workshop in Conjunction with the International World Wide Web Conference*, Edinburgh, UK, May 2006.

[9] Agbinya, J.I., "Fast interpolation algorithm using fast Hartley transforms," *Proc. IEEE*, 75(4) April 1987, 523–524.

[10] Agbinya, J.I., "Two-dimensional interpolation of eeal sequences using the DCT," *Electronics Letters*, 29(2), 1993, 204–205.

[11] Agbinya, J.I. "Interpolation using the discrete cosine-transform," *Electronics Letters*, 28(20), 1992, 1927–1928.

Index

Milton Keynes UK
Ingram Content Group UK Ltd.
UKHW031125141024
449569UK00006B/424